IET ENERGY ENGINEERING SERIES 245

Agrivoltaics

Other volumes in this series:

Volume 1	**Power Circuit Breaker Theory and Design** C.H. Flurscheim (Editor)
Volume 4	**Industrial Microwave Heating** A.C. Metaxas and R.J. Meredith
Volume 7	**Insulators for High Voltages** J.S.T. Looms
Volume 8	**Variable Frequency AC Motor Drive Systems** D. Finney
Volume 10	**SF_6 Switchgear** H.M. Ryan and G.R. Jones
Volume 11	**Conduction and Induction Heating** E.J. Davies
Volume 13	**Statistical Techniques for High Voltage Engineering** W. Hauschild and W. Mosch
Volume 14	**Uninterruptible Power Supplies** J. Platts and J.D. St Aubyn (Editors)
Volume 15	**Digital Protection for Power Systems** A.T. Johns and S.K. Salman
Volume 16	**Electricity Economics and Planning** T.W. Berrie
Volume 18	**Vacuum Switchgear** A. Greenwood
Volume 19	**Electrical Safety: A guide to causes and prevention of hazards** J. Maxwell ADAMS
Volume 21	**Electricity Distribution Network Design, 2nd Edition** E. Lakervi and E.J. Holmes
Volume 22	**Artificial Intelligence Techniques in Power Systems** K. Warwick, A.O. Ekwue and R. Aggarwal (Editors)
Volume 24	**Power System Commissioning and Maintenance Practice** K. Harker
Volume 25	**Engineers' Handbook of Industrial Microwave Heating** R.J. Meredith
Volume 26	**Small Electric Motors** H. Moczala *et al.*
Volume 27	**AC-DC Power System Analysis** J. Arrillaga and B.C. Smith
Volume 29	**High Voltage Direct Current Transmission, 2nd Edition** J. Arrillaga
Volume 30	**Flexible AC Transmission Systems (FACTS)** Y-H. Song (Editor)
Volume 31	**Embedded Generation** N. Jenkins *et al.*
Volume 32	**High Voltage Engineering and Testing, 2nd Edition** H.M. Ryan (Editor)
Volume 33	**Overvoltage Protection of Low-Voltage Systems, Revised Edition** P. Hasse
Volume 36	**Voltage Quality in Electrical Power Systems** J. Schlabbach *et al.*
Volume 37	**Electrical Steels for Rotating Machines** P. Beckley
Volume 38	**The Electric Car: Development and future of battery, hybrid and fuel-cell cars** M. Westbrook
Volume 39	**Power Systems Electromagnetic Transients Simulation** J. Arrillaga and N. Watson
Volume 40	**Advances in High Voltage Engineering** M. Haddad and D. Warne
Volume 41	**Electrical Operation of Electrostatic Precipitators** K. Parker
Volume 43	**Thermal Power Plant Simulation and Control** D. Flynn
Volume 44	**Economic Evaluation of Projects in the Electricity Supply Industry** H. Khatib
Volume 45	**Propulsion Systems for Hybrid Vehicles** J. Miller
Volume 46	**Distribution Switchgear** S. Stewart
Volume 47	**Protection of Electricity Distribution Networks, 2nd Edition** J. Gers and E. Holmes
Volume 48	**Wood Pole Overhead Lines** B. Wareing
Volume 49	**Electric Fuses, 3rd Edition** A. Wright and G. Newbery
Volume 50	**Wind Power Integration: Connection and system operational aspects** B. Fox *et al.*
Volume 51	**Short Circuit Currents** J. Schlabbach
Volume 52	**Nuclear Power** J. Wood
Volume 53	**Condition Assessment of High Voltage Insulation in Power System Equipment** R.E. James and Q. Su
Volume 55	**Local Energy: Distributed generation of heat and power** J. Wood
Volume 56	**Condition Monitoring of Rotating Electrical Machines** P. Tavner, L. Ran, J. Penman and H. Sedding
Volume 57	**The Control Techniques Drives and Controls Handbook, 2nd Edition** B. Drury
Volume 58	**Lightning Protection** V. Cooray (Editor)
Volume 59	**Ultracapacitor Applications** J.M. Miller
Volume 62	**Lightning Electromagnetics** V. Cooray

Volume 63 **Energy Storage for Power Systems, 2nd Edition** A. Ter-Gazarian
Volume 65 **Protection of Electricity Distribution Networks, 3rd Edition** J. Gers
Volume 66 **High Voltage Engineering Testing, 3rd Edition** H. Ryan (Editor)
Volume 67 **Multicore Simulation of Power System Transients** F.M. Uriate
Volume 68 **Distribution System Analysis and Automation** J. Gers
Volume 69 **The Lightening Flash, 2nd Edition** V. Cooray (Editor)
Volume 70 **Economic Evaluation of Projects in the Electricity Supply Industry, 3rd Edition** H. Khatib
Volume 72 **Control Circuits in Power Electronics: Practical issues in design and implementation** M. Castilla (Editor)
Volume 73 **Wide Area Monitoring, Protection and Control Systems: The enabler for smarter grids** A. Vaccaro and A. Zobaa (Editors)
Volume 74 **Power Electronic Converters and Systems: Frontiers and applications** A.M. Trzynadlowski (Editor)
Volume 75 **Power Distribution Automation** B. Das (Editor)
Volume 76 **Power System Stability: Modelling, analysis and control** A.A. Sallam and B. Om P. Malik
Volume 78 **Numerical Analysis of Power System Transients and Dynamics** A. Ametani (Editor)
Volume 79 **Vehicle-to-Grid: Linking electric vehicles to the smart grid** J. Lu and J. Hossain (Editors)
Volume 81 **Cyber-Physical-Social Systems and Constructs in Electric Power Engineering** S. Suryanarayanan, R. Roche and T.M. Hansen (Editors)
Volume 82 **Periodic Control of Power Electronic Converters** F. Blaabjerg, K. Zhou, D. Wang and Y. Yang
Volume 86 **Advances in Power System Modelling, Control and Stability Analysis** F. Milano (Editor)
Volume 87 **Cogeneration: Technologies, optimisation and implementation** C.A. Frangopoulos (Editor)
Volume 88 **Smarter Energy: from Smart Metering to the Smart Grid** H. Sun, N. Hatziargyriou, H.V. Poor, L. Carpanini and M.A. Sánchez Fornié (Editors)
Volume 89 **Hydrogen Production, Separation and Purification for Energy** A. Basile, F. Dalena, J. Tong and T.N. Veziroğlu (Editors)
Volume 90 **Clean Energy Microgrids** S. Obara and J. Morel (Editors)
Volume 91 **Fuzzy Logic Control in Energy Systems with Design Applications in MATLAB®/Simulink®** İ.H. Altaş
Volume 92 **Power Quality in Future Electrical Power Systems** A.F. Zobaa and S.H.E.A. Aleem (Editors)
Volume 93 **Cogeneration and District Energy Systems: Modelling, analysis and optimization** M.A. Rosen and S. Koohi-Fayegh
Volume 94 **Introduction to the Smart Grid: Concepts, technologies and evolution** S.K. Salman
Volume 95 **Communication, Control and Security Challenges for the Smart Grid** S.M. Muyeen and S. Rahman (Editors)
Volume 96 **Industrial Power Systems with Distributed and Embedded Generation** R Belu
Volume 97 **Synchronized Phasor Measurements for Smart Grids** M.J.B. Reddy and D.K. Mohanta (Editors)
Volume 98 **Large Scale Grid Integration of Renewable Energy Sources** A. Moreno-Munoz (Editor)
Volume 100 **Modeling and Dynamic Behaviour of Hydropower Plants** N. Kishor and J. Fraile-Ardanuy (Editors)
Volume 101 **Methane and Hydrogen for Energy Storage** R. Carriveau and D.S-K. Ting
Volume 104 **Power Transformer Condition Monitoring and Diagnosis** A. Abu-Siada (Editor)
Volume 106 **Surface Passivation of Industrial Crystalline Silicon Solar Cells** J. John (Editor)
Volume 107 **Bifacial Photovoltaics: Technology, applications and economics** J. Libal and R. Kopecek (Editors)

Volume 108 **Fault Diagnosis of Induction Motors** J. Faiz, V. Ghorbanian and G. Joksimović
Volume 109 **Cooling of Rotating Electrical Machines: Fundamentals, modelling, testing and design** D. Staton, E. Chong, S. Pickering and A. Boglietti
Volume 110 **High Voltage Power Network Construction** K. Harker
Volume 111 **Energy Storage at Different Voltage Levels: Technology, integration, and market aspects** A.F. Zobaa, P.F. Ribeiro, S.H.A. Aleem and S.N. Afifi (Editors)
Volume 112 **Wireless Power Transfer: Theory, technology and application** N. Shinohara
Volume 114 **Lightning-Induced Effects in Electrical and Telecommunication Systems** Y. Baba and V.A. Rakov
Volume 115 **DC Distribution Systems and Microgrids** T. Dragičević, F. Blaabjerg and P. Wheeler
Volume 116 **Modelling and Simulation of HVDC Transmission** M. Han (Editor)
Volume 117 **Structural Control and Fault Detection of Wind Turbine Systems** H.R. Karimi
Volume 118 **Modelling and Simulation of Complex Power Systems** A. Monti and A. Benigni
Volume 119 **Thermal Power Plant Control and Instrumentation: The control of boilers and HRSGs, 2nd Edition** D. Lindsley, J. Grist and D. Parker
Volume 120 **Fault Diagnosis for Robust Inverter Power Drives** A. Ginart (Editor)
Volume 121 **Monitoring and Control Using Synchrophasors in Power Systems with Renewables** I. Kamwa and C. Lu (Editors)
Volume 123 **Power Systems Electromagnetic Transients Simulation, 2nd Edition** N. Watson and J. Arrillaga
Volume 124 **Power Market Transformation** B. Murray
Volume 125 **Wind Energy Modeling and Simulation, Volume 1: Atmosphere and plant** P. Veers (Editor)
Volume 126 **Diagnosis and Fault Tolerance of Electrical Machines, Power Electronics and Drives** A.J.M. Cardoso
Volume 127 **Lightning Electromagnetics, Volume 1: Return stroke modelling and electromagnetic radiation, 2nd Edition** V. Cooray, F. Rachidi, M. Rubinstein (Editors)
Volume 127 **Lightning Electromagnetics, Volume 2: Electrical processes and effects, 2nd Edition** V. Cooray, F. Rachidi, M. Rubinstein (Editors)
Volume 128 **Characterization of Wide Bandgap Power Semiconductor Devices** F. Wang, Z. Zhang and E.A. Jones
Volume 129 **Renewable Energy from the Oceans: From wave, tidal and gradient systems to offshore wind and solar** D. Coiro and T. Sant (Editors)
Volume 130 **Wind and Solar Based Energy Systems for Communities** R. Carriveau and D. S-K. Ting (Editors)
Volume 131 **Metaheuristic Optimization in Power Engineering** J. Radosavljević
Volume 132 **Power Line Communication Systems for Smart Grids** I.R.S Casella and A. Anpalagan
Volume 134 **Hydrogen Passivation and Laser Doping for Silicon Solar Cells** B. Hallam and C. Chan (Editors)
Volume 139 **Variability, Scalability and Stability of Microgrids** S.M. Muyeen, S.M. Islam and F. Blaabjerg (Editors)
Volume 142A **Wind Turbine System Design, Volume 1: Nacelles, drive trains and verification** J. Wenske (Editor)
Volume 142B **Wind Turbine System Design, Volume 2: Electrical systems, grid integration, control and monitoring** J. Wenske (Editor)
Volume 143 **Medium Voltage DC System Architectures** B. Grainger and R.D. Doncker (Editors)
Volume 145 **Condition Monitoring of Rotating Electrical Machines** P. Tavner, L. Ran, C. Crabtree
Volume 146 **Energy Storage for Power Systems, 3rd Edition** A.G. Ter-Gazarian
Volume 147 **Distribution Systems Analysis and Automation, 2nd Edition** J. Gers
Volume 151 **SiC Power Module Design: Performance, robustness and reliability** A. Castellazzi and A. Irace (Editors)

Volume 152 **Power Electronic Devices: Applications, failure mechanisms and reliability** F. Iannuzzo (Editor)
Volume 153 **Signal Processing for Fault Detection and Diagnosis in Electric Machines and Systems** M. Benbouzid (Editor)
Volume 155 **Energy Generation and Efficiency Technologies for Green Residential Buildings** D. Ting and R. Carriveau (Editors)
Volume 156 **Lithium-ion Batteries Enabled by Silicon Anodes** C. Ban and K. Xu (Editors)
Volume 157 **Electrical Steels, 2 Volumes** A. Moses, K. Jenkins, Philip Anderson and H. Stanbury
Volume 158 **Advanced Dielectric Materials for Electrostatic Capacitors** Q Li (Editor)
Volume 159 **Transforming the Grid Towards Fully Renewable Energy** O. Probst, S. Castellanos and R. Palacios (Editors)
Volume 160 **Microgrids for Rural Areas: Research and case studies** R.K. Chauhan, K. Chauhan and S.N. Singh (Editors)
Volume 161 **Artificial Intelligence for Smarter Power Systems: Fuzzy logic and neural networks** M.G. Simoes
Volume 165 **Digital Protection for Power Systems, 2nd Edition** Salman K Salman
Volume 166 **Advanced Characterization of Thin Film Solar Cells** N. Haegel and M. Al-Jassim (Editors)
Volume 167 **Power Grids with Renewable Energy: Storage, integration and digitalization** A.A. Sallam and B. Om P. Malik
Volume 169 **Small Wind and Hydrokinetic Turbines** P. Clausen, J. Whale and D. Wood (Editors)
Volume 170 **Reliability of Power Electronics Converters for Solar Photovoltaic Applications** F. Blaabjerg, A.l Haque, H. Wang, Z. Abdin Jaffery and Y. Yang (Editors)
Volume 171 **Utility-scale Wind Turbines and Wind Farms** A. Vasel-Be-Hagh and D.S.-K. Ting
Volume 172 **Lightning Interaction with Power Systems, 2 Volumes** A. Piantini (Editor)
Volume 174 **Silicon Solar Cell Metallization and Module Technology** T. Dullweber (Editor)
Volume 175 **n-Type Crystalline Silicon Photovoltaics: Technology, applications and economics** D. Munoz and R. Kopecek (Editors)
Volume 178 **Integrated Motor Drives** Xu Deng and B. Mecrow (Editors)
Volume 180 **Protection of Electricity Distribution Networks, 4th Edition** J. Gers and E. Holmes
Volume 182 **Surge Protection for Low Voltage Systems** A. Rousseau (Editor)
Volume 184 **Compressed Air Energy Storage: Types, systems and applications** D. Ting and J. Stagner
Volume 186 **Synchronous Reluctance Machines: Analysis, optimization and applications** N. Bianchi, C. Babetto and G. Bacco
Volume 190 **Synchrophasor Technology: Real-time operation of power networks** N. Kishor and S.R. Mohanty (Editors)
Volume 191 **Electric Fuses: Fundamentals and new applications, 4th Edition** N. Nurse, A. Wright and P.G. Newbery
Volume 193 **Overhead Electric Power Lines: Theory and practice** S. Chattopadhyay and A. Das
Volume 194 **Offshore Wind Power Reliability, Availability and Maintenance, 2nd Edition** P. Tavner
Volume 196 **Cyber Security for Microgrids** S. Sahoo, F. Blaabjerg and T. Dragicevic
Volume 198 **Battery Management Systems and Inductive Balancing** A. Van den Bossche and A. Farzan Moghaddam
Volume 199 **Model Predictive Control for Microgrids: From power electronic converters to energy management** J. Hu, J.M. Guerrero and S. Islam
Volume 201 **Design, Control and Monitoring of Tidal Stream Turbine Systems** M. Benbouzid (Editor)
Volume 202 **Hydrogen from Seawater Splitting: Technology and outlook** A Ray and M Patel (Editors)

Volume 204	**Electromagnetic Transients in Large HV Cable Networks: Modeling and calculations** Ametani, Xue, Ohno and Khalilnezhad
Volume 208	**Nanogrids and Picogrids and Their Integration with Electric Vehicles** S. Chattopadhyay
Volume 210	**Superconducting Magnetic Energy Storage in Power Grids** M.H. Ali
Volume 211	**Blockchain Technology for Smart Grids: Implementation, management and security** H.L. Gururaj, K.V. Ravi, F. Flammini, H. Lin, B. Goutham, K.B.R. Sunil and C. Sivapragash
Volume 212	**Battery State Estimation: Methods and models** S. Wang
Volume 215	**Industrial Demand Response: Methods, best practices, case studies, and applications** H.H. Alhelou, A. Moreno-Muñoz and P. Siano (Editors)
Volume 213	**Wide Area Monitoring of Interconnected Power Systems, 2nd Edition** A.R. Messina
Volume 214	**Distributed Energy Storage in Urban Smart Grids** P.F. Ribeiro and R.S. Salles (Editors)
Volume 217	**Advances in Power System Modelling, Control and Stability Analysis, 2nd Edition** F. Milano (Editor)
Volume 225	**Fusion–Fission Hybrid Nuclear Reactors: For enhanced nuclear fuel utilization and radioactive waste reduction** W.M. Stacey
Volume 227	**Processing and Manufacturing of Electrodes for Lithium-Ion Batteries** J. Li and C. Jin (Editors)
Volume 228	**Digital Technologies for Solar Photovoltaic Systems: From general to rural and remote installations** S. Motahhir (Editor)
Volume 235	**Clean Energy Past and Future** P. Tavner
Volume 236	**Green HV Switching Technologies for Modern Power Networks,** K. Niayesh (Editor)
Volume 237	**High Voltage Power Transformers: State of the art and technological innovations** M. Sforna, S. Zunino and F. Ferrari
Volume 238	**AI for Status Monitoring of Utility Scale Batteries** S. Wang, K. Liu, Y. Wang, D. Stroe, C. Fernandez and J.M. Guerrero
Volume 239	**Intelligent Control of Medium and High Power Converters** M. Bendaoud, Y. Maleh and S. Padmanaban (Editors)
Volume 241A	**Power Electronic Converters and Systems, 2nd Edition, Volume 1: Converters and machine drives** M. Simões and T. Busarello (Editors)
Volume 241B	**Power Electronic Converters and Systems, 2nd Edition, Volume 2: Applications** M. Simões and T. Busarello (Editors)
Volume 247	**Power System Strength Evaluation Methods, Best Practice, Case Studies, and Applications** H. Alhelou, N. Hosseinzadeh and B. Bahrani (Editors)
Volume 250	**Interactions of Wind Turbines with Aviation Radio and Radar Systems** Alan Collinson
Volume 256	**Wireless Power Transfer Technologies: Theory and technologies, 2nd Edition** N. Shinohara (Editor)
Volume 905	**Power System Protection, 4 Volumes**

Agrivoltaics

Technical, ecological, commercial and legal aspects

Edited by
Constantin Klyk and Stephan Schindele

The Institution of Engineering and Technology

Published by The Institution of Engineering and Technology, London, United Kingdom

The Institution of Engineering and Technology is registered as a Charity in England & Wales (no. 211014) and Scotland (no. SC038698).

First published 2024

The Institution of Engineering and Technology
Futures Place
Kings Way, Stevenage
Hertfordshire SG1 2UA, United Kingdom

www.theiet.org

British Library Cataloguing in Publication Data
A catalogue record for this product is available from the British Library

ISBN 978-1-83953-797-4 (hardback)
ISBN 978-1-83953-798-1 (PDF)

Typeset in India by MPS Limited

Contents

About the editors

Constantin Klyk is a proficient independent consultant and expert specializing in Agrivoltaic Systems. He acquired extensive experience as a project manager for Agri-PV at Zimmermann PV-Steel Group in Germany, where he crafted intricate solutions for this emerging field. Prior to this role, he contributed to research endeavors at ISC Konstanz, focusing on PV system design and pioneering research initiatives in Agri-PV. Additionally, he has played integral roles in numerous international development projects as a PV engineer and project manager.

Stephan Schindele is the head of product management Agri-PV at BayWa r.e, Germany. In this function he supports global Agri-PV project development and implementation of BayWa r.e. and leverages synergies of the agricultural and energy industry. Between 2011 and 2020 Stephan built up the new business field of Agri-PV at Fraunhofer Institute for Solar Energy Systems ISE, where he worked as a PhD student and project manager. He received several awards for his work.

Preface

Constantin Klyk[1] *and Stephan Schindele*[2]

Welcome, dear reader, to a comprehensive exploration of the interdisciplinary land use configuration: Agrivoltaics. The book you now hold, *Agrivoltaics: Technical, ecological, commercial, and legal aspects* is a pioneering endeavour, the first of its kind, offering a complete, in-depth examination of the multifaceted world of Agrivoltaics (Agri-PV).

Agri-PV represents the confluence of agriculture and photovoltaics, a unique synergy of food production and renewable energy generation, with the potential to reshape the future of solar energy generation in profound ways. This innovative field holds the promise of capitalising on the sun's energy – the world's major energy source, to its fullest potential. It perfectly combines photosynthesis, the process by which plants capture solar energy, with the photoelectric effect, the principle behind solar power generation. This blend of food and energy, both fundamental needs of our species, stands as a testament to human ingenuity and our constant strive for sustainability – acknowledging the historical significance and the future potential of farmers in energy production. Prior to the coal and oil age, farmers were the primary energy producers, nurturing the grass and grains that powered animals – the main mode of transport and work. This critical role is set to re-emerge, albeit in a different form. With the world now on the cusp of a green energy revolution, farmers are again pivotal contributors to energy production. Through the utilisation of their land for photovoltaic systems or more specifically, Agri-photovoltaic systems, they hold one of the keys to a sustainable food and energy secure future.

Our mission with this book is to compile the vast expanse of existing knowledge on Agri-PV and present it in a manner that's accessible and engaging. We cover a broad spectrum of topics from the underlying technology to economic considerations, from plant physiology to legal frameworks across various geopolitical contexts, and from computer-aided yield predictions to the social implications of this nascent field. We provide unique perspectives into the largely underexplored areas in current publications such as O&M considerations and the necessities for agricultural machinery operation within Agri-PV projects. Further,

[1]Agri-PV Consulting, Jörg-Zürn-Str. 22, 88339 Bad Waldsee Germany
[2]BayWa r.e. AG, Arabellastraße 4, 81925 München, Germany

to reduce ambiguity on the topic, we study existing definitions and classifications on the topic and propose a globally valid definition for Agri-PV.

This book is not merely a theoretical discourse. It is a practical guide enriched with real-world case studies that provide a holistic and in-depth perspective on Agri-PV. We've made a conscious effort to bridge the asymmetric knowledge gap between the agricultural and photovoltaic sectors, making the complex world of Agri-PV as straightforward and commonly acceptable as possible.

Our contributors are leading experts in their respective fields, working across the globe. They bring to the table a wealth of expertise, passion and diverse perspectives, weaving a pompously tapestry of knowledge and experience on Agri-PV.

Our goal is to empower you, our reader, with the most comprehensive and up-to-date information on Agri-PV, which sometimes is not easy, considering how fast this relatively new land use configuration is evolving – technically, commercially, socially and legally. We hope this book will serve as a definitive reference, fostering a deeper understanding of Agri-PV and inspiring further breakthroughs in this field.

Designed for a wide audience, this book is for anyone keen on exploring the world of Agri-PV. Whether you are a student, an industry expert or a policymaker, we are confident that you will find this book insightful and illuminating. We aim to offer actionable recommendations that will help drive the future evolution of Agri-PV.

We invite you to join us on this journey discovering the fascinating world of Agri-PV. We hope that, armed with the knowledge from this book, you too can contribute to this diverse field and help shaping a sustainable future. We have tried to paint an all-encompassing picture of the genesis of Agri-PV. If we have missed a pioneer or a lighthouse project in Agri-PV, please let us know so we can continuously improve the book.

Remember, no matter how complex it may seem, learning is a journey, not a destination. So, let's embark on this exciting adventure together – harvesting the sun simultaneously for food and energy production!

Chapter 1

Agrivoltaics, the future of utility-scale photovoltaics on farmland

Stephan Schindele[1] *and Constantin Klyk*[2]

'Whoever causes two corn stalks to grow where one stalk grew before, has done more for a nation than a general who won a battle'.

Frederick II the Great (1712–1786), Prussian King, <The Old Fritz>

This chapter aims to answer two questions: First, why does it make sense to support Agri-PV diffusion? Second, what is the status quo of Agri-PV market development globally?

To answer the first question, basic knowledge, facts and trends in the different policy areas impacted by Agri-PV are outlined. This analysis gets to the heart of why it is reasonable for sustainable development to stimulate multifunctional and cross-cutting PV-solutions such as Agri-PV (or Floating-PV).

The second answer elucidates the genesis of Agri-PV between 1981 and today. Agri-PV pioneers from different regions of the world are introduced, some lighthouse projects are presented and an estimate on the current global Agri-PV market size is outlined.

1.1 Why agrivoltaics makes sense: policy areas impacted

At different stages of the Agri-PV innovation process, different policy areas are impacted. In principle, the following logic applies: the more advanced the Agri-PV market introduction and the higher the Agri-PV market share of the total PV market, the higher the socio-environmental risks and opportunities that Agri-PV can bring about. To better assess whether the benefits outweigh the cost of Agri-PV diffusion, this sub-chapter discusses basic knowledge, facts and trends in the following policy areas affected by Agri-PV: (I) Agriculture and Environment; (II) Climate and Energy; (III) Finance and Building/Permitting and (IV) National Sustainability Strategy.

[1]BayWa r.e. AG, Arabellastraße 4, Germany
[2]Agri-PV Consulting, Jörg-Zürn-Straße 22, 88339 Bad Waldsee, Germany

1.1.1 Agriculture, environment and the (missing) joy of farming

During the period from 1961 to 2018, the world's population rose from around 3.07 billion to 7.60 billion. This corresponds to a percentage increase of 148% [1]. During the same period, the global arable and permanent cropland area increased by 15% [2,3]. As the total land area remained almost constant over this period, but settlement area like traffic infrastructure, housing, industrial and recreation areas increased, it is assumed that the increase in arable land resulted mainly from logging of primary forests and the conversion of extensive pastures. Between 1990 and 2018, the global forest area decreased by 4%, which corresponds to about 163 million hectares or about 4.6 times the land area of Germany [4]. The conversion of forest and pastureland to arable land has a significant negative impact on biodiversity [5] (Figure 1.1).

The available arable land per person has decreased by 49% between 1961 and 2018 and statistically expresses the 'population trap' that Thomas Robert Malthus already warned about in the 18th century (Malthus 1798). While population growth exponential, the expansion of arable land and productivity growth can only be linear. Planetary boundaries seek population and economic growth to decouple from natural resource demand [6].

By 2050, it is estimated that 9.7 billion people will inhabit the planet Earth [7]. The Food and Agriculture Organization (FAO) estimates that the demand for food and biomass products will increase by more than 50% by 2050, compared to demand in 2021. At the same time, it is assumed that over a third of global arable land will be lost to climate change, desertification, water scarcity, settlement developments and overexploitation [5] and that annual global productivity increases due to the use of agrochemicals (fertilisers and pesticides) and modern agricultural machinery technology, as well as changes in agricultural structure are not sufficient to meet the growing demand for food [8]. Against this background, it is already foreseeable that the second Sustainable Development Goal of the United Nations – 'No Hunger by 2030' – will not be met. The number of undernourished people has risen from 604 million in 2014 to 768 million in 2020 and is very likely to continue to increase. The FAO report 'The State of the World's Land and Water Resources for Food and Agriculture. Systems at Breaking Point' [5] concludes that a 'business as usual' approach is not possible and disruptive adjustments in agricultural policy are needed to reduce the scale of the humanitarian crisis.

The agricultural sector is affected by climate change to the extent that it also contributes to causing it. In total, 34% of global CO_2 emissions are attributable to the agricultural sector [9]. This makes it the second largest CO_2 emitter after the energy sector. A large part of the CO_2 emissions in the agricultural sector is due to its energy requirements, whereby the energy-intensive production of pesticides and artificial fertilisers is methodically assigned to the industrial sector rather than the agricultural or energy sector. Dr. Josef Schmidhuber, FAO Deputy Director for Markets and Trade, judges: 'Modern agriculture is basically the conversion of energy via plants and animals into food' [10]. For both climate change adaptation

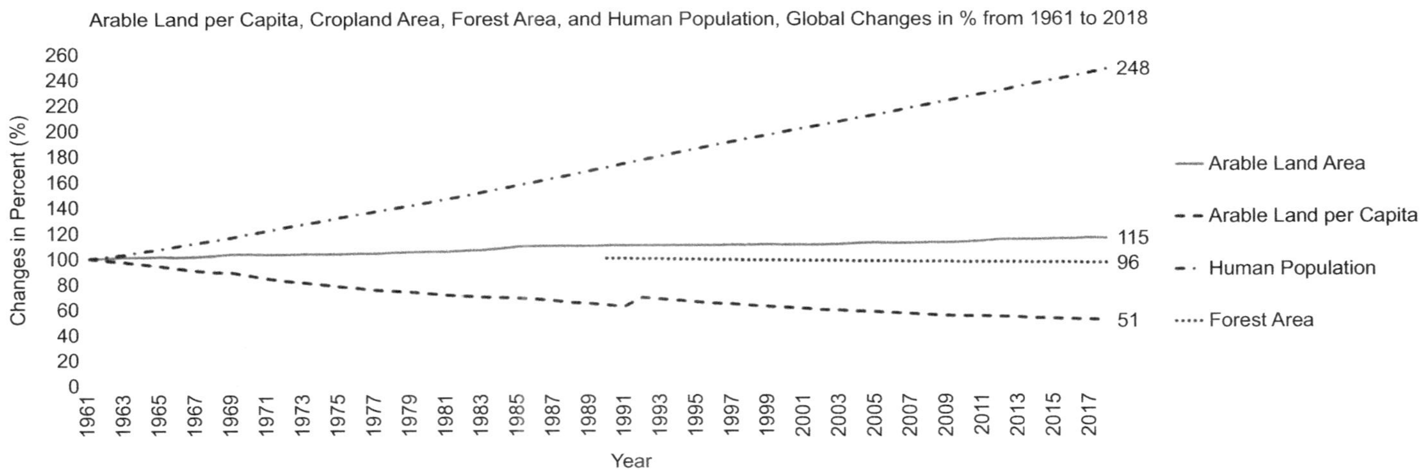

Figure 1.1 *Global change in per cent (%) from 1961 to 2018: Agricultural land arable land, arable land per person, forest area, population (Source: FAO; own representation)*

and defossilization of the food system, there is an urgent need for public and private investment in the agricultural sector. The demand for smart solutions that protect the climate and increase the resilience of farms is growing. The importance of the food-water-energy-biodiversity nexus and integrated food and energy systems is increasing [11]. The food retailers and several agribusinesses have recognised the challenges [12] and are taking it upon themselves (Scopes 1 and 2) and their suppliers (Scope 3) to increase the level of sustainability [13]. Farmers are faced with the challenge of ensuring economically viable agricultural production in times of extreme heat, drought, frost and heavy rain, while at the same time meeting the requirements of the food retailers and end-consumers on sustainable development. This combination of high operating risk, low profit expectations, high administrative requirements and great social responsibility makes the profession 'farmer' increasingly unattractive. Globally, the average age of farmers is above 60 years and rising – including developing countries where a vast majority of the population is below 25 years of age [14], and more and more young people are moving to the cities to pursue a career in industry, science or administration [15]. It is society's overall responsibility to make farming a joyful, trendy and economically lucrative passion. On a daily basis, the future farmer supplies our basic needs: healthy food, lively environment, clean energy and good social coexistence in rural areas.

1.1.2 Climate, energy and the ambivalence of (farm)land-take by ground-mounted photovoltaics

Climate policy is guided by the scientific findings of the Intergovernmental Panel on Climate Change (IPCC). In August 2021, via the organisation Scientist Rebellion, the third part of the IPCC's sixth report, which focuses on climate change mitigation, was leaked before it entered the political process within the United Nations climate policy regime [16]. The report shows that global CO_2 emissions are expected to peak before 2025 and should fall to zero between 2050 and 2075 to cost-effectively limit global warming to 1.5–2.0°C above pre-industrial levels [17]. Against this background, major global economies such as the USA, China and the EU signed the Paris Agreement in 2015 and have set themselves ambitious targets to become climate neutral by 2050 [18–20]. An example of such climate policy goal is the EU Green Deal: By the end of 2050, the EU wants to have established a circular economy emitting no more net greenhouse gases. To meet these targets, economic growth is to be decoupled from the use of resources. To guarantee high social acceptance during this transitions phase, no one, neither people nor regions, should be abandoned along the way. As an interim target, by the end of 2030 net greenhouse gas emissions in the EU are to be at least 55% below 1990 levels. In 2019, the EU-wide greenhouse gas emissions were 25.9% lower than in 1990. In the next eight years (beginning of 2023 to the end of 2030), the EU wants to achieve a higher greenhouse gas emission reduction compared to the past 32 years (1990 to the end of 2022). To meet the CO_2 reduction targets by 2030, the seven-year budget of the Green Deal allocates a total of 1.8 trillion Euro from the 'NextGenerationEU' package [21]. Globally, through the expansion of electromobility in the transport sector and of electric heat pumps or cooling systems in the heating/cooling industry, sector coupling between the power, heating/cooling and transportation markets will lead to an increase in gross electricity generation.

1.1.2.1 Germany as a case study to illustrate contradictory policy objectives

In Germany, for instance, the gross electricity generation is expected to increase by 13%, from about 655 TWh in 2020 to about 740 TWh in 2050. A large part of the gross electricity generation, approx. 86% or an average of 633 TWh, should then be generated from renewable energies. The remaining electricity generation is to be covered by biomass, other energy sources and electricity imports [22]. To ensure security of supply, in particular grid stabilisation and for balancing the fluctuating electricity generation from renewable energies, gas-fired power plants will serve as a so-called 'bridge' technology. The gas power plant infrastructure is part of the German hydrogen strategy [23]. For this reason, the German government demanded to include gas-fired power plant infrastructure as an allowed investment within the EU taxonomy regulation. The German government proposes that the fuel switch from natural gas to hydrogen should be completed by 2036 [24]. This target is remarkable, given that the coal phase-out was until recently planned for 2038 [25]. According to coalition agreement, the coal phase-out is now to be completed by 2030 and, according to the federal government, no more natural gas will be used to generate electricity for only six years after that [26]. In response to the Russian war of aggression in Ukraine in February 2022, the German government announced that it would reduce its dependence on Russian gas as quickly as possible. As early as 2035, Germany is to use 100% renewable energies in its power sector [27]. The energy transition must therefore be greatly accelerated, and the hydrogen infrastructure expanded as quickly as possible so that a high level of supply security is guaranteed even without Russian gas imports [28]. However, the expansion of renewable energies faces major challenges. In addition to the technical problems of system integration, for example the expansion of the electricity grid and hydrogen infrastructure, discussions on the social compatibility of the 'Energiewende' [29] and missing land availability for project implementation have become increasingly important [30–32]. Particularly onshore-wind and agrofuels are challenged by lacking social acceptance, for example for reasons of landscape and bird conservation regards onshore wind [33] or the 'Fuel or Food' debate in the context of agrofuels. To achieve the most cost-effective and rapid renewable energy expansion, offshore wind and solar PV are moving into the political and economic focus. The German government implemented policies to increase the offshore wind capacities by a factor of nine by 2030 [34] and the PV capacity by a factor of 2.5 from around 60 GWp (DC) in 2022 to more than 215 GWp in 2030, which corresponds to an annual PV expansion of 22 GWp [23,35].The PV expansion is to be achieved in equal parts by distributed rooftop PV systems and utility-scale conventional ground-mounted PV (GM-PV) on farmland. Assuming a ratio of 1 MWp (DC) per hectare (ha), the total demand of farmland resulting from GM-PV diffusion is roughly 77,000 ha by 2030 or 11,000 ha per year or 30 ha per day. Under the current legal framework, GM-PV implemented in rural areas is permitted as 'special areas', which allows electricity generation in underdeveloped outskirt areas. These special areas 'Solar PV' are considered an industrial area, due to the power generation aspect, and the farmland status of the project area is lost, but accounting for daily build up settlement area [36].

The expansion of GM-PV on farmland is thus in stark contradiction to the goal of introducing a circular land economy by 2050, which aims to reduce the daily increase in settlement areas – in Germany to less than 30 ha from 2030. Taking into account the limited availability of arable land and the goal of reducing the daily increase in residential, traffic, recreation and industrial settlement areas [26] it is foreseeable that the expanding land-take of GM-PV will lead to competition for land use and thus conflicts between agricultural, environmental and civil society stakeholders on local level.

1.1.2.2 Ground-mounted PV is the largest PV segment: more land-take expected globally

By end of 2022, a total of 1,300 TWh solar PV power was generated globally, accounting for 4.5% of total global power generation. The International Energy Agency (IEA), in alignment with its Net Zero Scenario, assumes that solar PV will reach a level of approximately 8,300 TWh in 2030. GM-PV represented just over 50% of installed PV capacity in 2022, followed by roof-top installations commercial and industrial (25%) and residential (23%) segments [37] (see Figure 1.2).

The IEA is recognising that the share of GM-PV was at its lowest since 2012 as governments generously stimulate distributed PV on rooftops. And although GM-PV has become the cheapest source of electricity generation in many

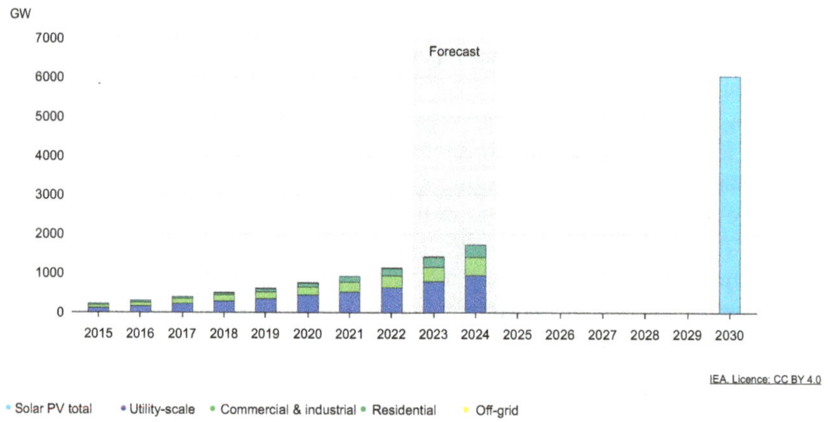

Figure 1.2 Solar PV power capacity in the Net Zero Scenario, 2015–2030 (IEA, 2023b)

regions of the world and can be implemented without subsidies, it is becoming increasingly difficult to find suitable sites and obtain a building permission for GM-PV projects [38].

To further accelerate PV implementation while at the same time diminish land-use conflicts, science recommends Integrated-PV solutions, where areas, spaces, surfaces and farmland are dual used, for example Building-integrated PV, traffic infrastructure and vehicle integrated PV, Carport- and Floating-PV as well as Agri-PV [39]. Simultaneously, integrated-PV solutions create cross-sectorial synergies and make the energy transition a positive experience.

In Germany alone, the technical potential of integrated-PV is estimated to be 3,200 GWp. Of all integrated-PV applications Agri-PV has the largest technical potential with a share of 53% or 1,700 GWp. The sooner governments support the implementation of multifunctional PV applications, the less farmland is taken by GM-PV, the more food production is available to meet an increasing demand. The relative advantages of Agri-PV against GM-PV are the following: (I) farming activity on project area is obtained and thus status of farmland remains unchanged. Therefore, Agri-PV is not accounting to daily increase build-up settlement area and the area potential for PV development is increased; (II) techno-ecological synergies are enabled, for example increase in land use efficiency, less evaporation and transpiration in agricultural soils and crops due to shading, wind protection, and lower temperatures and water demand on site. Therefore, the resistance within the authorisation process should be lower, social acceptance of the project higher and permitting faster and less risky; finally (III) farmer is included into the operation of the Agri-PV project, diversifying and increasing the farm's income and resilience against risk of climate change, while continuing food and biomass production in the most resource efficient way. The benefits mentioned must be compared with the costs of an Agri-PV diffusion and with a baseline scenario without Agri-PV market introduction.

1.1.3 Finance and permitting: how agrivoltaics defuses a tax bomb and widens the area potential for PV development in rural areas

In the following, three policy examples from finance and permitting are provided to illustrate the importance of these two policy areas in promoting Agri-PV diffusion.

Example 1 – how the land-use-change tax defines no-go-areas and hinders Agri-PV implementation: In some member states of the EU, for example in Poland and Romania, solar PV project developers must pay a land-use-change tax if farmland is converted to special area 'Solar PV' or 'Industrial Area'. Depending on the quality of soil that is converted, the tax fees are increasing (in the case of high-quality soil) and decreasing (in the case of low-quality soil). This governance regime is economically incentivising the land-take of lower quality soils for PV developments, while at the same time allowing immediate financial benefits for the local community impacted by the project. A strong downside of this tax-policy is the exclusion of solar PV on the highest quality soils. These high productivity soils are governed as no-go-areas for utility-scale PV as the main purpose of the farmland – the production of food – must be obtained. The European solar sector is criticising the definition of go-to and no-go areas within the EU Renewable Energy Directive (RED), as the PV sector knows best how to identify suitable land for solar PV project development including grid access. Regarding building permits, each building application should be balanced against public concerns and decided upon individually whether to grant permission or reject. Looking back when the land-use change tax and the definition of no-go-area for utility-scale PV development was introduced, policymakers very likely were not aware of the techno-ecological feasibility to produce food and energy via Agri-PV solutions on the same farmland area. Hence, one cannot blame policymakers for imbalanced action. Today, however, just as industry must constantly adapt and incrementally improve its services and products to be competitive and attractive for investment, good policy making also requires being innovative, able to learn and adapt to evolving surroundings. The goal of Agri-PV is to enable farming activity on the project site and therewith a simultaneous coexistence of food and energy production on farmland. To widen the area potential for the energy transition, to include as much rural areas and farmers as possible, and to increase the chances of free power grid capacities, it is necessary for policymakers to create enabling policy frameworks for multifunctional PV solutions such as Agri-PV, Floating-PV or Carport-PV. It simply does not make sense to exclude Agri-PV development from high-quality soils, as the idea of Agri-PV is to maintain these soils for food production. The challenge in Poland and Romania is that local authorities and communities oppose Agri-PV as the farmland remains farmland. Accordingly, there is no land-use change and thus less tax income from Agri-PV compared to GM-PV. A reduced tax income is not in the interest of local authorities. In such a case, not the design of the technical solution is the challenge, but more the legal framing. There is plenty of scientific output that proves the techno-economic and socio-ecological feasibility of Agri-PV. Many evidence-based policymakers used knowledge from science-policy-interface and implemented legal regulations incentivising investments for Agri-PV diffusion, for example, governments in France, Italy, Germany, Massachusetts/USA, Israel, Japan, China, Croatia and Czech Republic. Often building and permitting regulations are closely related to finance and tax governance and need amendment to enable Agri-PV diffusion.

Example 2 – the inheritance tax bomb of GM-PV project sites: In Germany, a farmer that is inheriting its farmland to her or his descendants is not obliged by financial authorities to pay an inheritance tax for the transition of farmland ownership. This tax privilege shall keep farming economically attractive to new generations of farmers who are willing to face the risk of climate change and technological transformation due to defossilization targets. At the same time, if a farmer dedicates land for GM-PV project development, this project area will be changed to a special area 'SolarPV' or 'Industrial Area' in the local land use planning. With this land-use change, the project area is accounted similar like an industrial estate to build-up settlement. Therefore, on the day the landowner inherits the project area to the next generation, the financial authority will ask for inheritance tax, which in Germany can be up to 50% of the market value of the project area. The farming sector is calling this tax-regulation a 'tax bomb', as many landowners who leased their farmland to project developers did not seek advice from a tax adviser before signing the lease agreement [40]. In Agri-PV, where farmland remains farmland, this tax-bomb is defused and tax-privileges for farmers and landowners remain in place. In July 2022, all 16 Ministries of Finance of the 16 German federal states passed a joint decree that the tax assessment law (in German 'Bewertungsgesetz') will be extended to the effect that if agricultural land is cultivated by means of Agri-PV that adheres to the DIN SPEC 91434 Agri-PV pre-standard, it will continue to belong to agricultural land and thus retain all tax-privileges [41]. Considering that the 16 state governments are led by different party affiliations, this agreement was not an easy one, and yet, it certainly is an important incentive for farmers to make their farmland available for climate protection and energy transition. Furthermore, this decree is a positive example of how important cross-sectorial cooperation is for a successful policy implementation: in this case, the diffusion of Agri-PV related to tax governance.

Example 3 – overwriting public interest for small-scale Agrivoltaics: In many markets, PV project developers complain that the approval process takes too long. Depending on the scope of the project, it can take two to three years before a building permit is issued [42]. First, the permitting request flows into the public bodies, for example, municipal council, mayor's office and public hearings in municipal participation procedures. If the project request is supported by the public bodies, the local community will ask the local or regional authorities to grant a building permit and update their spatial planning accordingly. In this permitting process, the specific departments of the lower building authority, for example, the water department, chamber of agriculture, finance, environmental authority, telecommunications and network infrastructure, transport and many others check the impact of the project on their respective specialist area. Parallelly, the public bodies evaluate the process, publish preliminary results asking for public feedback before initiating the second and final stage of the permitting process. Whereas the permitting process in the local authorities is shouldered on fact-based studies, such as biodiversity-monitoring, soil studies and visual impact analysis, the debate within the politically elected bodies is often non-rational, informal and behind the curtain. To obtain a permit, both pathways must be accomplished successfully: a majority support in favour of the project in public bodies and the official permitting process of the project. Far too often it is the very local informal restrictions and political nuances in the public bodies that cause the project to fail – and not the restrictions opposed by administrative authorities. Therefore, political objectives of the federal government, such as climate protection and PV-diffusion goals, are challenged or even fail by local political bias. 'Not in my back Yard' (NIMBY)-effects, delayed procedures or planned errors overly critical opinions in the involvement of public stakeholders, political grudges and a whole other bouquet of risks that bring down a single project and thus on the national level the political climate policy house of cards. Against this backdrop, there are efforts, especially in federalist democracies such as Germany, the USA or India, to circumvent the local political source of danger by enabling project developers submit their building application directly to the civil servants in the approval authority rather than to the elected local council, mayor's office or other public body. This kind of drastic restriction of political power at the local level by overwriting public interest is due to the climate emergency, which requires a rapid expansion of renewable energies on the ground for rapid decarbonisation. In Germany, there has been a debate on whether to introduce such a principle of privileges under the building regulation for PV applications developed on farmland in rural areas. The PV sector has discussed controversially whether such a privilege for utility-scale solar parks is advantageous or disadvantageous. Concerned that such a privilege under building law would only have a short-term positive effect, but that in the long term it would possibly impede and slow down the expansion of PV in rural areas due to a rebellious rural population and local politicians, a political compromise was found by introducing a privilege only for conventional ground-mounted PV systems along transport routes and for Agri-PV projects on agricultural land smaller than 2.5 ha.

At the end of 2022, there were 270,000 farmers in Germany [43]. About 1.5 MWp of Agri-PV capacity can be installed on 2.5 ha – depending on the Agri-PV application. If every farmer used this privilege, the entire PV expansion targets (435,000 MWp by 2045) could be achieved by farmers. However, project sizes between 1 MWp and 2 MWp are often uneconomical because they already must feed into the medium-voltage grid. Thus, the grid connection costs for these privileged project sizes are relatively high, disabling the economic viability for the overall investment. In a reality check, it would have been wiser if the restriction criterion for eligibility to this privilege had been raised from 2.5 ha to 10 ha, because then the chances of successful project implementation would have increased, and the expansion potential would have been easier and faster to develop. The average size of a farm in Germany is 60 ha [43]. Each farmer would therefore be able to set up an Agri-PV project covering between one-sixth and no more than half of his or her farmland, thus diversifying and increasing the farm's income, increasing the farm's economic and ecological resilience, and actively participating in the energy transition. Yet, to minimise the risk of policy failure, lawmakers should take care to reflect the needs of both groups of investors in the legislation: those investments in Agri-PV directly from agriculture by farmers and those from outside agriculture by institutional investors through Agri-PV investments. In both cases, the main interest of the capital invested is return and in both cases the farmer benefits – because without additional economic benefit for the farming activity within the project area, farmers would not participate in Agri-PV projects.

1.1.4 Added value of agrivoltaics for sustainable development compared to other PV technologies

Agrivoltaics in the context of Germany's National Sustainability Strategy: The German Sustainability Strategy (DNS) was prepared in the Schröder I cabinet and first published in 2002. Since then, the DNS has been updated every four years. In 2016, the DNS was fundamentally revised and adapted to the 17 Sustainable Development

Goals (SDGs) of the United Nations (UN) by 2030, known as the 2030 Agenda. The DNS was last updated in March 2021 to reflect developments from the European Green Deal and the corona pandemic. The current DNS version comprises 66 key performance indicators (KPI). Each KPI is assigned to one of the six transformation areas: (I) Pollution-free environment, (II) human well-being and capabilities: social justice; (III) Energy transition and climate action; (IV) Circular Economy; (V) Sustainable Building and transformation of transport; and (VI) Sustainable agriculture and food systems, as well as to one of the 17 UN SDGs [44] (Figure 1.3).

The grouping of KPIs in transformation areas supports two things: First, it helps identifying responsibilities for KPIs in the government and parliament. Second, it makes agenda setting and the coordination process between the impacted

Figure 1.3 Sustainable Development Strategy for Germany, areas of transformation, off-track indicators and measures [44]

policy areas easier. Results of the assessment what impact Agri-PV diffusion has on KPIs of the German Sustainability Strategy are presented in the following. This exercise is also a good methodology to identify added value of Agri-PV for sustainable development compared to other PV technologies such as building integrated PV (BiPV) or GM-PV on farmland.

Diffusion of interdisciplinary innovations such as Agri-PV demand cross-sectorial collaboration amongst ministries and industries. The more stakeholders are impacted by the innovation, the more complex the policy coordination becomes, as the chance of internal and external factors challenging the agenda setting and policy work on Agri-PV is increasing over time. Internal factors are, for example, the number of governmental actors that need to become engaged to synchronise the legal framework for Agri-PV, political path dependencies, existing decision-making rules, formal and informal contacts and networks, party driven interests, information asymmetries, forms of negotiations and political positions of individuals and associated institutions. External factors influencing the cross-sectoral policy coordination are, for example, international shocks such as the Russian aggression against Ukraine or the nuclear catastrophe in Fukushima, effects from climate change like draughts or flood disasters, party political calculations, new agenda settings, new technologies or unforeseen market developments. As a rule, the larger the number of participants in the negotiation system, the higher the transaction costs in the coordination process, and the more internal and external factors jeopardise the success of the negotiations and the effectiveness of the market launch.

Just as important as the number of negotiators is the impact of the negotiation item on the individual existing departmental goals. 'Generally speaking, a policy that gives to all and takes away from none can rely on broad public support' [45]. If this logic is applied to the Agri-PV market introduction, according to theories from Mayntz [46] and Scharpf, two aspects are decisive for the impact on departmental goals.

First, how big is the potential for conflict? This is measured by the extent whether and to what extent established political goals, KPIs and disciplinary interests are positively or negatively affected by the Agri-PV diffusion.

Second, how big is the socio-political pressure to solve the problem? Is there a crisis awareness among the population of the challenges that a political solution is trying to address?

In the context of Agri-PV diffusion in the EU, the war in Ukraine was of great importance, as it increased the societal awareness in multiple ways, for example how little room there is for (political) manoeuvre in global food supply chains, how relevant it is to protect and utilise existing fruitful soil to battle the global food security crises, and how essential it is to become more independent from fossil fuel supplies out of autocratic regimes. The war in the Ukraine has increased the socio-political pressure to use the limited farmland resources available in Germany (and EU) as efficiently as possible. Agri-PV, as a multifunctional land use option considering the food-energy-water-biodiversity nexus, was debated in media, society and federal parliaments and the political willingness to dedicate a budget for increasing the level of sustainability by introducing Agri-PV into the farming and PV markets enhanced.

The parties of the current government, the so-called 'traffic lights coalition' (Red, Yellow, Green = SPD, FDP, The Greens) expressed support for Agri-PV in their coalition agreement. Within the reassignment of the ministries, The Green Party concentrated their governing power on climate policy. Only two months after coming into office, the Green-led Ministry of Agriculture, Ministry of Economy and Climate, and Ministry of Environment published a joint cornerstone paper entitled: 'Expansion of photovoltaics on farmland in harmony with agricultural use and nature conservation'. The paper concretised the claim from the coalition agreement ('Innovative solar energy such as Agri- and Floating-PV shall be supported') and had a symbolic effect on the Green Party base, the relevant ministerial administrations, civil servants in local authorities, society and the market participants in the agricultural and PV sectors. The three ministries acted as a competence cluster and showed that they can think and act together and that complex, interdisciplinary challenges can be solved if sectoral silos are removed, and cross-collaboration enabled. The negotiating arena for the Agri-PV diffusion was thus informally moved to the Green Party headquarters, which made agenda-setting, agreements on responsibilities and the policy implementation easier. The coordination concentration of the issues of climate and energy, agriculture, soil protection and the environment on the Green Party is a prime example of Scharpf policy theory that follows a similar approach of grouping policy targets and KPIs from the German Sustainability Strategy into transformation areas and reduces the transactions costs of policy negotiation. Figure 1.4 illustrates this

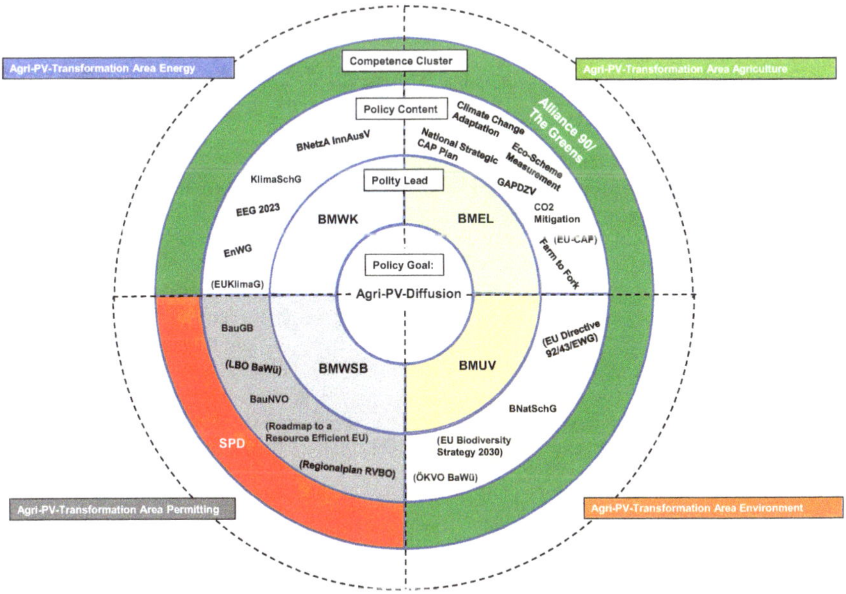

Figure 1.4 Leadership of the Agri-PV transformation areas, policy and competence-clusters of the Agri-PV market introduction in Germany (own illustration)

coordination concentration on gross-collaboration on Agri-PV policymaking on the Green Party in Germany.

However, the coordination process under the Green-Party leadership failed to involve the SPD-led Ministry of Building, Construction, and Permitting in the formulation of the cornerstone paper. Potential conflicts and ambiguities could have been avoided by integrating the SPD as a partner of the initiative and the policy-paper would have looked less like a solo effort of the Green Party, but more like a cross-party cooperation and coordination on Agri-PV diffusion within the traffic light coalition. The challenge of coordinating Agri-PV policymaking was not an issue of conflict, but more a question on how to define Agri-PV and differentiate the innovation from existing PV technologies, such as ground-mounted PV on farmland. This ambiguity was reduced with the publication of the pre-standard 'DIN SPEC 91434 – Requirements for primary agricultural use in Agri-Photovoltaic Systems' in May 2021 [47]. Today, the German government supports Agri-PV diffusion with a holistic regulatory regime, which is outlined in Chapter 10 of this book, Legal aspects of Agri-PV.

Results of the impact assessment of agrivoltaics diffusion on Germany's National Sustainability Strategy: Table 1.1 summarises the impacts that Agri-PV diffusion in Germany could have on the German Sustainability Strategy.

Table 1.1 Impact of Agri-PV diffusion on the German National Sustainability Strategy 2021 (own illustration)

SDG / KPI	SDG / Area of Transformation	Objectives and Postulate
SDG 2	**Zero Hunger / Sustainable Agriculture and Food Systems**	**End hunger, achieve food security and promote sustainable agriculture**
2.1.b	Increase share of organic farming	
SDG 7	**Affordable and Clean Energy / Energy transition and climate action**	**Ensure access to affordable, reliable, sustainable energy for all**
7.2.a	Increase share of renewable energies in gross final energy consumption	
7.2.b	Increase share of renewable energies in gross electricity consumption	
SDG 8	**Decent Work and Economic Growth / Circular Economy**	**Promote sustainable economic growth and full and productive employment for all**
8.4	Increase economic performance in an environmentally and socially responsible manner	
8.5.a	Increase level of employment	
SDG 9	**Industry, Innovation, and Infrastructure / Circular Economy**	**Promote sustainable industrialization and support innovation**
9.1	Shape the future with new solutions	
SDG 11	**Sustainable Cities and Communities / Sustainable Building and the Transformation of Transportation**	**Make cities and settlements resilient and sustainable**
11.1.a	Sustainable land use	
SDG 13	**Climate Action / Energy Transition and Climate Protection**	**Take immediate action to combat climate change and its effects**
13.1.a	Reduce greenhouse gases	

Six SDGs, eight KPIs and four DNS-transformation areas are directly affected by an Agri-PV diffusion. Out of the eight KPIs, the indicated four KPIs are unique propositions of Agri-PV compared to conventional GM-PV applications on farmland.

KPI 11.1.a sustainable land use: Agri-PV does not lead to an increase of daily build-up area, but the status of farmland is remained for the project. In contrast, in GM-PV project areas there is no farming activity and thus the status of farmland is lost. Accordingly, solar power generation is the preliminary production in GM-PV, which is considered an industrial production; thus the project area is changed to industrial area, and the daily build-up area of the region is increasing. The article 'Sustainable Land Use with Agri-Photovoltaics: Photovoltaic expansion in harmony with food production' addresses the conflicting legislation and DNS-targets in the Germany and was published in May 2021 in the GAiA-Journal [48]. A scenario analysis compares the farmland taken by GM-PV till 2050 and compares this to a scenario which introduces Agri-PV. Comparing results of both scenarios (without Agri-PV vs. with Agri-PV) shows that the KPI 11.1. 'Reduction of daily settlement land increase' to below 30 hectares per day by 2030 and to net zero by 2050 can only be achieved if the expansion of GM-PV on agricultural land is gradually substituted by Agri-PV diffusion until a circular land economy is implemented. Since the Ukraine War the PV expansion goals of the German government have increased. Latest impact assessment finds a total of 108,000 hectares agricultural land preservation with Agri-PV diffusion in Germany between 2021 and 2050 [49] (Figures 1.5–1.7).

KPI 8.4 Increase economic performance in an environmentally and socially responsible manner: by end of year 2050, the EU wants to establish a circular land economy. This objective has been integrated into Germany Climate Action Plan for 2050. From 2051 no additional agricultural land can be re-zoned in spatial planning to traffic development, industrial or residential area, and according to current legal status not for implementation of GM-PV. Therefore, the EU is

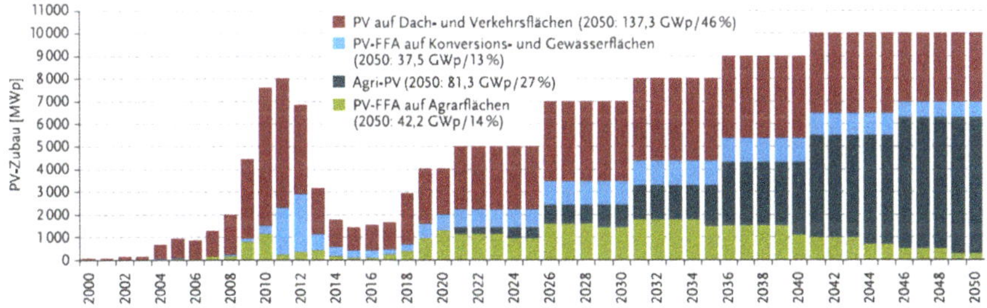

ABBILDUNG 5: *Szenario 2:* Jährliche Marktsegmententwicklung von Photovoltaik (PV): PV-Zubau in Deutschland bis 2050 mit Agri-PV (Daten aus Bundesnetzagentur 2016, ZSW 2018, ab 2021 Prognose aufgrund eigener Annahmen). Gesamte installierte PV-Leistung 2050: 298,3 GWp.

Figure 1.5 Scenario 2: annual market segment development of photovoltaics (PV): PV expansion in Germany by 2050 with Agri-PV market launch from 2021 to replace conventional ground-mounted PV capacity on agricultural land. Total installed PV capacity in 2050: 298.3 GWp.

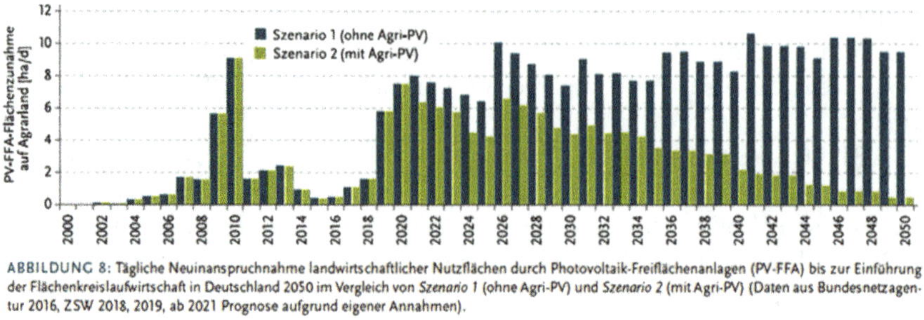

Figure 1.6 Daily new use of agricultural land by conventional ground-mounted PV projects (GM-PV) up to the introduction of the circular land economy in Germany in 2050 in comparison of Scenario 1 (without Agri-PV) and Scenario 2 (with Agri-PV) [48]

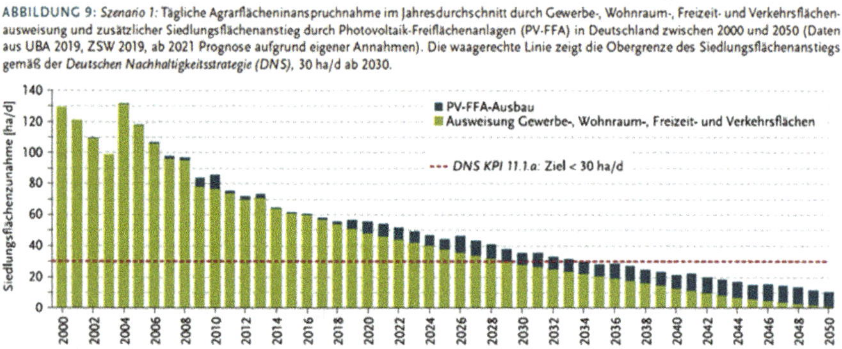

Figure 1.7 Scenario 1: Annual average daily agricultural land consumption due to commercial, residential, leisure and transport land use and additional increase in settlement area due to conventional ground-mounted PV systems (ground-mounted PV) in Germany between 2000 and 2050. The horizontal line shows the upper limit of the daily increase in settlement area of 30 hectares/day by 2030 according to the German Sustainability Strategy [48].

asking its members states to use their fruitful farmland in the most resource efficient way. Agri-PV is a multifunctional land use solution and compared to monofunctional land use the land use efficiency is higher. Agri-PV is economically and ecologically reasonable, if the agricultural output in a combined system is higher than in two separate land uses. Land use efficiency is measured by the parameter Land Equivalent Ratio (LER). Especially for permanent grassland and permanent cropland, an increase in land use efficiency is relatively easy to achieve by dual-using the land through Agri-PV. For arable land with crop rotation, the LER calculation is more challenging and the assumptions on the LER increase for agricultural yield should be justified in a detailed agricultural concept in the project approval process. Socially, Agri-PV benefits from a higher acceptability in the local community, if food and energy is produced in a project area, compared to PV only. Therefore, with high confidence, Agri-PV is supporting economic performance in rural areas in greater ecological and social manner compared to GM-PV.

KPI 9.1 Shaping the Future with New Solutions: in recent years Agri-PV became a new field of research. Globally, from 2018 till end of July 2023 over 120 scientific articles have been published in established international research journals. PV associations have been collaborating with the farming associations and published joined position papers and statements, new Agri-PV associations have been founded in the USA (The American SolarGrazing Association; ASGA), France (two associations), Italy, Korea and Germany. Agri-PV is a dynamic and innovative PV solution that is technically, legally and commercially developed together with the farming sector. Successful Agri-PV projects integrate a farmer and a farming process into the project area and record a higher agricultural yield in multifunctional land use compared to GM-PV as a monofunctional land use option. Local authorities on regional or local level have the interest in using their limited land resources as efficiently as possible. Agri-PV brings new income streams to farms, keeps food production higher compared to monofunctional land use, and receives higher acceptance by surrounding neighbourhoods. It thereby shapes the energy transition particularly in rural areas.

KPI 2.1.b increase the share of ecological farming: the goal of KPI 2.1.b is to increase the share of ecological farming in Germany from approx. 9% of the total farmland by end of 2022 to 30% by 2030. The difference of 21% of Germany's farmland is thus not available for the development of GM-PV, as on GM-PV there is no farming activity feasible. Yet, Agri-PV projects can be combined with organic farming and is thus compatible with KPI 2.1.b. In general, organic food is more costly compared to conventional farming products and the organic food sector is hit hard by lower consumption since higher inflation rates and increasing cost of living. Consumers keep their spendings low and often consume less expensive agricultural products. Yet, organic farming records a higher quality of biodiversity on local level, show a larger diversity of agricultural production on farm level, even though average organic farm size tends to be lower than the average farm size in general. Synchronising the growth of organic farming with the growth of Agri-PV diffusion could support the achievement of KPI 2.1.b.

Climate change adaptation of the farming sector: Besides the impact assessment of Agri-PV on KPIs of the German Sustainability Strategy, there is an increasing risk of climate change to the farming sectors. However, the adaptation of the farming sector to climate change, is no DNS-KPI. Yet, Agri-PV solutions such as Fruitvoltaics and Cowvoltaics show a relative advantage against GM-PV, for example if the micro-climate on the project area is improved to favour the agricultural production. For example, the heat stress of livestock is decreased if cows or sheep graze under Agri-PV systems [50] or less water is demanded in Fruitvoltaics compared to conventional plastic-based rain covers in berry production [51]. In summary, the following added values and distinguishing features of Agri-PV compared to conventional GM-PV can be identified: (I) land use efficiency is greater in the Agri-PV project because the agricultural yield in multifunctional land use is higher than in monofunctional land use; (II) agricultural activity on the project area allows the status of agricultural land to be maintained and the project is not counted in the daily increase in settlement area; (III) the income of farmers is increased and diversified, which increases the resilience of farms to climate change; (IV) the expansion of organic agriculture can be combined with Agri-PV, while land use by means of GM-PV in combination with organic agriculture is excluded; (V) the compatibility of Agri-PV with agriculture increases the agricultural land potential of PV expansion in rural areas and thus the probability of finding a suitable grid connection for the PV generator; (VI) Agri-PV is still a relatively young and innovative PV application with a steep learning curve and the potential to shape the future of rural areas. These benefits of Agri-PV seem to justify the inclusion of Agri-PV into existing support schemes of renewable energies, common agricultural policy, permitting and tax in Germany. At the beginning of the market entry phase, two aspects should be considered according to Rogers' diffusion theory [52]: First, the pioneers and early adopters of the innovation should be identified to keep the cost-benefit ratio as efficient and effective as possible. Second, policymakers and funding agencies with an interest in diffusion of the technology should have the courage to be inefficient and take experimental risk in policy implementation so that the market introduction succeeds, and a learning process of technical and political nature is initiated.

According to Ortwinn Renn, only the technology with the highest net-benefit will prevail in the long term [53]. Against this background, it can be assumed that Agri-PV will only assert itself if awareness of the global food crisis in society is high and policymakers recognise the net-benefit of Agri-PV towards efficient farmland use and its contribution to increase the level of sustainability. Market introduction is assumed to be a success if only incremental policy adjustments are necessary for steering the continuation of further market penetration of Agri-PV.

1.1.5 In a nutshell: future of utility-scale PV on farmland is agrivoltaics

When taking a holistic view of global trends in the agricultural, energy, environmental and climate protection sectors, and considering the planetary boundaries, an Agri-PV market introduction at national and local level is inevitable. Farmland is declining globally due to over-use, effects from global heating, and economic development. At the same time, the population and demand for food are increasing. Energy transition and decarbonisation demands a fast PV diffusion, also on farmland. Conventional ground-mounted PV (GM-PV) contributes to land-take from farmers and increases land use competition where relief is needed. The objective of soil protection by establishing a circular land use economy by end of 2050 cannot be achieved without substantially substituting monofunctional GM-PV with multifunctional Agri-PV applications.

The two main advantages of Agri-PV against GM-PV are, first, Agri-PV obtains the agricultural activities along the food-energy-water-biodiversity nexus on the project area and includes a farmer as the prime land user. The PV sector learns about farming processes and is adjusting the PV generator in such a way that land use efficiency with regard agricultural yield increases. Thus, the status of farmland is maintained despite the simultaneous secondary PV power generation on the project area. The landowner and farmer continue to benefit from financial and tax privileges in the agricultural sector. The second relative advantage of Agri-PV over GM-PV is the possibility of exploiting the many techno-ecological synergies between the sectors. Agri-PV can become a strategy for the farmer to adapt parts of its productive farmland climate change, decrease CO_2 emissions on its land use, reduce water demand due to micro-climate changes on the project area, and/or protecting the crops from strong weather events.

On the one hand, and in all cases, the participating farmer benefits from income diversification and a higher income, which in sum increases the economic resilience of the farm and thus the rural area. On the other hand, the solar sector benefits from avoiding land use conflicts, less restrictions in permitting, higher social acceptability, and enlarging the area potential in rural areas, which increases the probability to power grid access.

Agri-PV enables a win-win-win-situation for the farming, environment and solar sector. An effective and efficient Agri-PV diffusion in harmony with these sectors demands intensive cross-sector collaboration in policymaking. Particularly the farming, energy, climate, environmental, building and finance ministries represents the transformation area to synchronise their political actions to promote an all-embracing policy regime supporting Agri-PV diffusion and with this increasing the national level of sustainability.

1.2 The genesis of agrivoltaics: from idea to diffusion

Globally, the Agri-PV market is in the early stages of market development. In some markets, such as China, Japan and France, market development is more advanced. In other markets, the regulatory course is only just being set, for example in Germany, Italy and Israel. Then there are countries that are only very initially involved with Agri-PV, for example the Netherlands, Austria, the Czech Republic, Croatia and some states in the USA. All markets have in common that they have relatively little agricultural land available per person, the pressure on land is correspondingly high, and public institutions support piloting and Agri-PV demonstration projects to enable scientific knowledge increase and evidence-based policymaking.

It is recognised worldwide that the 1,000 roofs program, followed by the 100,000 roofs program, and then the implementation of the feed-in tariff within the Renewable Energy Sources Act (EEG) in Germany between 1995 and 2015 were the door openers for today's global PV industry. Those who have witnessed these beginnings of PV market development notice a parallel between the implementation of the first rooftop PV systems and the first Agri-PV systems today. At that time, questions regarding structural engineering, building integration, protection of historical monuments, approval procedures, fire protection, economic efficiency, legal security, installation technology, technical management, insurance and guarantees, among others, had to be clarified. In short: PV technology learned to integrate itself into the building envelope. Today, building-integrated PV (BiPV) projects are a market standard and only minor adjustments to the regulatory framework are necessary to further expand this PV market segment. Governments around the globe managed to bridge with public funding the so-called 'Valley of Death' [54]. In the conventional GM-PV market segment, contradictory policies continue to exist and resistance from landowners and authorities threatens to slow down the market growth in this important sector. With more than 50% of the global installed PV capacity implemented as GM-PV, the PV sector will need to learn how to integrate PV into agricultural areas and farming processes. Just as the years between 1995 and 2015 have proven that PV can become an integrated part of buildings, the years 2015 to 2035 will prove that Agri-PV can become an integrated part of agriculture. Again, to bridge the valley of death in innovation process, public support is needed to answer technical and legal questions, scientifically accompanied demonstration projects document the empirical values and prove the techno-economic feasibility. On the long-run, public support can be reduced as the private sector attracts investment and drives diffusion (Figure 1.8).

Existing legislation will be adapted to stimulate Agri-PV market development or at least not stand in the way of market development. From the private contributions and investments, it is recognisable that the first users are once again

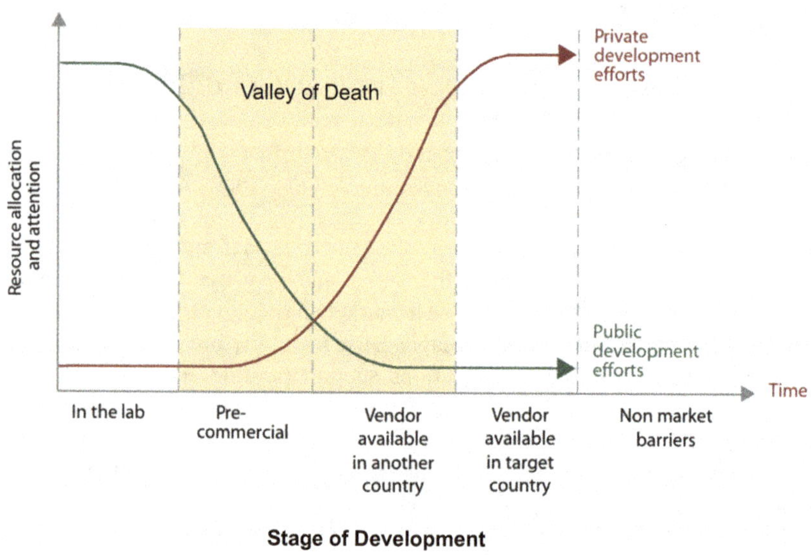

Figure 1.8 Overview of stages of development of a technology from research to market implementation [54]

farmers, which were already among the most important investor groups for BiPV due to their wealth of experience in dealing with funding programme, affinity for new technology, willingness to innovate and invest, and availability of surface – in the case of BiPV roofs of barns, stables, and other farm buildings, and in the case of Agri-PV their farmland. The major differences compared to the market development 30 years ago are the following: climate change and global warming have progressed strongly, the world population, food demand and land pressure on agricultural land have risen sharply, conventional GM-PV is the cheapest source of electricity generation worldwide and can be implemented without subsidies, while roof-top PV is in many markets still dependent on regulatory support due to the higher electricity generation costs. In the bouquet of PV technologies, Agri-PV is already cost-competitive, less costly than BiPV, sometimes at similar or slightly higher costs compared to GM-PV, but in all cases, showing the largest area potential of all PV technologies. Compared to BiPV the construction of Agri-PV is utility-scale, offering lower levelised cost of electricity, and a higher economic benefit for farmers and landowners. For a successful Agri-PV diffusion the agricultural, environmental and energy sectors must jointly learn how to progress and innovate in the field of Agri-PV – technically, legally and commercially. Figure 1.9 describes the Agri-PV innovation process from idea to imitation. Each activity along this process generates a result and enables Agri-PV to climb up the technology-readiness-levels within a certain time frame. At different times and stages of development, different policy actors are responsible for driving Agri-PV market development. Today, Agri-PV is moving from innovation to diffusion globally, which is indicated by the red frame. The purpose of this book is to (I) support Agri-PV on its way from innovation to global diffusion, (II) present the current state of the science from its many facets and (III) highlight the technological, ecological, social, legal and political challenges through case studies.

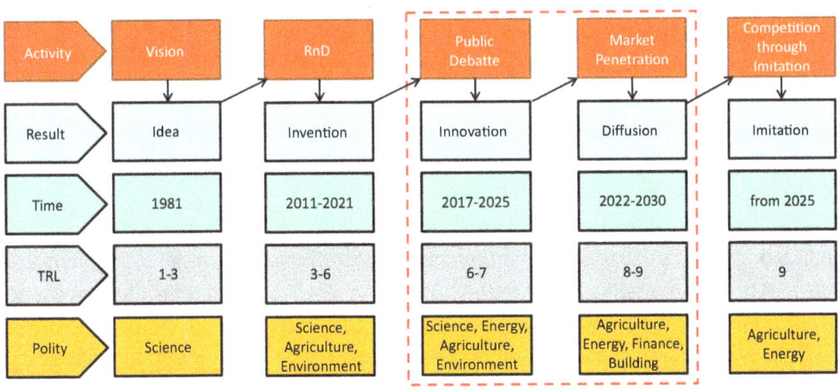

Figure 1.9 Agri-PV innovation process and political responsibilities in federal government departments in Germany (own illustration)

In the following sub-chapters, the individual Agri-PV innovation steps from idea to diffusion are examined in more detail on a global level.

1.2.1 From idea to invention: 1981–2011

Pioneering countries in the field of Agri-PV are Germany, Japan, France and Italy.

1981, Germany: In 1981, Fraunhofer ISE was founded by Adolf Goetzberger. In the same year, together with Zastrow, he published the research article 'Potatoes under the collector' [55]. This paper predicted that the manufacturing costs for photoelectric systems would become competitive with other types of energy. In contrast to studies at that time, which considered the first attractive application was the production of photoelectric energy on the roofs of buildings in connection with electric battery systems, the authors proposed 'photoelectric agricultural farms' built on average-sized farmland connected to the power grid. 'This type of solar farm would generally not be connected directly to a consumer but would produce for the grid. For example, it could be owned and maintained by the power company and maintained by it' [55]. With an optimal arrangement of the collectors and taking into account the distances between the rows of collectors, it should be possible to use the farmland simultaneously for solar energy production and for agricultural purposes. Zastrow was responsible for the calculations and simulations of the direct and diffuse light reaching the ground. Goetzberger had the idea of the 'double harvest' and conceptualised the systemic embedding of his idea with a view to grid integration, ownership and economic viability. Goetzberger recognised over 40 years ago that utility-scale GM-PV would become the most affordable of energy production and that this would create a competitive situation between energy and agricultural production. He noted that when solar panels are installed in temperate climatic zones, the growth of weeds can be kept under control. Simultaneous use of agricultural land and the co-existence of crop cultivation and solar power generation would even create synergy effects. Goetzberger and Zastrow mentioned in their visionary article that if the PV system and the crop cultivation 'are economical on a given piece of land by themselves, the only additional costs of combining the two uses are those for raising the collectors to a sufficiently stable frame'. Finally, the two scientists thought about what plants could be grown under the collectors. Plants whose growth is limited by the availability of light, for example C4 metabolism, should be excluded, but shade tolerant crops such as oats, barley, potato and sugar beet were identified as suitable. 'Very attractive in some areas could be sheep, deer or even cattle grazing, if only the collector structure is made stable enough' [55]. Parts of this visionary article was translated into English and published in 1982 with the title 'On the Coexistence of Solar-Energy Conversion and Plant Cultivation' in the internationally renowned journal *Solar Energy* [56] (Figure 1.10).

2004, Japan: According to his own statements, without prior knowledge of Goetzberger's publications, Akira Nagashima developed and installed the world's first 'solar sharing' system in Japan in 2004. Nagashima was a retired agricultural machinery engineer, who later became biologist and thus got to

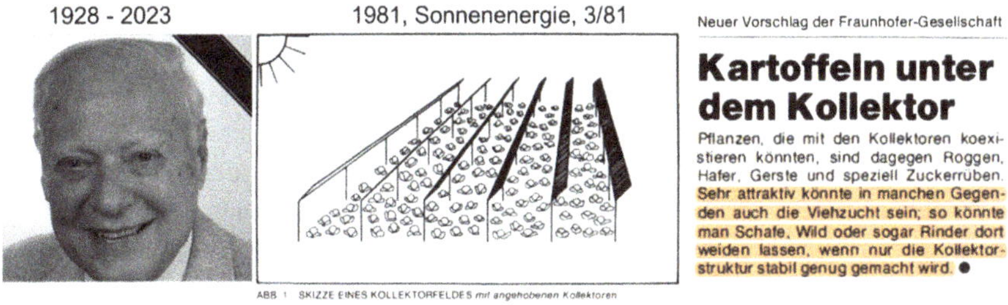

Figure 1.10 Adolf Goetzberger, originator of the Agri-PV idea, the world's first scientific publication on dual harvesting in 1981, in which it was proposed to combine photovoltaics with arable farming and livestock keeping, for example with sheep and cows

Figure 1.11 Typical solar sharing project in Japan, where the Agri-PV idea was first implemented by Akira Nagashima in 2004

know the plant-physiological 'light saturation point'. This gave him the idea of combining PV systems with agriculture, because for certain crops, too much light can lead to a decrease photosynthesis rate. He installed several test fields with different degrees of shading for different crops and thus improved the double harvest [57] (Figure 1.11).

2006, Germany/Japan/USA: In 2006, Goetzberger, together with the solar entrepreneurs Swanson and Werner, back then managing directors of California headquartered SunPower Corporation, and with Masafumi Yamaguchi, then Head of the Toyota Technological Institute, Department of Materials Science and Engineering, Japan, published the article 'Energy Farming' [58]. This was the first global initiative to promote the idea of Agri-PV and should act as 'a call to the PV industry to come together to make this vision a reality'. The authors proposed that 1% of the approximately 300 billion US dollars spent annually by industrialised countries in the in the EU, USA and Japan should be paid to farmers as a direct subsidy supporting the implementation of 'SunFarming' projects. Benefits arising from such a SunFarming subsidy program for the agricultural and solar sectors have been summarised as follows: farmers can benefit financially through the expansion of renewable energy and rural development is promoted, Kyoto Protocol climate policy targets are strengthened, food production is preserved, negative publicity against GM-PV on farmland prevented, and finally, trade deficits caused by the import of fossil fuels would be reduced. The idea to support business development and commercialisation of Agri-PV was to utilise the agricultural subsidies to compensate for the cost difference between a GM-PV on farmland without food production and an elevated SunFarming project with food production.

2007, Italy: In 2007, Hans Günter Czaloun realised Europe's first Agri-PV project based on a suspension cable structure above an apple orchard near Bolzano in South Tyrol [59] see Figure 1.12. Czaloun was not aware of the Goetzberger publication and the SolarSharing idea from Japan, but he worked in the manufacture of ski lifts in Austria and the spanning of long distances with steel cables led him to make double use of the scarce and expensive agricultural land in South Tyrol, Italy.

2009, Germany: In 2004, Manfred Guggenmos, electrician by profession with master craftsman's examination, installed a 70 kWp PV system on the foundation of an old crane in Warmisried, Bavaria. PV modules were mounted at a height of over 3 m. In the years that followed, Guggenmos noticed that grass and herbs grow better under the elevate PV system than in the meadow next to it without partial shading. Guggenmos began to grow vegetables and cereals without the use of machinery below the PV system. In 2009, he approached the local public authority,

Figure 1.12 The first overhead Agri-PV project in Europe implemented by Hans Günter Czaloun in northern Italy in 2007

Figure 1.13 Agri-PV system by Manfred Guggenmos, who planted winter wheat under the solar modules in the fall of 2008, enabling the first state-certified agricultural yields in an Agri-PV project in Germany in 2009

Chamber for Food, Agriculture, and Forestry in Mindelheim to verify the yield of the winter wheat that was sown under his PV system in autumn 2008. The authority confirmed to Guggenmos that the winter wheat yield is estimated at 60 decitons per hectare and a higher yield would not have been expected under comparable conditions outside the PV system [60] (Figure 1.13).

2011, France: 30 years after Goetzberger published his idea of double harvesting, a research group from the field of agroforestry under the leadership of Christian Dupraz re-called Goetzberger's dual harvest idea and implemented the world's first 'Agrivoltaics' research facility at the INRA, Montpellier. Since then, Dupraz supervised several PhD students in plant physiology and transferred his competence from crop cultivation in shading conditions in agroforestry systems to Agri-PV [61] (Figure 1.14).

2011, Germany: After several failed attempts to convince the German government to fund an Agri-PV research pilot project, Eicke R. Weber, then Director of Fraunhofer ISE, decided to use own financial resources of the institute to support an in-house project on the subject of 'double harvesting'. The objective of the project was to simulate and optimise the cultivation of crops by means of photosynthesis with and energetic use of sunlight by means of the photo-electrical effect on the same agricultural area. The process of light- and photon management was registered as an EU patent in 2012 [62] and presented at the EU PV-SEC [63] For the consideration of the boundary effects in this cultivation process, a second patent was filed by Goetzberger in 2014 [64] Stephan Schindele, who from 2011 worked as a doctoral student on the topic 'Solar Technologies within the United Nations Framework Convention on Climate Change (UNFCCC)' was the contact person and coordinator of the Fraunhofer ISE activities in UNFCCC. In November 2011, Schindele represented Fraunhofer ISE at 'The Water, Energy and Food Security Nexus – Solution for the Green Economy' conference in Bonn and met Joachim Sauerborn, back then Dean of the Faculty of Agriculture at University of Hohenheim. Sauerborn was convinced of the usefulness and interdisciplinarity of

Figure 1.14 Agri-PV pilot project from INRA, Montpellier, which was implemented under the direction of Christian Dupraz in 2011

the double-harvesting results and together with Fraunhofer ISE, University of Hohenheim, Institute for Technology Assessment and System Analysis (ITAS) at Karlsruhe Institute of Technology (KIT), BayWa r.e., EWS-Schönau, Demeter-Farming Community, and Regional Association Lake Constance Upper Swabia (RVBO) the APV-RESOLA project in Heggelbach was initiated in 2012 [65]. Between 2012 and 2020 Schindele coordinated Agri-PV activities of Fraunhofer ISE and established the new Agri-PV research area at the institute (Figure 1.15).

2011, Italy: The first utility-scale Agri-PV project was installed by REM TEC Energy in Northern Italy, 2011. The project area is 11.42 hectares and the total PV capacity is 2.15 MWp. REM TEC uses a two-axis tracking system and elevated the PV modules 4.5 m above the ground in this pioneering project [66] (Figure 1.16).

1.2.2 From invention to innovation: 2012–2021

Besides the pioneering countries Germany, Japan, France and Italy, the following markets became early adaptor countries in the field of Agri-PV: China, the USA, the Netherlands and South Korea.

2012–2021, EU and UK: The idea of combining PV projects on agricultural land with sheep grazing, published in their original paper on dual harvest in 1981 by Goetzberger and Zastrow, became a widespread practice in EU and UK between 2012 and 2021. Alone the project developer BayWa r.e. implemented 47 so-called 'SolarGrazing' projects in France, Germany, the Netherlands, the UK and Spain during this period, totalling over 1.211 MWp of PV capacity installed. Unlike in the USA, Japan or China, there is unfortunately no centrally recorded overview of Agri-PV projects in the EU (Figure 1.17).

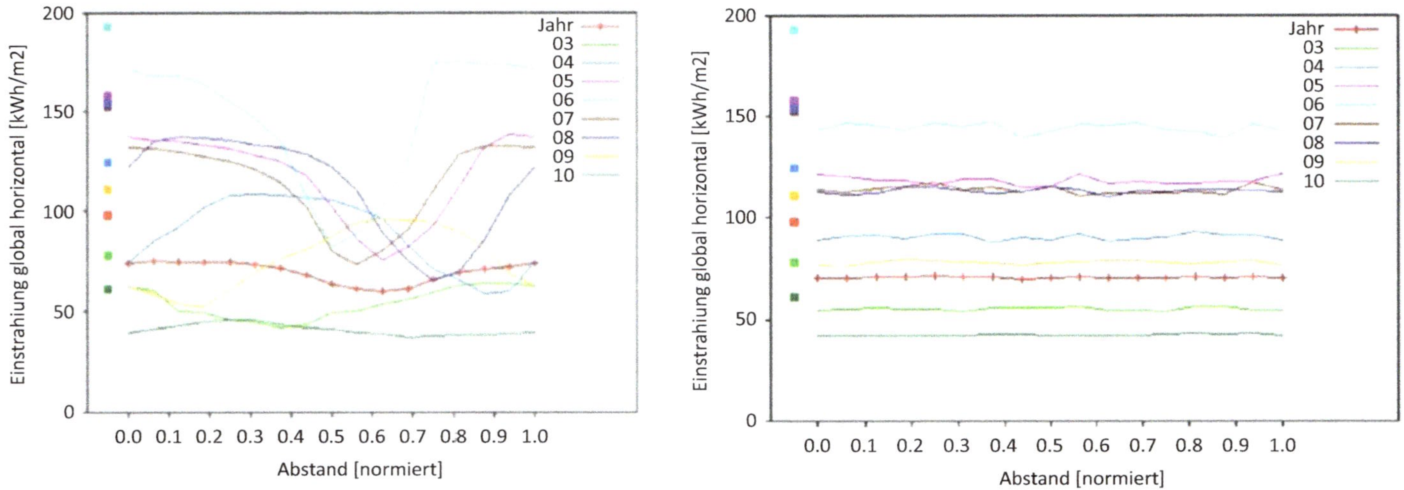

Figure 1.15 Simulation results of the irradiation conditions taking into account different configurations with regard to the sky orientation of the PV modules, their tilt angle, installation height, PV module row spacing for obtaining sufficient and uniform light irradiation in the Agri-PV project. The method for simultaneously cultivating crop plans and utilising the energy of sunlight was registered as an EU patent by Fraunhofer ISE in 2012.

Figure 1.16 The first elevated/overhead multi-megawatt Agri-PV project on arable land installed by the company REM Tec in Italy in 2011

Figure 1.17 Solargrazing projects with sheep in the EU implemented by BayWa r.e. The sheep are looked after by trained shepherds.

2013–2021, Japan: In April 2013, the Japanese Ministry of Agriculture, Forestry and Fisheries approved the installation of PV systems on farmland. Previously, PV installation on farmland was prohibited [67]. Until now, the prerequisite for approval to install a PV system on farmland has been that agricultural activity under the PV system must be guaranteed. The regionally responsible chambers of agriculture require the farmers who farm under the PV system to report their agricultural yields. If, after ten years, the agricultural yield has decreased by more than 20% in comparison with the time before the installation of the PV system, the Agri-PV system must be dismantled, and the remaining ten years of the subsidy period will expire. The law was introduced in response to the nuclear disaster in Fukushima in 2011 and enables farmers to diversify their income by participating in the SolarSharing project. This was to compensate for financial losses in agriculture resulting from the decline in Japanese agricultural exports as a result of the Fukushima disaster. Furthermore, the SolarSharing promotion was intended to counter the rural exodus and the declining number of farms in Japan. Many farmers gave up their farm and sought work in cities. In total, more than 3,000 Agri-PV projects were implemented between 2013 and 2021, representing an estimated total Agri-PV capacity of about 320 MWp [68]. With an average size of approx. 100 kWp per project, Agri-PV or SolarSharing projects in Japan are relatively small and not utility-scale. Yet, the Japanese Agri-PV market has very likely the largest number of projects and with over 150 different crops cultivated in these 3,000 projects, Japanese authorities have a great knowledge base on what crop works well in shading conditions and which crop not so well. So far, only 60 projects did not meet the 80% minimum agricultural yield target.

2014–2021, China: In 2022, the Sino-German Energy Partnership published 'An Overview of the Current Situation of Agrivoltaic Development in China and Some Recommendations' [69]. Between 2014 and 2021 a series of supporting Agri-PV policies were introduced, creating a favourable market environment for Agri-PV. In October 2015, the 'Modern Industrial Alliance on Agrivoltaics' was initiated by the Ministry of Science and Technology, with the objective to integrate agriculture and PV and to implement Agri-PV projects on arable cropland, combined with vegetable planting, fruit and melon planting, herb and flower planting, animal breeding, fishery and greenhouses (glass, helio- and open-type greenhouses). The Agri-PV market has grown since then and even accelerated with the Central Committee of the Communist Party of China and the State Council introducing the Rural Revitalisation Strategy in February 2018. The goal was to combine agriculture, PV and tourism in rural areas and support was granted from the National Energy Administration (NEA) and the Ministry of Agriculture and Rural Development (MARD). Jointly a Plan for Promoting Agricultural and Rural Modernization and the Plan for the Innovation and Development in the Smart PV Industry during the 14th Five-Year Plan Period was introduced. By end of 2021, the 23 provinces and regions in China received a total of 350 fillings for Agri-PV projects, representing 37.6 GWp PV-capacity. The average size of the filled Agri-PV project is thus approximately 107.4 MWp. The largest installed Agri-PV capacity is registered in South China (nearly 19 GWp),

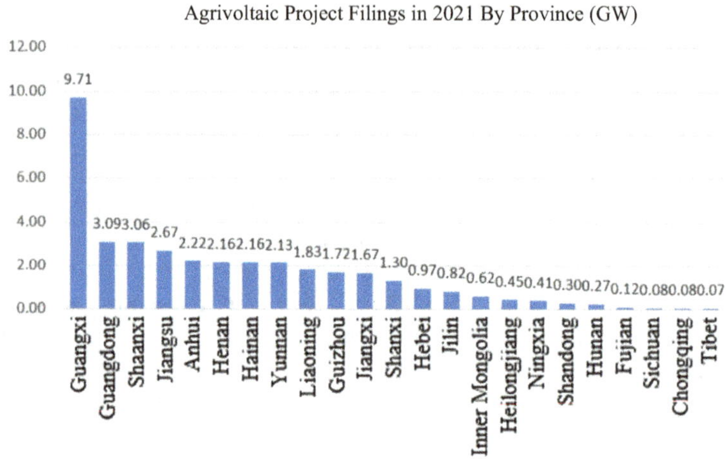

Figure 1.18 Agri-PV project fillings in China by province in GWp by end of 2021

Figure 1.19 The world's largest Fruitvoltaics project with an installed capacity of over one GWp in combination with goji berry production in China

followed by East China (6.9 GWp). Figure 1.18 shows the Agri-PV project filings in 2021 by province in GWp.

Precisely, 22.26 GWp, or 54.6% of the total filled Agri-PV capacity, are hybrid solar farm projects that include vegetable, flower, mushroom, melon or herb planting within the project area. The second largest sub-segment of Agri-PV is hybrid PV-fishery projects, totalling 10,77 GWp (31,4%). The largest Fruitvoltaics project in China is cultivating goji berries below the PV modules. 640 MWp of the PV capacity is already grid-connected and upon completion the project size will be 1 GWp [70,71], see Figures 1.19 and 1.20.

By the end of 2021, a total of 309 GWp of PV capacity was registered in China [72]. The market share of Agri-PV with 37.6 GWp was therefore around 12.1%.

2018–2021, USA: The most widespread Agri-PV application in the USA are so-called 'SolarPollinator' projects, which create a habitat for pollinators and other beneficial insects on the project area while at the same time restoring soils and ecosystems as well as generating PV power. The second largest sub-segment of Agri-PV is SolarGrazing with sheep. NREL is the leader of the Agri-PV research project 'Inspire', through which an Agri-PV project map for the USA is published. This includes 369 Agri-PV projects, of which 293 are SolarPollinator and 73 are SolarGrazing projects. The regional distribution of Agri-PV projects is clustered to North-East and along the east coast, see screenshot of the map [73] (Figures 1.21 and 1.22).

Figure 1.20 Fish production on land (not in the sea) with PV production above the water body in China

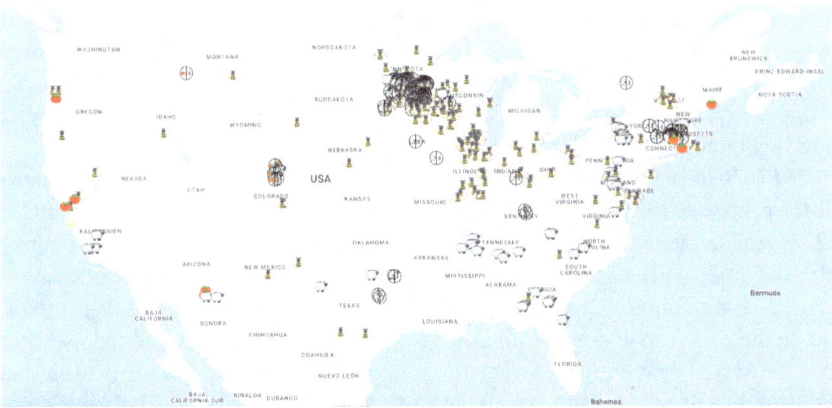

Figure 1.21 Agri-PV Map from NREL in the USA. Agri-PV projects are mainly implemented in the north and east of the USA

Figure 1.22 Solar pollinator projects are considered Agri-PV in the USA because they benefit agriculture and beekeeping is an agricultural activity

The higher Agri-PV market share of SolarPollinator projects is due to legislation promoting pollinator-friendly solar PV projects in 7 US-states by end of 2020 [74,75], showing how important Agri-PV policymaking is on market development. The Agrisolar Clearing House publishes factsheets on SolarPollinator Habitats and is a good source of further information on the topic [76]. The ASGA provides also a map of SolarGrazing projects in the USA [77] and informs publicly and its members on different grazing concepts, vegetation management and health and safety regulations ensuring high animal welfare while at the same time diminish the risk of negative impact from livestock on solar PV yields generated by the project [78]. The first US-state promoting Agri-PV projects with a higher power off take price was Massachusetts in 2018. Since then, the Agricultural Solar Tariff Generation Units Guideline was updated regularly and is active till today [79]. Other Agri-PV lighthouse projects in the US include Jack's Solar Garden [80] and the Agrivoltaics: Food, Energy, and Water Solutions research group within the Biosphere 2 project [81–83] (Figures 1.23 and 1.24).

2017–2020, France: In each of the years 2017, 2019 and 2020, an innovation call for proposals for Agri-PV projects was carried out by the French Government. In 2017, the tender was only open for Agri-PV projects and comprised 15 MWp. In 2019 and 2020, other innovative PV technologies were permitted in addition to Agri-PV, for example, Carport-PV, and the tenders comprised 80 MWp in each case. In the first Agri-PV-tender in 2017, 57 MWp of bids were submitted, of which only 15 MWp were awarded. In the following years, the company Sun'Agri established itself as the main beneficiary of the tenders. In 2020, 31 Agri-PV projects were awarded through the tender, of which 22 projects (71%) with a total

Figure 1.23 Agri-PV lighthouse project 'Jack's Solar Garden' in Colorado, USA

Figure 1.24 Agri-PV research project at the University of Arizona, USA

capacity of 58 MWp were awarded to Sun'Agri. The French innovation tender was aimed at elevated so-called overhead Agri-PV solutions that included a 'dynamic shading' function, which allows turning the PV modules away from the optimum power generation but enables a higher solar irradiation reaching the crop if needed in a certain time of the vegetation period [84]. The number of solar grazing projects are not recorded in France. Today, local agricultural champers have a veto right in the approval process of PV projects on agricultural land. Hence, in France almost all future PV projects in rural areas on farmland must be implemented as Agri-PV projects with the participation of a farmer and an agricultural activity on the project area (Figure 1.25).

Figure 1.25 Agri-PV viticulture project of the Agri-PV pioneer Sun Agri in France

2015–2021, Germany: In 2016, Fraunhofer ISE was finally able to inaugurate its first Agri-PV research and demonstration project on agricultural land belonging to the Demeter farming community of Heggelbach in southern Germany. On this day, Goetzberger said that it sometimes takes a long time for an idea to be realised and that he is glad that the double harvest, as his last unrealised vision, is now finally being tested in a real environment. The need for an Agri-PV quality standard arose in one of the group discussions that Schindele executed with representatives of the Ministry of Agriculture, the Ministry of Education and Research, the University of Hohenheim and Fraunhofer ISE in Berlin on 26 June 2018. The APV-RESOLA project turned into a lighthouse project in Germany and EU and received several awards. Before the APV-RESOLA project was terminated in July 2021, Fraunhofer ISE published in October 2020 Agri-PV guidelines with the title 'Agri-Photovoltaics: Opportunity for agriculture and the energy transition'. This guideline contains the most important scientific results for practitioners from the target groups of farmers, energy project developers, politics and research and development [65] (Figure 1.26).

The APV-RESOLA project seemed to have struck a chord and received a lot of attention in society and science. Supported by the federal state of Rhineland-Palatinate and the Federal Ministry of Agriculture, Fraunhofer ISE was able to implement the first Agri-PV research facility in the field of special crops in Gelsdorf in 2021. Eight different apple varieties are being analysed under three different crop protection systems: Hailnet, plastic cover and Fruitvoltaics, which at the same time supplies the electric Fendt tractor and the water irrigation system with solar power [85] see Figure 1.27.

While the two Fraunhofer ISE research projects were elevated 'overhead' Agri-PV projects, the solar industry was already implementing the first lower-cost

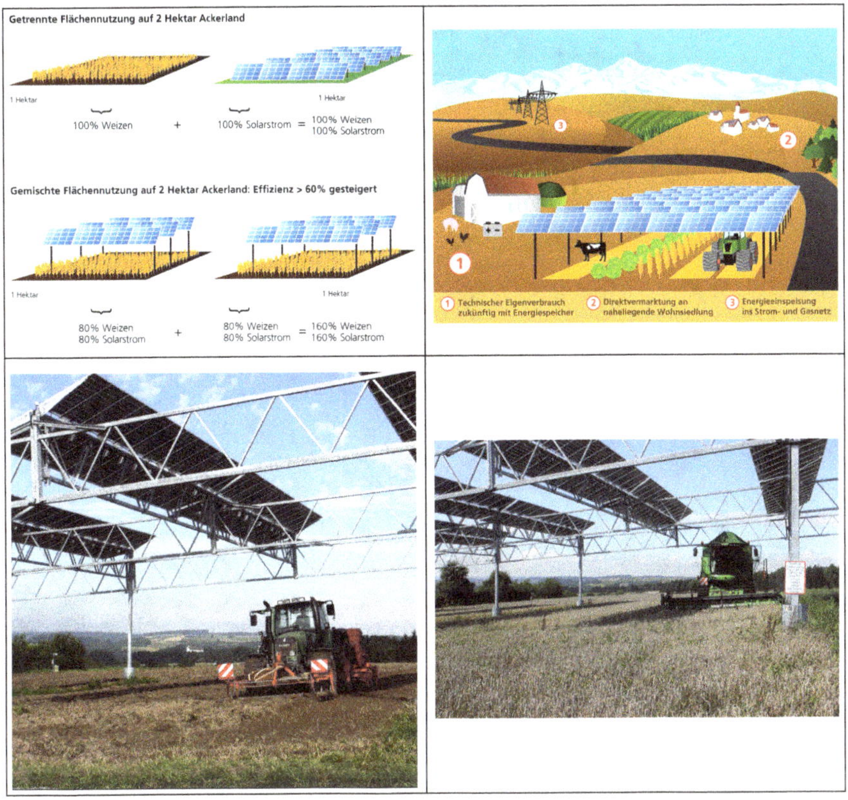

Figure 1.26 Graphics and impressions of the APV-RESOLA project of Fraunhofer ISE in Germany

'inter-row' Agri-PV projects in Germany. The company Next2Sun developed vertical, fixed-tilt installed inter-row Agri-PV systems in which the PV module rows run north-south, and the bifacial modules are oriented east-west [86], see Figure 1.28.

In 2019, BayWa r.e. set up the first subsidy-free utility-scale PV project in northern Germany as SolarGrazing with sheep [87], see Figure 1.29.

From 2021 onwards, Agri-, Floating- and Carport-PV were included as an innovation tender within Germany's Renewable Energy Act (EEG), resulting in 12 Agri-PV projects and a total of 21 MWp, which will need to be implemented in Germany by 2024. There was not a single Floating-PV, and only one Carport-PV project also receiving funding from this tender [88].

2017–2021, Netherlands: From 2017 to 2021, a generous support mechanism called 'Stimulation of Sustainable Energy Production and Climate Transition (SDE)' enabled the implementation of several innovative Agri- and Floating-PV projects in the Netherlands. Floating-PV technology made its breakthrough in the EU in the Netherlands. The Netherlands also achieved great success in the research

Figure 1.27 Fruitvoltaics project with apples, where the electric tractor of Fendt/ AGCO is used and the fruit farm is supplied exclusively with solar power in Gelsdorf, Germany

and commercialisation of several Agri-PV projects in combination with permanent crops such as berries, stone and pome fruit. In particular, the Fruitvoltaics project in Babberich received worldwide attention because it tested the optimisation of light transmitting PV modules in practice for the first time in 2019. The results were promising, and the berry producer decided to convert half of its fruit farm to an Agri-PV/Fruitvoltaics system. The Dutch government supports the Sunbiose project, which conducts scientific research to accompany several Agri-PV projects in the Netherlands, including the site in Babberich [89] (Figure 1.30).

2015–2021, new projects in new markets: In addition to the significant Agri-PV market growth in China, Japan, the USA and the EU, the first Agri-PV projects were implemented in Australia, Chile, South Korea and Taiwan between 2015 and 2021. In Australia, the first solar grazing project with sheep was implemented in

Figure 1.28 *Inter-row Hayvoltaics project of the Agri-PV Pioneer Next2Sun in Germany*

Figure 1.29 *Germany's first subsidy-free PV project was implemented by Agri-PV pioneer BayWa r.e. as solar grazing with sheep on permanent grassland in northern Germany in 2019*

2015. Since then, over a dozen such utility-scale projects have been installed [90]. Fraunhofer Chile together with Fraunhofer ISE tested three small-scale (13 kWp each) Agri-PV system in the Chilean communities El Monte, Curacavi and Lampa. The research project was sponsored by the Santiago de Chile Metropolitan Regional Government [91] As far as we know, these were the first Agri-PV systems in South America (Figure 1.31).

Figure 1.30 First commercial Fruitvoltaics project with berries implemented by GroenLeven/BayWa r.e. in the Netherlands in 2020

Figure 1.31 First Agri-PV project in South America installed by Fraunhofer Chile in 2018

Between 2016 and 2019, 14 Agri-PV projects were implemented in South Korea. Most of the projects were test and demonstration projects by energy companies or universities. The technical implementation, the social circumstances of an ageing society and high average age of farmers and very low availability of agricultural land per person as well as the project sizes between 70 and 100 kWp are very similar to the SolarSharing approaches from Japan [92,93] (Figure 1.32).

1.2.3 From innovation to diffusion: 2022–2041 and beyond

Future market growth of PV globally and its demand for agricultural land by the end of 2100: At the end of 1994, 0,4 GWp of PV capacity was installed worldwide, representing 0,0001% of the global share of electricity production. By the end of 2022, this figure had already risen to 1 TWp and a share of 4,5% of

Figure 1.32 Solar sharing project in Korea above a rice field

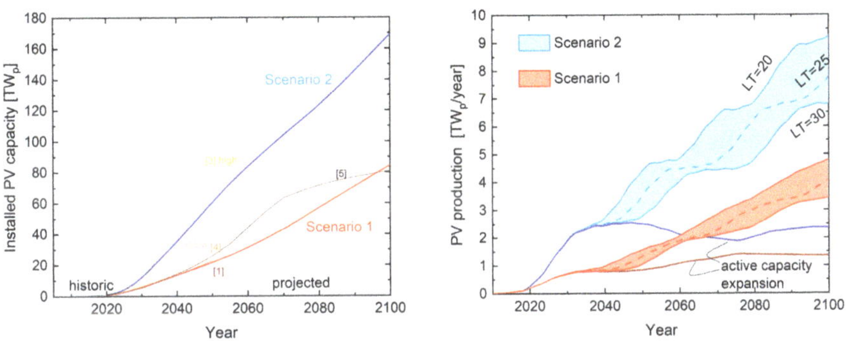

Figure 1.33 Modelled installed PV capacity at end of year for two scenarios in comparison to scenarios from literature. Depending on the scenario assumptions, the global actively operating PV capacity is reaching 80–170 TWp in the year 2100 [95].

global power production [37]. The rapid growth of PV implementation will continue in the coming decades if the 1,5-degree targets of the Paris Climate Agreement are to be met. The International Renewable Energy Agency (IRENA) assumes that around 14 TWp of PV capacity will be installed worldwide in 2050 [94]. Scientists at Fraunhofer ISE and the Potsdam Institute for Climate Impact Research (PIK) assume that between 20 and 80 TWp of PV capacity will be installed in 2050 and between 80 and 170 TWp in 2100 [95] (Figure 1.33).

In political scenarios, PV expansion targets were often underestimated. Current trends of sector coupling by electrifying the transport and heat sectors lead to an increasing electricity demand and thus require a larger PV expansion than assumed in the IRENA publication. The lower assumption of 20 TWp by 2050 from the Fraunhofer ISE and PIK study is therefore used as a conservative assumption for PV expansion by 2050, representing 40–50% of global power production. Furthermore, it is assumed that only 50% of the PV power is implemented on agricultural land by means of conventional utility-scale GM-PV projects, which corresponds to 10 TWp. The other 50% or 10 TWp will be implemented as BiPV on rooftops, as Floating-PV or on conversion and non-agricultural land, for example desert areas. Although there will certainly be a technical learning process in the following years, we still assume that 1 ha is needed for the implementation of 1 MWp GM-PV. 10 TWp corresponds to 10,000 GWp or 10,000,000 MWp or thus 10,000,000 ha or 100,000 km^2 of agricultural land, which is roughly equivalent to the total land area of the three countries Netherlands, Belgium and Switzerland. Assuming that half of the PV capacity at the end of 2022 has been installed on agricultural land, this would be ceteris paribus approx. 0.5 TWp, or 500 GWp or 500,000 MWp or 500,000 ha. The difference between the end of 2022 and the end of 2050 is therefore 9,500,000 ha. Including leap years, 10,227 days are available for implementation. The daily agricultural land use by GM-PV is therefore about 930 ha. The average utility-scale GM-PV project size is assumed to be 30 ha. Each day for 28 consecutive years, about 31 GM-PV projects are thus built.

The impact of Agri-PV diffusion on farmland preservation can also be assessed by how many of these 31 GM-PV projects can be replaced by Agri-PV, thereby preserving farming activity on the project area.

The EU's goal is to establish a circular land economy by 2050 [96]. This means that from 2051 onwards, no further agricultural land may be used for the designation of new traffic infrastructure, industrial land, residential land and GM-PV expansion on agricultural land. To achieve this goal, the daily increase in settlement area must be reduced by 2050 so that it falls to net zero in 2050. The German government has introduced this goal of a circular land economy in its national sustainability strategy with the KPI 11.1.a Sustainable Land Use, and in its climate action plan [27,44]. Sooner or later, estimated from 2040 onwards, the expansion of GM-PV on agricultural land would therefore have to be scaled down, at least in the EU. From 2051 onwards, only Agri-PV projects could be installed on agricultural land and the GM-PV market segment would come to a standstill on agricultural land in the EU.

Goldschmidt *et al.* from Fraunhofer ISE and PIK assume that at least 80 TWp of PV capacity will be needed by the end of 2100, representing more than 60% of global power production. Starting from 2023 until the end of 2100, ceteris paribus, about 40 TWp or 400,000 km^2 of agricultural land would be taken up by GM-PV. This would roughly correspond to the total land area of the six countries Netherlands, Belgium, Switzerland, Tunisia, South Korea and Costa Rica. The agricultural land demand would then amount to 1.386 ha per day for 78 consecutive years from 2023 onwards, or the implementation of 46 GM-PV projects per day. Table 1.2 summarises these findings.

The 46 utility-scale PV projects that would have to be installed on agricultural land every day for the next 78 years always mean balancing public interests in rural

Table 1.2 Global demand for agricultural land to install PV capacity until the end of the century

		Year			
		1994	**2022**	**2050**	**2100**
Global PV capacity	GWp	0,4	1.000	20.000	80.000
Global GM-PV capacity on farmland	GWp	0	500	10.000	40.000
Global farmland area demanded by GM-PV	Ha	0	500.000	10.000.000	40.000.000
Share of power production globally	%	0,001	4,5	40-50	>60

Figure 1.34 EWS Sonnenfeld, Cropvoltaics project in Bruck an der Leitha, Austria

areas at the individual and municipal level. Because the world's population and prosperity, and thus the demand for food, will continue to grow over the same period, it is to be expected that the potential for conflict in PV implementation on agricultural land will most likely increase rather than decrease. Governments in the current key markets for Agri-PV, including China, the USA and major EU agricultural producers such as France, Germany, Italy and the Netherlands, are therefore very likely to expand rather than restrict their Agri-PV support regimes. At the same time, new Agri-PV markets will be opened. Since the beginning of 2022, the following governments have been discussing the adoption of Agri-PV support instruments or have already implemented them: Israel [97,98], Luxemburg [99], Croatia [100], Czech Republic [101] and Austria [102]. In addition, further Agri-PV pilot projects have been implemented in Kenya [103], Spain [104] and Switzerland [105]. In France [106], Germany [107] and Austria [108,109]), the first inter-row 'Cropvoltaics' projects have been successfully tested, see Figures 1.34–1.36.

In Bavaria, Germany, the first 'Cowvoltaics' project has been implemented by PV developer 'OneSolar' [110], see Figure 1.37.

Figure 1.35 Total Energy, Cropvoltaics project in France

Figure 1.36 Doppelernte, Cropvoltaics project in Germany

Agri-PV is in the early stages of diffusion and continues to require a great deal of policy learning, policy coordination and intersectoral cooperation. It is not the technical and economic frameworks that stand in the way of Agri-PV diffusion, but often the legal ones. Whether policymakers recognise the contribution of Agri-PV to the preservation of agricultural activity in rural areas and thus show a willingness to promote it depends on the public's crisis awareness regarding food security and climate protection. Good political communication on why Agri-PV can contribute to increase the level of national sustainability is therefore of great importance. At

Figure 1.37 OneSolar, First Cowvoltaics project in Germany

the same time, the GM-PV sector must accept that it is not enough to produce only green energy. In a complex environment, thinking and acting should be cross-sectoral. The societal demands on PV projects in their immediate environment will increase and so will the demand for multifunctionality. The food-energy-water-biodiversity nexus in the Agri-PV project increases land-use efficiency and acceptability among affected citizens. Agri-PV is as diverse as agriculture is. For PV project developers, this diversity is both a challenge and an opportunity. For a successful Agri-PV diffusion, capacity building in PV companies is necessary and industry decision makers need to work hand in hand with the capacity builders to ensure that Agri-PV projects are successfully implemented in harmony with agriculture and environmental protection. The introduction of Agri-PV project implementation requires the adaptation of previous technical and legal standards and the individual functions in project development and project implementation of the operational business must learn to meet the requirements of agricultural activity in the project area. Change management, organisational learning and WeQ concepts combined with new tools that support the Agri-PV project implementation process will be crucial for Agri-PV diffusion. Agri-PV market research assumes the global market value of Agri-PV at 3.6 billion US dollars in 2021. This market value is projected to increase to 9.3 billion US dollars in 2031, growing at a CAGR of 10.1% from 2022 to 2031 [111]. These figures underpin that the future of utility-scale PV on agricultural land is Agri-PV.

References

[1] Food and Agriculture Organization of the United Nations. Land use statistics and indicators: Global, regional and country trends 1990–2019. 2019 [cited 2023 Nov 7]. Available from https://www.fao.org/3/cb6033en/cb6033en.pdf.

[2] FAO. *The State of Food and Agriculture: Making Agri-Food Systems More Resilient to Shocks and Stresses*. Rome: Food & Agriculture Org; 2021 [cited 2022 May 17]. Available from https://www.fao.org/3/cb4476en/cb4476en.pdf.

[3] FAO. *FAOSTAT: Land Use*: Food and Agriculture Organization of the United Nations; 2022 [cited 2022 May 17]. Available from https://www.fao.org/faostat/en/#data/RL.

[4] FAO. *The State of the World's Forests 2020: Forest, Biodiversity and People*. Rome: FAO; 2020. (*State of the World's Forests*; Vol 2020) [cited 2020 May 17].

[5] FAO. *The State of the World's Land and Water Resources for Food and Agriculture: Systems at Breaking Point*; Synthesis Report. Rome; 2021 978-92-5-135327-1 [cited 2022 May 16]. Available from https://www.fao.org/documents/card/en/c/cb7654en.

[6] Rockström, J., Steffen, W. L., Noone, K., *et al. Planetary Boundaries: Exploring the Safe Operating Space for Humanity*. In Ecology & Society Vol. 14 (No. 2), Article Art. 32. [cited 2024 April 7] Available from htps://www.ecologyandsociety.org/vol14/iss2/art32/

[7] World Bank. *Population, Total*. Data; 2022 [cited 2022 May 17]. Available from https://data.worldbank.org/indicator/SP.POP.TOTL.

[8] Steensland A. *Global Agricultural Productivity Report: Strengthening the Climate for Sustainable Agricultural Growth*; 2021 GAP Report. 2021 [cited 2022 May 16]. Available from https://globalagriculturalproductivity.org/wp-content/uploads/2021/10/2021-GAP-Report.pdf.

[9] European Commission. *EDGAR - The Emissions Database for Global Atmospheric Research. EDGAR-FOOD [A global emission inventory of GHGs from the food systems];* 2022 [cited 2022 May 16]. Available from https://edgar.jrc.ec.europa.eu/edgar_food.

[10] Schmidhuber J. *Die Rückkehr der Missernten. Der Spiegel;* 2021 [cited 2022 May 16]. Available from https://www.genios.de/presse-archiv/artikel/SPIE/20211224/die-rueckkehr-der-missernten/PMG4SPIEGEL-Heftimport-SP20211224-38831_b7a6ed78-6334-412e-a59e-716a644df5f1.html.

[11] FAO. *Making Integrated Food-Energy Systems Work for People and Climate: An Overview*. FAO Books; Rome, Italy, 2014.

[12] EIT Food. *New Coalition Announces Bold Plan to Decarbonize Europe's Food System*. EIT Food iVZW; 04.05.2021; 2021 [cited 2022 May 16]. Available from https://www.eitfood.eu/news/post/new-coalition-announces-bold-plan-to-decarbonize-europes-food-system.

[13] Handley L. Walmart has a grand plan to help suppliers club together to buy green energy. *CNBC* 2020 Oct 23 [cited 2022 Sep 2]. Available from https://www.cnbc.com/2020/10/23/walmart-wants-suppliers-to-buy-renewable-energy-collectively.html.

[14] Vos R. Is global food security jeopardised by an old age timebomb? *The Guardian* 2014 Feb 4 [cited 2023 Nov 7]. Available from https://www.theguardian.com/global-development-professionals-network/2014/feb/04/global-food-security-old-age-timebomb.

[15] Ritchie H and Roser M. *Urbanization: Our World in Data*. 2018 [cited 2023 Nov 7]. Available from https://ourworldindata.org/urbanization#what-share-of-people-will-live-in-urban-areas-in-the-future.

[16] Spiegel D. *Geleakter Teil des IPCC-Berichts: Die reichsten zehn Prozent verursachen mehr als ein Drittel der Treibhausgase*. Der Spiegel 2021 Aug 13 [cited 2022 May 17]. Available from https://www.spiegel.de/wissenschaft/mensch/geleakter-teil-des-ipcc-berichts-die-reichsten-zehn-prozent-verursachen-mehr-als-ein-drittel-der-treibhausgase-a-6d2e8438-7f3e-49bc-8967-4e3aeb6be585.

[17] Scientist Rebellion. We leaked the upcoming IPCC report!. 2021 [cited 2022 May 17]. Available from https://scientistrebellion.com/we-leaked-the-upcoming-ipcc-report/.

[18] IEA. *An Energy Sector Roadmap to Carbon Neutrality in China*. International Energy Agency; 2021 [cited 2023 Nov 7]. Available from https://iea.blob.core.windows.net/assets/9448bd6e-670e-4cfd-953c-32e822a80f77/AnenergysectorroadmaptocarbonneutralityinChina.pdf.

[19] European Commission. National energy and climate plans: EU countries' 10-year national energy and climate plans for 2021–2030. 2022 [cited 2022 May 17]. Available from https://ec.europa.eu/info/energy-climate-change-environment/implementation-eu-countries/energy-and-climate-governance-and-reporting/national-energy-and-climate-plans_de.

[20] United States Executive Office of the President. *The Long-Term Strategy of the United States: Pathways to Net-Zero Greenhouse Gas Emissions by 2050*. United States Department of State; United States Executive Office of the President; 2021 [cited 2023 Nov 7]. Available from https://www.whitehouse.gov/wp-content/uploads/2021/10/US-Long-Term-Strategy.pdf.

[21] European Commission. *A European Green Deal: Striving to be the First Climate-Neutral Continent*. 2022 [cited 2022 May 16]. Available from https://ec.europa.eu/info/strategy/priorities-2019-2024/european-green-deal_en.

[22] Prognos, GWS and Fraunhofer ISI, IINAS. *Energiewirtschaftliche Projektionen und Folgeabschätzungen 2030/2050: Gesamtdokumentation der Szenarien; Bericht im Auftrag des Bundesministeriums für Wirtschaft und Energie*. 2021 [cited 2022 May 17]. Available from https://www.bmwk.de/Redaktion/DE/Publikationen/Industrie/energiewirtschaftliche-projektionen-und-folgeabschaetzungen-2030-2050.pdf?__blob=publicationFile&v=32.

[23] Die Bundesregierung. *Entwurf eines Gesetzes zu Sofortmaßnahmen für einen beschleunigten Ausbau der erneuerbaren Energien und weiteren Maßnahmen im Stromsektor: Drucksache 20/1630*. Deutscher Bundestag 20. Wahlperiode; 2022 May 2 [cited 2022 Nov 14]. Available from https://dserver.bundestag.de/btd/20/016/2001630.pdf.

[24] BMF. *Stellungnahme zur Taxonomie: Eingangsbemerkung*. 2022 Jan 21 [cited 2022 May 17]. Available from https://www.bundesfinanzministerium.de/Content/DE/Downloads/Europa/stellungnahme-zur-taxonomie.pdf?__blob=publicationFile&v=4.

[25] BMWK. 'Kommission Wachstum, Strukturwandel und Beschäftigung'. BMWi; 2019 [cited 2022 May 17]. Available from https://www.bmwk.de/Redaktion/DE/Downloads/A/abschlussbericht-kommission-wachstum-strukturwandel-und-beschaeftigung.pdf?__blob=publicationFile&v=4.
[26] SPD, Bündnis 90/Die Grünen und FDP. *Mehr Fortschritt wagen: Bündnis für Freiheit, Gerechtigkeit und Nachhaltigkeit*. 2021 [cited 2022 May 13]. Available fromhttps://www.spd.de/fileadmin/Dokumente/Koalitionsvertrag/Koalitionsvertrag_2021-2025.pdf.
[27] BMWK. *Klimaschutzplan 2050*. Bundesministerium für Wirtschaft und Klimaschutz; 2022 [cited 2022 May 16]. Available from https://www.bmwk.de/Redaktion/DE/Artikel/Industrie/klimaschutz-klimaschutzplan-2050.html.
[28] Pepe J M. Der Ukraine-Krieg und seine Folgen: Deutschland muss seine Energietransformation neu austarieren. 2022 [cited 2022 May 18]. Available from https://www.swp-berlin.org/publikation/der-ukraine-krieg-und-seine-folgen-deutschland-muss-seine-energietransformation-neu-austarieren.
[29] Goldammer K, Mans U and Rivera M. *Beiträge zur sozialen Bilanzierung der Energiewende*. Potsdam; 2013 [cited 2022 May 18]. Available from https://www.iass-potsdam.de/sites/default/files/files/report_beitraege_zur_sozialen_bilanzierung_der_energiewende_0.pdf.
[30] Kühne O and Weber F, eds. *Bausteine der Energiewende*. Wiesbaden: Springer Fachmedien Wiesbaden; Imprint; Springer VS; 2018. (RaumFragen: Stadt - Region - Landschaft).
[31] Böhm J and Tietz A. *Abschätzung des zukünftigen Flächenbedarfs von Photo-voltaik-Freiflächenanlagen*. Thünen Working Paper 204. Braunschweig; 2022.
[32] Böhm J. Wie viele landwirtschaftlichen Flächen sind bereits durch PV-Freiflächenanlagen aus der Produktion genommen? 2022 Jan 4 [cited 2022 May 27]. Available from https://www.dlg-mitteilungen.de/fileadmin/img/content/cover/heft/2022/22-02/DLG0222_Flaechennutzung_PV-FFA_Boehm.pdf.
[33] Eichenauer E, Reusswig F, Meyer-Ohlendorf L and Lass W. Bürgerinitiativen gegen Windkraftanlagen und der Aufschwung rechtspopulistischer Bewegungen. In: Kühne O and Weber F, eds. *Bausteine der Energiewende*. Wiesbaden: Springer Fachmedien Wiesbaden; Imprint; Springer VS; 2018. pp. 633–651 (RaumFragen: Stadt - Region - Landschaft) [cited 2022 May 18]. Available from https://link.springer.com/chapter/10.1007/978-3-658-19509-0_32.
[34] Die Bundesregierung. *Beschleunigter Ausbau: Mehr Windenergie auf See*. 2022 [cited 2022 May 18]. Available from https://www.bundesregierung.de/breg-de/themen/klimaschutz/windenergie-auf-see-gesetz-2022968.
[35] BMWK. Photovoltaik-Strategie: Handlungsfelder und Maßnahmen für einen beschleunigten Ausbau der Photovoltaik; 2023 May 5.
[36] Die Bundesregierung. Antwort der Bundesregierung: auf die Kleine Anfrage der Abgeordneten Stephan Protschka, Peter Felser, Franziska Gminder, weiterer Abgeordneter und der Fraktion der AfD – Drucksache 19/11956 –;

Direkte und indirekte Landnutzungsänderungen. 2019 [cited 2020 Feb 3]. Available from https://kleineanfragen.de/bundestag/19/12697.

[37] IEA. Solar PV Power Capacity in the Net Zero Scenario, 2015–2030. Paris; 2023 [cited 2023 Sep 7]. Available from https://www.iea.org/data-and-statistics/charts/solar-pv-power-capacity-in-the-net-zero-scenario-2015-2030.

[38] IEA. *Solar PV: Recommendations, Policy Makers, Facilitate Permitting for Utility-Scale Systems*. International Energy Agency; 2023 [cited 2023 Nov 17]. Available from https://www.iea.org/energy-system/renewables/solar-pv.

[39] Fraunhofer ISE. *Integrated Photovoltaics – Areas for the Energy Transformation*. Fraunhofer Institute for Solar Energy Systems ISE; 2019 [cited 2023 Oct 31]. Available from https://www.ise.fraunhofer.de/en/key-topics/integrated-photovoltaics.html#expand-all.

[40] Rolink D. Steuerfalle: 'Die Freiflächenanlage wird für uns zur Steuerbombe'. *top agrar online* 2021 Sep 22 [cited 2022 May 27]. Available from https://www.topagrar.com/energie/news/steuerfalle-die-freiflaechenanlage-wird-fuer-uns-zur-steuerbombe-12691115.html.

[41] Finanzbehörden der Länder. Zurechnung und Bewertung von Agri-Fotovoltaik-Anlagen: Bewertungsgesetz; BStBl I 1226; 2022 Jul 15 15.

[42] SolarPower Europe. AgriSolar. 2023. Available from https://agrisolareurope.org/.

[43] BMEL. *Daten und Fakten: Land-, Forst- und Ernährungswirtschaft mit Fischerei und Wein- und Gartenbau [Datum01.07.2022]*. Bundesministerium für Ernährung und Landwirtschaft; 2022 [cited 2023 Nov 17]. Available from https://www.bmel.de/SharedDocs/Downloads/DE/Broschueren/daten-fakten-2022.pdf.

[44] Die Bundesregierung. German Sustainable Development Strategy: Update 2021; 2020 [cited 2023 Nov 20]. Available from https://www.bundesregierung.de/resource/blob/974430/1940716/4bdf89ceea3b1e4367918384b8839a37/2021-07-26-gsds-en-data.pdf?download=1.

[45] Scharpf F W. *Planung als politischer Prozess: Aufsätze zur Theorie der planenden Demokratie*. 1. Aufl. Frankfurt am Main: Suhrkamp; 1973. (Theorie) [cited 2022 May 18].

[46] Mayntz R and Scharpf F, eds. *Planungsorganisation: Die Diskussion um die Reform von Regierung und Verwaltung des Bundes*. München: R. Piper & Co. Verlag; 1973. (Piper-Sozialwissenschaft Reader zur Politologie, Soziologie und Ökonomie; vol 17).

[47] DIN. DIN SPEC 91434:2021-05: Agri-Photovoltaik-Anlagen - Anforderungen an die landwirtschaftliche Hauptnutzung. DIN; 2022 [cited 2022 Apr 26]. Available from https://www.beuth.de/de/technische-regel/din-spec-91434/337886742.

[48] Schindele S. Nachhaltige Landnutzung mit Agri-Photovoltaik: Photovoltaikausbau im Einklang mit der Lebensmittelproduktion: Szenarioanalyse zur Inanspruchnahme landwirtschaftlicher Nutzflächen durch Photovoltaik in Deutschland bis 2050. *GAIA - Ecological Perspectives for Science and Society*. 2021 [cited 2022 Apr 27];30(2):96–105. Available from https://www.oekom.de/_files_media/zeitschriften/artikel/GAIA_2021_02_96.pdf.

[49] Schindele S. *Nachhaltige Landnutzung Dank Doppelernte: Eine Mehrdimensionale Politikanalyse der Agri-Photovoltaik-Diffusion in Deutschland [Dissertation]*. Tübingen, Germany: Universität Tübingen; 2023.
[50] World Economic Forum. *Putting Solar Panels On Farms Makes Sheep Happier*. 2023.Geneve, Switzerland, [cited 2024 April 11]; Available from htps://www.weforum.org/videos/putting-solar-panels-on-farms-makes-sheep-happier/
[51] Schindele S. Feldfrüchte und Strom von Agrarflächen: Was ist Agri-Photovoltaik und was kann sie leisten? *GAIA – Ecological Perspectives for Science and Society*. 2021 [cited 2022 Apr 26];30(2):87–95. Available from https://www.ingentaconnect.com/content/oekom/gaia/2021/00000030/00000002/art00007.
[52] Rogers E M. *Diffusion of Innovations. 5th edn*. New York: Free Press; 2003. Available from https://ebookcentral.proquest.com/lib/kxp/detail.action?docID=4935198.
[53] Ortwin R. *Methoden und Verfahren der Technikfolgenabschätzung und der Technologiebewertung*. undefined 1982.
[54] UNDP. *Handbook for Conducting Technology Needs Assessment for Climate Change*. 2010. New York, USA, [cited 2024 April 11]; Available from htps://www.undp.org/sites/g/files/zskgke326/files/publicaons/Technology_Needs_Assessment_Handbook.pdf
[55] Goetzberger A and Zastrow A. Kartoffeln unter dem Kollektor. *Sonnenenergie*. 1981. Nürnberg, Germany, Publisher Präsidium der DGS, Berlin, Germany, [cited 2022 Apr 25];(3):19–22. Available from https://produktdatenbank.innovationsgruppen-landmanagement.de/kartoffeln-unter-dem-kollektor.
[56] Goetzberger A and Zastrow A. On the coexistence of solar-energy conversion and plant cultivation. *International Journal of Solar Energy*. 1982 [cited 2022 Apr 25];1982(1):55–69. Available from DOI:10.1080/01425918208909875.
[57] Movellan J. Japan next: Generation farmers cultivate crops and solar energy. *Renewable Energy World*. 2013 Oct 10 [cited 2023 Oct 31]. Available from https://www.renewableenergyworld.com/solar/japan-next-generation-farmers-cultivate-agriculture-and-solar-energy/#gref.
[58] Goetzberger A, Swanson R, Werner T and Yamaguchi M. Energy farming: Für einen intelligenteren Gebrauch von Agrarsubventionen. *Solarzeitalter*. 2006;2006(4):44–48.
[59] Leitner H and Czaloun G. *Innovatives Seil-Tragsystem für Anwendungen in der Agro-Photovoltaik*. 2018. (15. Symposium Energieinnovation) [cited 2022 Apr 27]. Available from https://www.tugraz.at/fileadmin/user_upload/Events/Eninnov2018/files/lf/Session_C6/461_LF_Leitner.pdf.
[60] AELF Krumbach (Schwaben)-Mindelheim. Versuchsanbau (Schauversuch) von Winterweizen unter der Photovoltaik-Anlage in Warmisried; Besichtigung am 02.09.2009; Brief vom 04.09.2009 [Brief]; 2009 [cited 2022 Apr 25]. Available from http://elektro-guggenmos.de/uploads/media/Bescheinigung_Landwirtschaftsamt_Mindelheim_-_Versuchsanbau_Doppelernte.pdf.
[61] Dupraz C, Marrou H, Talbot G, Dufour L, Nogier A and Ferard Y. Combining solar photovoltaic panels and food crops for optimising land use:

Towards new agrivoltaic schemes. *Renewable Energy*. 2011 [cited 2022 Apr 27];36(10):2725–2732. Available from https://www.sciencedirect.com/science/article/abs/pii/S0960148111001194#!

[62] Bopp G, Goetzberger A, Obergfell T and Reise C. *Verfahren zur simultanen kultivierung von nutzpflanzen und energetischen nutzung von Sonnenlicht: Patent Nr. EP2811819B1*. Fraunhofer-Gesellschaft; 2012. Available from https://patents.google.com/patent/EP2811819B1/de.

[63] Beck M, Bopp G, Goetzberger A, Obergfell T, Reise C and Schindele S. Combining PV and Food Crops to Agrophotovoltaic – Optimization of Orientation and Harvest [5 pages / *27th European Photovoltaic Solar Energy Conference and Exhibition*; 4096-4100] 2012.

[64] Goetzberger A. *Solarmodulanordnung mit reduzierten Randeffekten und Verwendung der Solarmodulanordnung zur simultanen Kultivierung von Nutzpflanzen und energetischer Nutzung von Sonnenlicht: Patent Nr. DE102014218458A1*. Deutsches Patent- und Markenamt: Fraunhofer-Gesellschaft; 2014 [cited 2022 Apr 25]. Available from https://patents.google.com/patent/DE102014218458A1/ja.

[65] Fraunhofer ISE. Agri-Photovoltaik: Chance für Landwirtschaft und Energie: Ein Leitfaden für Deutschland. 2020 [cited 2022 Apr 26]. Available from https://www.ise.fraunhofer.de/content/dam/ise/de/documents/infomaterial/brochures/photovoltaik/APV-Leitfaden_2020_web_neu.pdf.

[66] REMTEC. *CASTELVETRO Agrovoltaico plant*. Piacenza, Italy; 2011 [cited 2020 Jan 7]. Available from http://www.remtec.energy/en/agrovoltaico-2/castelvetro-plant/.

[67] PRI. *Japanese Farmers Producing Crops and Solar Energy Simultaneously: Revitalizing the Farming Sector for the Next-Generation and Could Provide more than Enough Electricity for the Entire Country*. [Restriction to solar PV installation on farmlands removed.]. Permaculture Research Institute; 2013 [cited 2022 May 30]. Available from https://permaculturenews.org/2013/10/16/japanese-farmers-producing-crops-solar-energy-simultaneously/.

[68] Sugibuchi K. *Solar Sharing: Symbiosis of PV and Agriculture in Japan*. Munich; 2019. (Agrovoltaics – Sharing Resources for Multiple Benefits), Presented at Intersolar Munich, May 15 2019 [cited 2022 May 25]. Available from https://www.rts-pv.com/en/news/201905_7655/.

[69] GIZ. *An Overview of the Current Situation of Agrivoltaic Development in China and Some Recommendations: Sino-German Energy Partnership*. The Sino-German Energy Partnership; 2022.

[70] Bellini E. Giant agrivoltaic project in China. *PV Magazine*. 2020 Sep 3 [cited 2022 May 13]. Available from https://www.pv-magazine.com/2020/09/03/giant-agrivoltaic-project-in-china/.

[71] Bellini E. Chinese fish pond hosts 550 MW solar farm. *PV Magazine*. 2022 Jan 7 [cited 2023 Oct 31]. Available from https://www.pv-magazine.com/2022/01/07/chinese-fish-pond-hosts-550-mw-solar-farm/.

[72] Shaw V and Hall M. Chinese PV industry brief: China added 53 GW of new PV capacity in 2021. *PV Magazine*. 2022 [cited 2023 Nov 7]. Available from

https://www.pv-magazine.com/2022/01/21/chinese-pv-industry-brief-china-added-53-gw-of-new-pv-capacity-in-2021/.

[73] NREL. Agrivoltaics Map: InSPIRE Project. 2023 [cited 2023 Nov 19]. Available from https://openei.org/wiki/InSPIRE.

[74] Terry G. *State Pollinator-Friendly Solar Initiatives*. Clean Energy States Alliance; 2020 [cited 2023 Nov 7]. Available from https://www.cesa.org/wp-content/uploads/State-Pollinator-Friendly-Solar-Initiatives.pdf.

[75] Davis R. Bee-Friendly Solar is a Win-Win-Win for People, Planet and Pollinators. 2019 [cited 2023 Nov 7]. Available from https://www.organicvalley.coop/blog/bee-friendly-solar-is-win-win-win-for-people-planet-pollinators/.

[76] Walston LJ and Hartmann H. *The Science of Solar-Pollinator Habitat: Fact Sheet*. Agri Solar Clearing House; 2023 [cited 2023 Nov 7]. Available from https://www.agrisolarclearinghouse.org/the-science-of-solar-pollinator-habitat-fact-sheet-2/.

[77] ASGA. Solar Grazing Map USA; 2020 [cited 2023 Nov 7]. Available from https://solargrazing.org/map/#.

[78] ASGA. Solar-Multiuse-Farming-Practices-Fact-Sheet-2019-v3 2019.

[79] DOER MA. *Dual-Use Shading Analysis Tool*. Massachusetts Department of Energy Resources; 2022 . Boston, Massachusetts, United States of America, [cited 2022 Sep 1]. Available from https://massgov.github.io/DOER/doer.html.

[80] Jack's Solar Garden Developments, LLC. Jack's Solar Garden. 2022 [cited 2023 Nov 7]. Available from https://www.jackssolargarden.com/.

[81] Brean H. *University of Arizona Researchers Unveil New Model for Desert Farming in Warming World*. Tucson; 2020 [cited 2023 Nov 19]. Available from https://tucson.com/news/local/university-of-arizona-researchers-unveil-new-model-for-desert-farming-in-warming-world/article_2c9ecfdf-9128-53c2-b3c9-dd08eba3a832.html.

[82] University of Arizona. *Biosphere 2 Honors Teams: Vertically Integrated Projects*. 2022 [cited 2023 Nov 7]. Available from https://uavip.arizona.edu/teams/biosphere-2-honors-teams.

[83] Allen G and Hickman T. Why the ground under Colorado solar panels is ripe for growing food: A Republican lawmaker has joined the previously Democrat-only push for solar agriculture, or agrivoltaics, priming Colorado to become a national leader. *The Colorado Sun*. 2023 [cited 2023 Nov 19]. Available from https://coloradosun.com/2023/01/22/colorado-solar-agriculture-agrivoltaics-lawmakers/.

[84] Sun'Agri. *The World's First Agrivoltaic Power Plant*. Domaine de Nidolères; 2021 [cited 2022 Apr 27]. Available from https://sunagri.fr/en/project/nidoleres-estate/.

[85] BayWa r.e. Erste Agri-PV-Anlage für Äpfel in Deutschland errichtet. 2021 [cited 2022 Apr 27]. Available from https://www.baywa-re.com/de/news/details/erste-agri-pv-anlage-fuer-aepfel-in-deutschland-errichtet.

[86] next2sun. Agriculture and Livestock Farming. 2022 [cited 2022 Nov 14]. Available from https://www.next2sun.de/en/our-concept/#agriculture.

[87] BayWa r.e. First subsidy-free solar park in Germany: 05.09.2019. 2019 [cited 2022 May 31]. Available from https://www.baywa-re.com/en/news/details/first-subsidy-free-solar-park-in-germany.

[88] Neumann H. Innovationsausschreibung: 12 Agri-PV-Projekte erhalten Zuschlag. *top agrar online* 2022 [cited 2023 Nov 19]. Available from https://www.topagrar.com/energie/news/innovationsausschreibung-12-agri-pv-projekte-erhalten-zuschlag-13091209.html.

[89] Sunbiose. Sunbiose is a 4 year project for the development of Agri-PV solutions. 2023 [cited 2023 Nov 7]. Available from https://sunbiose.nl/.

[90] Clean Energy Council. Solar and agriculture could be a perfect match. 2023 [cited 2023 Nov 7]. Available from https://www.cleanenergycouncil.org.au/news/agrisolar.

[91] Fraunhofer ISE. *Agrophotovoltaik goes global: von Chile bis Vietnam.* Fraunhofer-Institut für Solare Energiesysteme ISE; 2018. Freiburg im Breisgau, Germany, [cited 2022 May 27]. Available from https://www.ise.fraunhofer.de/de/presse-und-medien/presseinformationen/2018/agro-photovoltaik-goes-global-von-chile-bis-vietnam.html.

[92] Minsu K, Oh S-Y and Jung J H. *History and Legal Aspect of Agrivoltaics in Korea*. AIP; 2022 [cited 2023 Nov 19]. Melville, New York, United States of America. Available from https://pubs.aip.org/aip/acp/article/2635/1/050001/2830610/History-and-legal-aspect-of-agrivoltaics-in-Korea.

[93] Yoon-seo L. Growing crops under solar panels: Korea tests agrivoltaic farm. *The Korea Herald*. 2022 [cited 2023 Nov 7]. Available from https://www.koreaherald.com/view.php?ud=20220509000749.

[94] IRENA. *World Energy Transitions Outlook 2023*. International Renewable Energy Agency; 2022. Abu Dhabi, United Arab Emirates, [cited 2023 Nov 8]. Available from https://www.irena.org/Digital-Report/World-Energy-Transitions-Outlook-2023.

[95] Goldschmidt J C, Wagner L, Pietzacker R and Friedrich L. Technological learning for resource efficient terawatt scale photovoltaics. Energy and Environmental Science 2021 Royal Society of Chemistry [cited 2023 Nov 7];14:5147–5160. Available from htps://publications.pik-potsdam.de/rest/items/item_26513_1/component/file_26533/content.

[96] European Commission, Communication from the Commission to the European Parliament, The Council, The European Economic and Social Committee and the Committee of the Regions, Roadmap to a Resource Efficient Europe, SEC 2011, 1067 and 1068 final, 15–16, Brussels, Belgium, Sept. 20 2011, [cited 2024 April 11]. Available from htps://eur-lex.europa.eu/legal-content/EN/TXT/PDF/?uri=CELEX:52011DC0571.

[97] Bellini E. Israeli government wants to boost development of agrivoltaics. *PV Magazine*. 2021 [cited 2023 Nov 7]. Available from https://www.pv-magazine.com/2021/02/01/israeli-government-wants-to-boost-development-of-agrivoltaics/.

[98] Bellini E. Israel launches tender for 100 MW of agrivoltaics. *PV Magazine*. 2022 Jan 14 [cited 2022 May 13]. Available from https://www.pv-magazine.com/2022/01/14/israel-launches-tender-for-100-mw-of-agrivoltaics/.

[99] Enkhardt S. Luxembourg allocates 52.7 MW in agrivoltaic tender. *PV Magazine*. 2023 [cited 2023 Nov 7]. Available from https://www.pv-magazine.com/2023/10/05/luxembourg-allocates-52-7-mw-in-agrivoltaic-tender/.

[100] Maisch M. Croatia adopts legal framework for agrivoltaics. *PV Magazine*. 2023 [cited 2023 Nov 7]. Available from https://www.pv-magazine.com/2023/07/28/croatia-adopts-legal-framework-for-agrivoltaics/.

[101] Ceenergy News. Czechia to further support agrivoltaics. *Ceenergy News*. 2022 [cited 2023 Nov 7]. Available from https://ceenergynews.com/renewables/czechia-to-further-support-agrivoltaics/.

[102] Mariacher A, Musch R and Zapfl W. Agri-PV: der Vorteil liegt im Doppelpack. *Ikonline*. 2023 [cited 2023 Nov 20]. Available from https://stmk.lko.at/agri-pv-der-vorteil-liegt-im-doppelpack+2400+3853040.

[103] Kamadi G. Kenya to use solar panels to boost crops by 'harvesting the sun twice'. *The Guardian*. 2022 Feb 22 [cited 2023 Nov 7]. Available from https://www.theguardian.com/global-development/2022/feb/22/kenya-to-use-solar-panels-to-boost-crops-by-harvesting-the-sun-twice.

[104] Sanchez Molina P. BayWa, Velux sign Europe's first agrivoltaics-linked PPA. 2022 [cited 2023 Nov 20]. Available from https://www.pv-magazine.com/2022/11/16/baywa-velux-sign-europes-first-agrivoltaics-linked-ppa/.

[105] Diermann R. Schweizer Pilotversuch mit Agri-Photovoltaik bringt gute Himbeer-Ernte. 2023 [cited 2023 Nov 20]. Available from https://www.pv-magazine.de/2023/07/05/schweizer-pilotversuch-mit-agri-photovoltaik-bringt-gute-himbeer-ernte/.

[106] Gwenaelle D. TotalEnergies announces results of vertical agrivoltaic pilots in France. *PV Magazine*. 2023 [cited 2023 Nov 20]. Available from https://www.pv-magazine.com/2023/01/16/totalenergies-announces-results-of-vertical-agrivoltaic-pilots-in-france/.

[107] Steinhart C. Erste Agri-Photovoltaik-Anlage steht in Merching. *Augsburger Allgemeine*. 2022 [cited 2023 Nov 7]. Available from https://www.augsburger-allgemeine.de/themenwelten/leben-freizeit/Oekostrom-Ackerbau-Merchinger-Betrieb-entwickelt-erste-Agri-Photovoltaik-Anlage-id65601471.html.

[108] EWS Consulting GmbH. *EWS Sonnenfeld – Bruck/Leitha (NÖ), Agri-PV Referenzanlage*. EWS Consulting GmbH; 2022. Munderfing, Austria, [cited 2023 Nov 7]. Available from https://www.ews-sonnenfeld.com/referenzprojekte/detail/ews-sonnenfeld-bruck-an-der-leitha.

[109] RWA. Öko-Solar-Biotop Pöchlarn: Biodiversität und Photovoltaik vereint auf einer Fläche. 2021 [cited 2023 Nov 20]. Available from https://www.rwa.at/oeko-solar-biotop-poechlarn-biodiversitaet-und-photovoltaik-vereint-auf-einer-flaeche+2500+1014863.

[110] OneSolar International GmbH. *Spatenstich Freiflächenanlage Pfeffenhausen/AGRAR-PV*. OneSolar International GmbH; 2021. Eching, Germany, [cited 2023 Nov 7]. Available from https://www.onesolar.de/spatenstich-freiflaechenanlage-pfeffenhausen-agrar-pv/.

Chapter 2

To farm or not to farm, that is the question: definition and potentials of agrivoltaics

Stephan Schindele[1], Alice Lefort[1], Luca Schler[1], Maximilian Tegtmeyer[1], Erion Bousi[1] and Constantin Klyk[2]

> "In the end, only the technology that promises the highest net benefit will prevail."
>
> —Ortwin Renn [1]

Basically, this chapter aims to answer two questions: First, what is a globally and universally valid definition for Agricultural-Photovoltaics (Agri-PV or APV)? Second, which types of Agri-PV applications have how much area and PV capacity potential worldwide?

To answer the first question, a brief theoretical introduction to the international standards for the development of a definition or new terminology is given as well as information on why a clear definition of Agri-PV is of importance to the industry, society, and policymakers alike. This is followed by a literature review of existing scientific contributions and national Agri-PV standards to the classification and definition of Agri-PV. As a result, some key indicators and parameters are evaluated that have been used as Agri-PV quality attributes and differentiation criteria from conventional ground-mounted Photovoltaics (GM-PV) GM-PV applications. Finally, a proposal for a globally and universally valid Agri-PV definition is made, and the vast diversity of Agri-PV applications is highlighted.

The answer to the second question is worked out by breaking down the agricultural land potential into hectares (ha) worldwide and per region. The respective agricultural land use is linked to the installed PV capacity per hectare of the specific Agri-PV application. Finally, the results for both are presented: the hectare and the Agri-PV capacity potential in GWp (DC) worldwide and per region. These results are put into relation with the currently installed PV capacity globally and the international PV expansion targets until 2050, which are necessary for climate protection and an energy supply system based entirely on renewable energies.

[1]BayWa r.e. AG, Arabellastraße 4, München, Germany
[2]Agri-PV Consulting, Jörg-Zürn-Straße 22, 88339 Bad Waldsee, Germany

2.1 Definition of agrivoltaics and its distinction from conventional ground-mounted photovoltaics

The basic idea of Agri-PV is already recognizable in its terminology: Agri- and Photovoltaics. Agri-PV combines agricultural production (Agri-) with the generation of solar power (Photovoltaics or PV) on one and the same agricultural land area. Even though the basic idea is simple, there are several international terminologies for Agri-PV: the originator of the Agri-PV idea, Adolf Goetzberger, called it "dual harvest," "sun farming," or "energy farming" [2,3]. Later, Fraunhofer ISE, which was founded by Goetzberger, initiated the terminology "Agro-Photovoltaics" [4], which was based on the already existing terminologies agroforestry, agrofuels (such as E5, E10, biodiesel and biogas), and agroecology. In France, Christian Dupraz and his team established the terminology "agrivoltaics," which is today's terminology-standards in science [5]. In Japan, the idea of Agri-PV was socially and politically implemented as "SolarSharing" [6], and in China the regulatory framework dealing with Agri-PV describes the dual harvest approach as "PV+" [7]. In Italy the terms "Agrofotovoltaico," "Agrivoltaici," or "Agrivoltaico" [8] are common.

Not only the difference in terminology is remarkable, but also the variety and differences in the indicators of how Agri-PV is defined in national quality standards that are enshrined in laws and regulations, for example in Germany [9], Italy [10], and France [11]. The included quality criteria in these national standards, e.g. of expected minimum agricultural yield varies from 66% (in Germany) to 80% (in France and Japan), maximum land loss varies from 15% (in Germany) to 30% (in Italy), and some standards categorically distinguish between highly elevated Agri-PV applications, so-called "Overhead," where the agricultural activity is carried out under the PV modules, and so-called "Inter-row" Agri-PV applications, which are installed close to the ground and the agricultural production takes place between the PV module rows. The intensity of the respective criteria is subject to a certain political arbitrariness and seems to depend on how technology-open, innovation-friendly, or interest-driven the standardization and legislation processes were. The uncertainty of a common concept and understanding of Agri-PV leads to the expression of doubts and concerns in society [12], which in turn leads to ambiguity and hesitant political action [13–15]. Before entering the substantive debate on which indicators and terminology are considered most useful, the following section looks at the theoretical basis of standards in the definition of terminologies and why a clear definition of Agri-PV is of importance—especially for policymakers.

2.1.1 Theoretical background on development of international standards and why a clear definition of Agri-PV is of importance for cross-sectorial policy-making

Background on development of international standards: Most nations have a standardization organization. These national standards organizations control and coordinate official standardization processes. In Germany, standardization work is coordinated by the German Institute for Standardization (DIN). In France, standardization work is organized by the Association Française de Normalization (AFNOR) and in the USA by the American National Standard Institute (ANSI). The standardization process itself is a joint effort by a group of experts and competence carriers from industry, science, authorities, associations, consumers, or other groups of interest. The goal of the standardization process is to develop a new standard or to check if a new standard on the same topic has been developed elsewhere, e.g. on regional level like the EU or internationally. In case a standard had been developed already, the standard committee will evaluate if the existing standard can be adopted nationally, too, or to what extend the existing standards needs adjustments. Tangible and intangible products and services can be standardized. Standardization is often utilized if identical or similar objects are used in many markets by different groups of people. The new standard aims to avoid technical obstacles within the circle of interested parties, both nationally and internationally. Unification supports the exchange of goods and services, increases the usability, compatibility, and safety in the use of the products and services. The standardization procedure is structured in several stages. First, the object of standardization is described and evaluated if a similar standard already exists. If yes, the initiating group of experts is linked to the already established standard committee to further proceed in a joint effort. If the object can be sufficiently distinct from already available standards, a new standardization process on the topic is initiated. Second, a broad participation of all interested and affected parties is recommended as this will ensure the acceptance and applicability of the new standard. The standard committee will start drafting the new standard. Third, once a final draft of the new standard is available it is usually put up for public comment and objection procedure. Fourth, incoming suggestions will be considered and analyzed by the steering committee, and if necessary, incorporated into a new version of the draft standard. These steps are repeated until a satisfactory status is reached and no more significant objections are received. Fifth, the result of the standard procedure is documented as a new "standard" or "norm" and is made available to the public. All standards are developed in a multi-stage procedure and in a democratic manner. A new standard is finalized in accordance with consensus principle. It is not the standardization organization that is standardizing, but the participating experts. The introduction and establishment of a new standard is a lengthy process and can take up to multiple years. Formal, international settings have therefore come under criticism [16]. National standardization organizations, such as the British Standards Institution (BSI) have answered with the introduction of a new "pre-standardization" process called "Publicly Available Specification (PAS)." The PAS-system provides a standardizing framework, which supports mainly emerging technologies. It offers a "fast-track" definition phase and decisions are taken by absolute majority amongst the group of experts, rather than by the principle of consensus. The pre-standard offers an "effective means of quickly introducing standardization," allows lower transaction costs for all stakeholders involved, and can serve as a basis "for further standardization at [. . .] a European or International level" [17].

Standardization is driven by interested parties, mainly by industry and trade organizations. It is therefore considered a field of industrial policy. The intrinsic motivation of participating parties in the standardization committee is to set technical standards faster, more flexible, and with more expertise than a governmental body could do it. It therefore has a liberalization and deregulatory effect and relieves the burden on the state government of policy coordination. Standardization and pre-standardization make the economy function more efficiently, supports the free movement of products, services, and markets, and encourages companies to be innovative. Standards and pre-standards themselves have no legal binding force. They are an output of voluntary work and have more the character of recommendations which everyone is free to adapt and observe. However, they might become legally binding if legislators include them into laws and regulations. Especially when policymakers govern new technologies and lack the competence to define terms and conditions of the object to be regulated, referencing the standard or pre-standard reduces the ambiguity of the law [18].

Reducing ambiguity: why a clear definition of agrivoltaics is of importance for cross-sectorial policy-making: In Chapter 1, it was outlined why an effective and efficient Agri-PV diffusion in harmony with farming and nature conservation depends on interdisciplinary collaboration in policymaking. Particularly the farming, energy, climate, environment, building, and finance ministries are asked to coordinate and synchronize their political actions to promote Agri-PV. Yet, at different stages in the innovation process, different departments in the executive ministerial administration are responsible for the cross-cutting issue of Agri-PV. In the early phase of market introduction, research, pilot and demonstration projects support evidence-based policymaking and the ministry of education and science is responsible for the innovation steps from idea to invention. In the market introduction phase, when the techno-economic feasibility and proof of concept for Agri-PV has been established and the public visibility of the technology rises, demand for Agri-PV projects increases and responsibility for the coordination of the Agri-PV innovation processes is handed over from ministry of education and science to the leading ministries in agriculture, environment/soil protection, and energy.

As the Agri-PV market evolves, the number of actors engaged in cross-collaboration in policy making increases, too. This increases the complexity of negotiations as the interests of more stakeholders must be satisfied. At this stage, there is a risk that the regulatory framework is not in balance and the efficiency of market introduction is decreased due to over-controlling, too many reporting obligations, and the introduction of new quality control criteria that make it unattractive to invest into Agri-PV. Policymakers are challenged to find a compromise that is liberal enough to attract investment while at the same time avoiding the risk of misuse of incentives. For the transfer of knowledge from Agri-PV research and pilot projects to the market introduction with the help of a holistic Agri-PV policy-regime, an Agri-PV quality standard is needed to reduce the ambiguity and risk of misguided incentives. Figure 2.1 illustrates the Agri-PV innovations process and illustrates where an Agri-PV standardization process is needed in policy

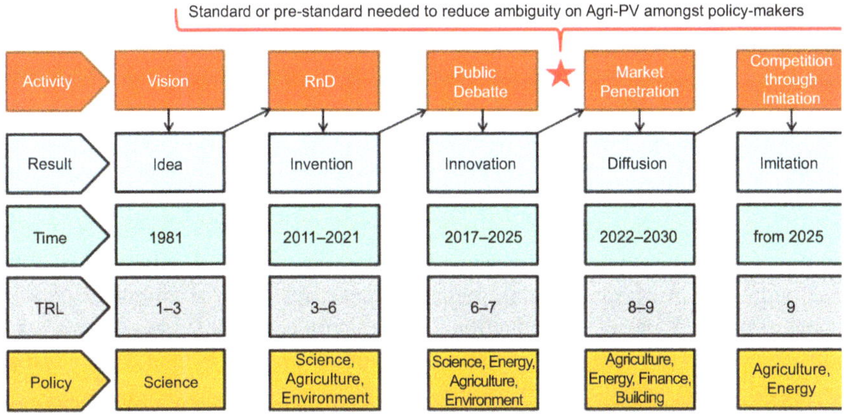

Figure 2.1 Agri-PV innovation process and indication at what stage a standard or pre-standard is needed to reduce ambiguity on Agri-PV amongst policy-makers (own illustration)

coordination. Figure 2.1 also captures the outcomes of each innovation stage, the corresponding timeline, technology-readiness level (TRL), and the policy areas involved in Agri-PV diffusion.

In the context of the Agri-PV market introduction, the "Ambiguity-Conflict Model" of Richard E. Matland [19] can be used to answer two questions: First, what is the level of ambivalence and the resulting level of conflict between the different departments/ministries in the federal administration? Second, is there a similar idea and understanding amongst political decision-makers on the topic that needs regulation—in this case the market introduction of the innovative multi-functional land use configuration "Agri-PV."

The **level of ambivalence** expresses whether the envisaged policy implementation in the impacted departments has a high or low risk of conflict with already existing targets. If the policy implementation leads to a worsening of other lines of action and objectives of the department, the ability to make a positive decision is assessed as conflictual. The decision-making power and willingness to compromise on policy implementation are thus largely limited. If no setbacks are expected, but a positive net-benefit for the ministry, the ideal-typical constellation assumes a low-conflict implementation and thus a positive decision.

The **level of ambiguity** expresses if policy-makers have a similar or different understanding related to the policy object and the means of how to achieve a certain goal. If there are contradictory perceptions in the coordination process in relation to the policy goals and the policy implementation, the inefficiency and the risk of a negative decision against the policy initiative increases. Hence, the reduction of ambiguity leads to more efficient coordination and increases the likelihood of a willingness to compromise. Matland's "ambiguity-conflict model" sees an increase in complexity of policy implementation as the number of actors involved increases.

To answer the first question on the level of ambivalence and conflict regards Agri-PV diffusion, it is important to politically scrutinize how big the socio-political pressure is to solve the problem? Is there a crisis awareness among the population of the challenges that political action is trying to address? Through the war in Ukraine, societal awareness has been increased in multiple ways, all addressing core Agri-PV topics. It has come ever clearer how little room there is for (political) maneuver in global food supply chains, how relevant it is to protect and utilize existing fruitful soil to battle the global food security crises, and how essential it is to become more independent from fossil fuel supplies—especially when supplied out of an autocratic regime. Hence, the level of ambivalence with respect to Agri-PV market introduction and diffusion amongst leading ministries is relatively low, as the topic of food security, climate protection, and energy transition are rather high on the global political agenda currently.

The second question focusing on the level of ambiguity and insecurity regarding Agri-PV and how Agri-PV can be distinct from GM-PV is not so easy to answer and is therefore expressing the need for an Agri-PV standard or pre-standard.

Matland subdivides his ambivalence-ambiguity hybrid model into four ideal-typical constellations of policy implementation: (I) Administrative implementation if there is low conflict and low ambiguity; (II) Political implementation if there is high conflict, but low ambiguity; (III) Symbolic implementation with high conflict and high ambiguity, and finally, (IV) Experimental implementation with low conflict, but high ambiguity is the fourth constellation we find for the Agri-PV market introduction.

In "experimental implementation" processes, there is extensive political agreement that policy implementation should take place and the policy objective is clear. The departments/ministries involved want to proactively contribute to policy implementation, but there are opposing views on the terminology and the subject matter to be regulated, as well as the availability of resources and the timetable have not yet been clearly defined. In this constellation, policy implementation requires support from science and industry to reduce ambiguity. In the negotiation system, the question is not whether the policy measure should be implemented, but rather answers are needed, what exactly is being regulated, how and when political action is needed. Standardization and consultation phases on policy drafts in the legislative process can close information gaps and make intersectoral compromises visible. In such context, ambiguity is seen as an opportunity to get to know one another, and the coordination process requires the awareness that shared learning is the goal.

Due to the relatively high political pressure to act regarding food security, climate and soil protection as well as the expansion of PV, there are only minor conflicts between the ministries involved in the coordination process for Agri-PV market introduction and diffusion. However, there is no common agreement within the solar industry, as well as between the solar and agricultural sectors, on what exactly Agri-PV is and how it should be introduced to the market in the most effective and efficient way. Legally binding standardization or pre-standardization can reduce the ambiguity between the interest groups and thus support the transition from an "experimental implementation" with low risk and high ambiguity to an "administrative implementation" with low risk and low ambiguity. If Agri-PV is to become the new industry standard, a strong political and intersectoral effort is needed at an early stage of market introduction. Once the Agri-PV market matures, the political efforts can be reduced, resulting in incremental policy-learning and path dependencies, leading to a continuous learning-by-doing on project and technical level, further decreasing cost of Agri-PV, thus increasing demand while at the same time lowering the risk of social dis-acceptance of the policy. Looking back to the background to the chapter on development of international standards, governments have increasingly requested experts from research, solar industry, and farming to develop their own national Agri-PV pre-standard according to the PAS-procedure, such as the three following:

1. Germany: DIN SPEC 91434:2021-05
2. Italy: Comitato Elettrotecnico Italiano (CEI), PAS 82-93 [10]
3. France: AFNOR, Label Projet Agrivoltaïque [11]

Agri-PV is therefore a perfect example of cross-collaboration, mutual learning, and why pre-standards developed within PAS-procedures are an effective tool to support Agri-PV market introduction.

2.1.2 Case study Germany: background information on the origination and making of the German DIN SPEC 91434 Agri-PV pre-standard

The first Agri-PV pre-standard that was developed according to the PAS-procedure was the DIN SPEC 91434 in Germany. In the following, some background information on this process will be explained. Details on some challenges will be given and topics explained on where more attention should have been paid during the development of the pre-standard. In addition, new findings and positions have been released since the publication of the pre-standard in May 2021 and some quality criteria should be updated to integrate these into the Agri-PV pre-standard in Germany. Suggestions for improvement also result from the fact that the number of Agri-PV stakeholders has increased significantly in recent years. Criteria for compromise and consensus should therefore be adapted so that the broadest possible support can be given to an updated version of DIN SPEC 91434. The idea to take the German DIN SPEC 91434 as a case study is to increase the transparency of how such a process works, to explain how complex and multilayered the standardization process is, and to allow other markets and regions to benefit from these lessons learned when initiating their own Agir-PV pre-standardization process.

The German DIN SPEC 91434 Agri-PV pre-standard was an output of the inter- and transdisciplinary "APV-RESOLA" project, which was funded by the German Federal Ministry of Education and Research (BMBF) between March 2015 and July 2021. The APV-RESOLA project had five main areas of work: Agriculture, Environment, Technology, Society and Politics and Economics. In the context of the Policy and Economy area, a policy analysis and several expert interviews and group discussions were conducted to evaluate the introduction of an Agri-PV market. The need for an Agri-PV standard arose in a group discussion between representatives of the Ministry of Agriculture, the Ministry of Education and Research, the University of Hohenheim and Fraunhofer ISE in Berlin on 26 June 2018. Schindele, who was the project coordinator of APV-RESOLA and leader of the Politics and Economics area initiated and moderated the discussion. When the topic of quality assurance of Agri-PV projects was analyzed, MinR Dr. Wolf Junker, Head of the Resources, Recycling Management, and Geo Research Division at BMBF suggested to initiate a DIN SPEC. MinR Niendieker, Head of the Energy Affairs and Bioenergy Division at the Ministry of Agriculture welcomed this approach. Two prerequisites were definite for the funding of the DIN SPEC on Agri-PV: first, the APV-RESOLA project consortium had to include interested experts from outside the APV-RESOLA project to widen the research scope from a case study with regional context to a national level. The aim was to transfer knowledge and research results into a national expert environment. Second, the DIN SPEC should be initiated as soon as possible. In summer 2019, the workplan of DIN SPEC 91434 was published. Sixteen organizations with voting rights decided to participate in the process. Other organizations and ministries decided to take part in the process but preferred to be without voting rights. On 19 December 2019, the inter- and transdisciplinary DIN SPEC 91434 consortium kicked-off the first Agri-PV pre-standardization process globally. In 2020, the German Government decided to implement support policies for the multifunctional PV applications Floating- and Agri-PV within the innovation auction that was part of the amendment of the Renewable Energy Act (EEG) 2021. The Federal Network Agency (BNetzA), which is coordinating the PV auctions, was asked to regulate and publish the definition of Agri-PV by June 2021. The DIN SPEC 91434 was published just a month prior to this date in May 2021, which seemed to be just the right time. The regulations on the innovation tender for special and innovative solar PV systems included a reference to DIN SPEC 91434, which made the fulfillment of the Agri-PV pre-standard a legally binding prerequisite to become an Agri-PV project eligible for funding within the innovation auction. Also in 2021, DIN SPEC 91434 was integrated into the Direct Payment Implementing Regulation, making the agricultural activity in the Agri-PV project eligible to receive 85% of the agricultural subsidies, 85% only because DIN SPEC 91434 stipulates as a quality feature of Agri-PV the mandatory requirement that a maximum of 15% of the agricultural area may be taken away from agricultural activity for the installation of the Agri-PV system on the project area.

The aim of DIN SPEC 91434 was to define quality criteria for Agri-PV projects and to distinguish Agri-PV from conventional ground-mounted PV. The avoidance of ambiguity regarding the Agri-PV technology was seen as a basic prerequisite for a political handling and consideration of Agri-PV in existing support schemes. DIN SPEC 91434 defines Agri-PV as follows: "Agri-Photovoltaics: combined use of one and the same area of land for agricultural production as the main use and for electricity generation by means of a PV system as secondary use" [9]. Hence, the PV-system is adapted to such a way that agricultural production is technically and economically feasible in the project area. DIN SPEC 91434 defined two Agri-PV application categories: Category 1, Overhead, is defined by elevated PV-modules and a clearance height equal or above 2.10 m, where the agricultural production takes place below the PV-modules, see Figure 2.2.

Category 2, Inter-row, projects are installed closer to the ground, where the agricultural production takes place in between the PV module rows and PV-modules are mounted with clearance height <2.10 m, see Figure 2.3.

For both DIN SPEC 91434 Agri-PV categories, different quality criteria were defined, e.g. the loss of farmland due to substructures, foundation, fences, transformer station, and other project related hardware must not exceed 10% of the total project area for Category 1 and a maximum of 15% for Category 2. Regardless of the Agri-PV category, at least 66% of the agricultural reference yield after the Agri-PV installation must be generated on the project area. An external expert must assess the possibility of an economically viable agricultural activity in the Agri-PV project. This confirmation can be requested during the building permit procedure by the lower building authority. At the federal level it is sufficient if in the application for funding within the EEG's innovation tender coordinated by BNetzA the proof of the municipal decision to set up the Agri-PV project according to the DIN SPEC 91434 pre-standard with a copy of the expert report confirming the

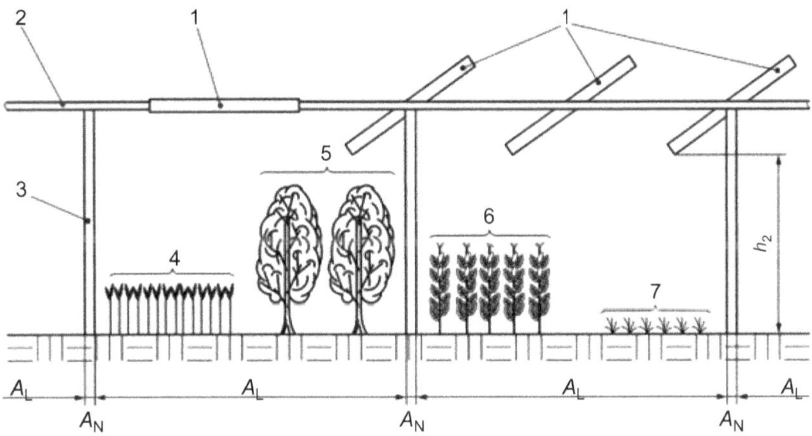

Key

A_L	agriculturally usable area
A_N	agriculturally usable area
h_2	clear height over 2, 10 m
1	examples of solar modules
2	strut
3	mounting structure
4 to 7	example of agricultural crops

Figure 2.2 DIN SPEC 91434, Illustration of Agri-PV Category 1, Overhead, and examples of different agricultural crops cultivated below the PV modules [9]

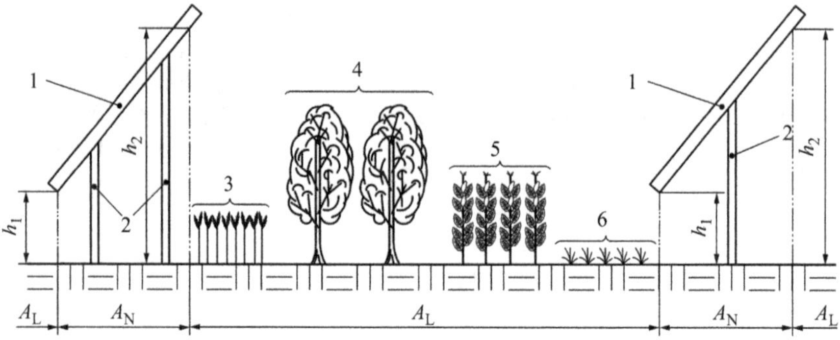

Key

A_L agriculturally usable area
A_N agriculturally unusable area
h_1 clear height under 2, 10 m
h_2 clear height over 2, 10 m
1 examples of solar modules
2 mounting structure
3 to 6 examples of agricultural crops

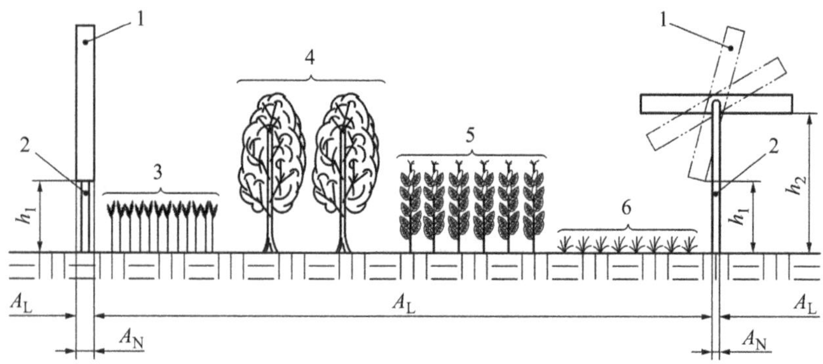

Key

A_L agriculturally usable area
A_N agriculturally unusable area
h_1 clear height under 2, 10 m
h_2 clear height over 2, 10 m
1 examples of solar modules
2 mounting structure
3 to 6 examples of agricultural crops

Figure 2.3 DIN SPEC 91434, Illustration of Agri-PV Category 2, Inter-row, and examples of different agricultural crops cultivated between the PV modules [9]

agricultural concept of the Agri-PV project accordingly. Detailed information and background on the Agri-PV definition and its regulatory integration, in particular the distinction between GM-PV and Agri-PV were provided by two scientific publications by Schindele published in German language in the *GAiA Journal* in May 2021 [13,14]. The publication of the DIN SPEC 91434 and its inclusion into legal frameworks has further increased the recognition of Agri-PV and with it the debate on Agri-PV standardization. Again, the number of actors involved in the industry shaping process has increased and different opinions and positions need consideration to develop the pre-standard to a consensus standard in future. Any pre-standard, also the DIN SPEC 91434 is valid for a maximum of three years. After three years, the same consortium participants must reconvene and take a decision on (i) if the DIN SPEC 91434 should be extended by another three years with no alternations being made; (ii) if the pre-standard should be updated and the opened to other participants; (iii) if the pre-standard should be taken as the base for a norming process; or (iv) if the pre-standard shall be terminated or replaced by another standard or pre-standard. Currently, there is a new DIN SPEC 91492 process open in Germany focusing on livestock keeping in Agri-PV projects [20]. This DIN SPEC 91492 might be used to extend and update DIN SPEC 91434, which needs an update in year 2024 anyhow. Looking back from the three years of DIN SPEC 91434 experience in Germany, the following learnings can be derived:

1. **Parity of interest between sectors increasing credibility of the (pre-) standard**
   DIN SPEC 91434 defined requirements for the main agricultural use in an Agri-PV project. Yet, Agri-PV is a power generating technology and on national level the national standardization organization in the field of electricity should be involved, e.g. VDE in Germany, at regional level like the EU the European Committee for Electrotechnical Standardization (CENELC) and at global level the International Electrotechnical Commission (IEC). For the engineering and construction of Agri-PV the companies must be registered as an electrical organization. For this reason, most participation in the DIN SPEC 91434 consortium came from the solar PV sector. Thus, most organizations from the PV sector have defined requirements for the main agricultural use in Agri-PV, without being (officially) registered as agricultural experts. From the 16 participating experts, only 4 had a purely agricultural background: AB-CERT, University of Hohenheim, Magdalena Farm, and the Competence Centre for Renewable Resources, Technology- and Funding Promotion TFZ as an institution of the Bavarian State Ministry of Food, Agriculture and Forestry. Of course, also the Agri-PV experts coming from the PV sector, e.g. at Fraunhofer ISE, BayWa r.e. or TubeSolar, have strong agricultural expertise and practical experience, but officially they belong to the PV sector and not to the agricultural sector. Associations and stakeholders from the agricultural sector participated in some meetings and commented on the drafts of the pre-standard, but were not authorized in their bodies to support an official decision and therefore did not make use of their voting rights. This passive participation from the agricultural sector reduces the credibility of DIN SPEC 91434 and it would be beneficial for policy makers if representatives from the agricultural sector could also officially participate in the standardization process with

decision-making power. Ultimately, however, the DIN SPEC process was able to facilitate a transfer of knowledge between the two sectors, enable joint learning and support cross-sector collaboration as well as a common basic understanding of the definition and meaningfulness of Agri-PV.

2. **Representing the entire diversity of Agri-PV**
The DIN SPEC 91434 was initiated from the APV-RESOLA project, which focused on cereal and vegetable production in an overhead Agri-PV system. Already during the standardization process, it became clear that an expert for animal welfare was missing in the consortium to specify the parameters that are relevant for the types of Agri-PV in combination with livestock farming. It was agreed that animal husbandry in Agri-PV projects should be included and thus enabled by the existing DIN SPEC 91434, but at a later stage more detailed regulations should be included into an Agri-PV standard. This process has started in 2023 with a new DIN SPEC 91492 on "Animal Husbandry Specific Requirements in Agri-PV Systems" [20]. Also, in 2023 the EU Biodiversity Directive became active and farmers in the EU are able to voluntarily set a site a minimum of 4% of their productive farmland for biodiversity measurements. The idea of "ecovoltaics" projects is to promote biodiversity in Agri-PV projects and have the project area registered as such a non-productive set-a-site area. The advantage of such biodiversity-supporting Agri-PV projects is that no additional farmland needs to be occupied for PV diffusion on farmland. This biodiversity-considerations had not been addressed in the DIN SPEC 91434 and the first publication of the workplan of DIN SPEC 91492 and further Agri-PV quality criteria need to be defined, especially biodiversity specific requirements, such as the implementation and maintenance of flower strips, bird and insect housing, sand bunkers and piles of stones for amphibians, insect friendly hay-cutting, and intercropping for soil restoration. External experts need to verify if the quality of flora and fauna are on similar or higher level compared to a set aside, non-productive biodiversity area that is included into the farmers crop rotating system. With a high probability the biodiversity quality in ecovoltaics/Biodiversity-supporting Agri-PV projects is higher than on set aside areas, as these projects remain 30 years on the same area and long-term measurements such as bushes and deadwood for insects can be included within the border area of the projects—which would not be included by the farmer in a crop rotation scheme. Nevertheless, the legislation of the EU member states does not reflect ecovoltaics on set-a-side area and the link to the EU's biodiversity directive. Especially the agricultural policy regime must guarantee to the farmer that the ecovoltaics project area remains the farmland status after 30 years and does not merge into nature protected area due to the relatively high biodiversity. In future, any pre-standard on Agri-PV must therefore include Agri-PV quality specifications on the following four land use categories: (1) arable land with temporary crops, pastures, or fallow; (2) land under permanent crops, e.g. horticultural use; (3) land under permanent cultivated pastures, e.g. for animal grazing and hay-cutting; and (4) Non-productive farmland for biodiversity increase. Only if cereal and

vegetable production, animal husbandry, and biodiversity measurements are considered in an Agri-PV quality standards, the entire variety of farming and thus the diversity of Agri-PV is represented.

3. **Suggested improvements to the must-have requirements in DIN SPEC 91424:**

(i) **Suggestion to rename requirement "land loss" to "non-productive farmland area" and increasing it to maximum 20% of project area**: to increase the competitiveness of a greater number of Agri-PV solutions and to have a greater flexibility with respect to the engineering and planning of the project area, the max. "land loss" criteria, which is a must-requirement, is suggested to increase form currently 15% to 20% in future. With this suggestion, 80% of the farmland on the project area stays active and the max. 20% "non-productive farmland area" is increasing the biodiversity and PV power yield on the project area. It is also recommended to re-name this requirement from "land loss" to "non-productive farmland," as this land is not lost, but used differently. It does not seem appropriate to speak about "land loss," if in fact the solar radiation is used more efficiently with the photo-electric effect compared to photosynthesis, and biodiversity measurements and improvements are demanded from the farmer anyhow.

(ii) **Suggestion to rename the requirement "land use efficiency" to "minimum agricultural yield" and to change from must- or shall-have to should-have**: the requirement 5.2.10 is called "Land use efficiency." Yet, it defines a minimum agricultural yield on the total project area after construction of the Agri-PV project to be at least 66%. The requirement should therefore be called "Minimum Agricultural Yield" and not "Land use efficiency" as it is today. The land use efficiency is measured by the land equivalent ration (LER). Once the name of the requirement is changed to "Minimum Agricultural Yield" it is recommended to change this criterion from a must-have criterion to a should-have. There are four arguments against a minimum requirement for agricultural yields. Firstly, the reference value can be very easily manipulated and is difficult to control. Second, external factors such as drought and disease can undermine the minimum requirement in some years. Third, in times of global heating it is close to impossible for agricultural experts to simulate if the agricultural yield will be 66% or higher in the upcoming 30 years of the project lifespan. Also, for a farmer it is impossible to predict his agricultural activity to be the same over a period of 30 years. Fourth, the German Government aims to increase the ecological/organic farmland area from 10% in 2022 up to 30% in 2030. Considering the agricultural yield gap between conventional farming using pesticides and chemical fertilizer and ecological farming of up to 50% per area, it makes it difficult to combine Agri-PV with the goal to convert farmland from conventional to organic farming. Today's regulation on this criterion is considered over-regulated and the risk of the investment that requires re-finance from the power sales and the confirmed bid-price in a PV auction is shifted to the farmer, as he or she must-have an agricultural yield output of 66% or larger. This criterion is perceived as too theoretical and not practical for the farming and the PV sector; thus, it should be adjusted from a hard- to become a soft-criterion accordingly. Even if the “tenor” of the EEG-regulation claims this requirement only as a benchmark, it still creates legal ambiguity and risk. Project developers are expecting a higher return against a higher risk to re-allocate internal and external resources and to shift their strategy from GM-PV to Agri-PV, and yet, the current financial regulation initiates higher risk, and lower returns, which makes the development of Agri-PV under these circumstances none-attractive for project developers. A member of the German Parliament from the SPD fraction urges the government to show a higher political courage, be more liberal and less restrictive and leave room for financial inefficiencies, especially at the early stage of market introduction.

(iii) **Suggestion to drop the “no land use change” requirement**: A permanent, economically viable agricultural activity on the project area must be obtained and confirmed by an external assessment of the agricultural concept. For this purpose, the requirement “agricultural usability of the area” must be in place. Yet, in the current DIN SPEC 91434 the requirement goes further and requests that the status quo of agricultural land use is unchanged and remains in place. It reads that the “previous agricultural usability of the area shall be maintained” [9] and continuous that there is no land use change feasible from permanent (1A and 2A) and Annual and multi-year cropland (1B and 2B) to Permanent grassland for mowing (1C and 2C) or Permanent grassland uses as pasture for animal husbandry (1D and 2D). Yet, farmers want to choose themselves on how the farm’s agricultural land is used in future. Any restriction on the freedom to adapt the agricultural structure of the farm limits the implementation and diffusion of Agri-PV. For sure, the 80% agricultural land within the Agri-PV project area shall be usable for agriculture, but the type of agricultural activity and the status of the future agricultural land category shall be chosen by the farmer and not be limited by a pre-standard. The recommendation is to drop the “no land use change” requirement entirely or at least change it from a must-have to a can-have criteria. Of course, the criteria “agricultural cultivation proposal” remains a must-have criterion, as this will allow a third-party expert to assess the agricultural feasibility of the permanent, economically viable agricultural activity on the project area.

4. **Communication and capacity building how to use and apply the Agri-PV pre-standard in the multi-level policy-regime:**

Besides the learnings and suggested improvements to the content of DIN SPEC 91434, there are also improvements to be made on the communication and capacity building within the multi-level policy-regime. The coordination of government action in a multi-level policy-system is distinguished between the horizontal and vertical level. The horizontal level is the intersectoral coordination between different departments/ministries impacted on a single policy level, e.g. coordination of Agri-PV diffusion on national level. The vertical level concerns coordination across policy levels, e.g. policies from the UN-level, e.g. Paris agreement, the regional level, e.g. the EU's Biodiversity Directive, national level, e.g. the EEG in Germany, and local level, e.g. state level and local authority level, e.g. spatial planning. Besides the challenge to coordinate a new policy or policy amendments on a single, horizontal policy-level, also Agri-PV diffusion must be coordinated across the higher and lower policy levels. An example of a higher policy-level integration is the EU's Common Agricultural Policy objective that every member state is publishing its own National EU CAP strategy plan. These strategies shall include the opportunity for farmers to develop Agri-PV on their farmland to increase and diversify their income and increase their financial resilience. An example for a lower policy-level is permitting and local planning. Many building departments at local authority level have never before worked on a Agri-PV permit procedure and are confronted for the first time with multifunctional land use options like Agri-PV (or Floating-PV). Fritz W. Scharpf speaks of policy interdependence on multi-level policy coordination. The interdependence is observed when either a public task cannot be carried out autonomously by the national state level, but cooperation involving actors at serval policy level is necessary, e.g. climate policy, or, if parallel to the hierarchical relationship in the multi-level policy-system informal cooperations are witnessed at one policy level, e.g. between several local state governments or local communities [21]. Arthur Benz coined the term "policy interdependence trap," which describes the inability of the federalist system to facilitate institutional changes to resolve certain challenges [22]. In Germany, the 16 federal states gave a best practice example on how policy-coordination in a multi-level policy-system can work. In Germany, tax-legislation privileges the farming sector as farmers do not need to pay inheritance taxes when handing over the farm and farmland to their ancestors. In Chapter 1, it was described how the change in land use category of GM-PV projects from farmland to industrial land detached a tax-bomb on farmers and how Agri-PV could defuse this tax-bomb for utility-scale PV projects. The inheritance tax incomes, however, become a governmental budget of the federal state level, e.g. Bavaria or Baden-Württemberg in Germany, and not to the national state level. Hence, the tax-privileges for Agri-PV had to be negotiated between the Finance Ministries of 16 federal states. In July 2022, the so-called "Bewertungsgesetz" that is assessing the assets of ownership and allocating the taxes accordingly was amended by a regulation confirmed by all 16 Ministries of Finance of Federal States in Germany, that the status of farmland is not changed to industrial land, if and only if the requirements of the pre-standard DIN SPEC 91434 are fulfilled on the project area [23]. To enable capital flows and continuous dept and equity investments into the Agri-PV diffusion it is important to involve the Finance and Building Ministries into the innovation process and to educate civil servants on all policy-levels regards the advantages of an available Agri-PV pre-standard such as the DIN SPEC 91434.

Conclusion in a nut-shell: Today, the DIN SPEC 91434 is perceived in the farming and solar sectors as a non-perfect, but good pre-standard that fostered cross-collaboration and reduced ambiguities in policy-making. Yet, some quality criteria are considered too strict and thus posing a risk for the financing of Agri-PV projects, e.g. the 66% minimum agricultural reference yield, the maximum 15% land loss or the no land use change criteria. All three criteria hinder the farming and solar sector alike to further optimize techno-ecological synergies, the increase of land use efficiency, and the economics of the farm and the project. By making DIN SPEC 91434 legally binding by integrating it into laws and regulations, more actors become involved into the standardization procedure. It became clear that the scope of the Agri-PV standard shall be technologically wider, and that animal husbandry and biodiversity aspects must be included. A more liberal approach, less hard-restrictions, and better communication in all levels of policy making are necessary to make Agri-PV a new industry standard on farmland.

2.1.3 Literature review on Agri-PV definitions, national Agri-PV pre-standards, key performance indicators of Agri-PV, and differentiation criteria from conventional GM-PV

In the years between 1981 and 2015, several terminologies were used for the Agri-PV idea, e.g. agrivoltaics, agrophotovoltaics, or solar sharing. However, there has been no scientific analysis of the different types of Agri-PV applications that are theoretically possible, how they can be distinguished from GM-PV, and how they can be classified and standardized. The first scientific publication that distinguishes between "inter-row" and "overhead" Agri-PV solutions was published in November 2015 by H. Dinesh and M. J. Pearce [24]. In relation to GM-PV, the PV module row spacing is extended so that agricultural activity can take place between the module rows (inter-row) or the PV modules are elevated so that agriculture can be carried out below them (overhead), see Figure 2.4.

The study by Dinesh and Pearce had a great influence on the categorization of the German pre-standard DIN SPEC 91434, which becomes clear when comparing the figures. While Dinesh and Pearce only presented the distinction in a schematic way, the two different categories (inter-row and overhead) were defined and made measurable in DIN SPEC 91434 by specifying key performance indicators (KPIs) and mandatory requirements, e.g. in category 1/overhead 90% of the Agri-PV project area must remain agricultural area and the clearance height between the ground and the lowest point of the PV-generator must be 2.10 m or larger.

In July 2019, Hernandez *et al.* published their results on how strategic engineering of solar energy technologies can enable synergistic outcomes across industrial and ecological boundaries. The developed framework includes 16 so-called solar energy "Techno-ecological Synergies (TES)" and characterized 20 TES-outcomes, see Figure 2.5.

Solar energy TES No. 3 represents PV generation "coupled with ecological restoration and/or pollinator habitat" and is pioneering the idea of ecovoltaics, in which the farmer implements ecological measurements to promote flora and fauna on the Agri-PV project area. TES No. 5 captures TES-outcomes for agrivoltaics co-located with crop production, and TES No. 6 establishes for the first time the terminology "Rangevoltaics," describing agrivoltaics combined with intensive animal husbandry such as SolarGrazing with sheep and cattle. These three Agri-PV

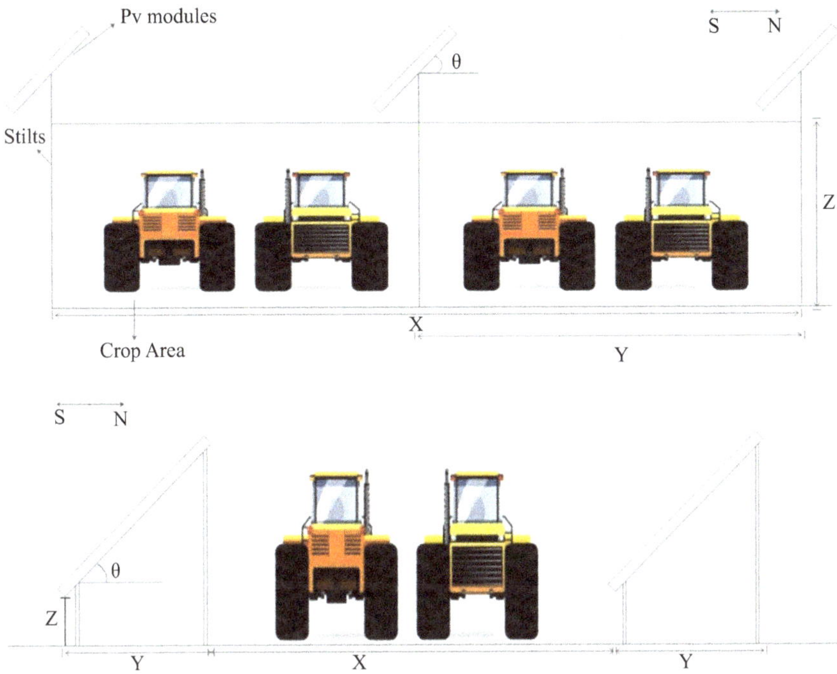

Figure 2.4 Differentiation between inter-row and overhead Agri-PV land use configuration by Dinesh and Pearce published in 2015

applications: ecovoltaics, agrivoltaics combined with crops or livestock include 17 of the 20 TES-outputs identified, highlighting the great potential of Agri-PV applications to increase the level of sustainability in rural areas and "represent a powerful frontier in sustainable engineering to minimize unintended consequences on nature associated with a rapid energy transition" [25]. The first comprehensive study that included the previous distinctions of inter-row and overhead (Dinesh and Pearce), Agri-PV applications combined with crops or animal husbandry (Hernandez *et al.*), as well as closed growing systems such as PV greenhouses, and additionally differed farming type, and flexibility was Willockx *et al.* in October 2020 [26], see Figure 2.6.

The inclusion of closed cultivation systems, such as Greenhouses, follows the rational that Agri-PV solutions manage light or more precisely the photons within the light spectrum reaching a certain area. In fact, there are solar greenhouse projects that follow the idea of Agri-PV and enable simultaneous food production and solar power generation on one site. It, therefore, seems consistent that this distinguishing feature is included in a classification compared to GM-PV. Yet, following the idea of "photon-management" in Agri-PV projects, it appears contractionary that the terminology "agrivoltaics" cancels the "photons" by eliminating the "photo" in its abbreviation of agri-*photo*voltaics. While in GM-PV projects no agricultural activity takes place, but the project site is used exclusively for solar power

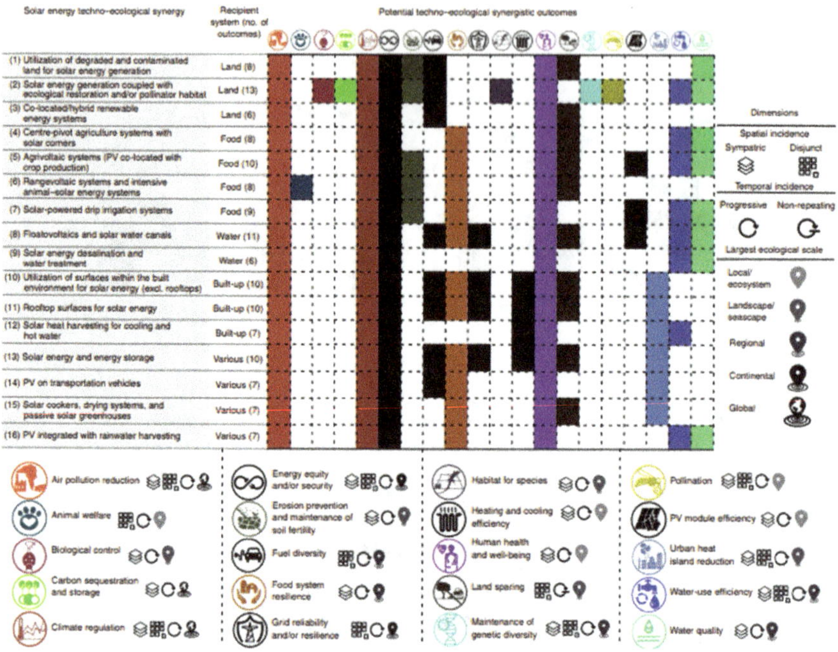

Figure 2.5 Framework for Techno-Ecological Synergies (TES) of Solar Energy Development by Hernandez et al. published in 2019

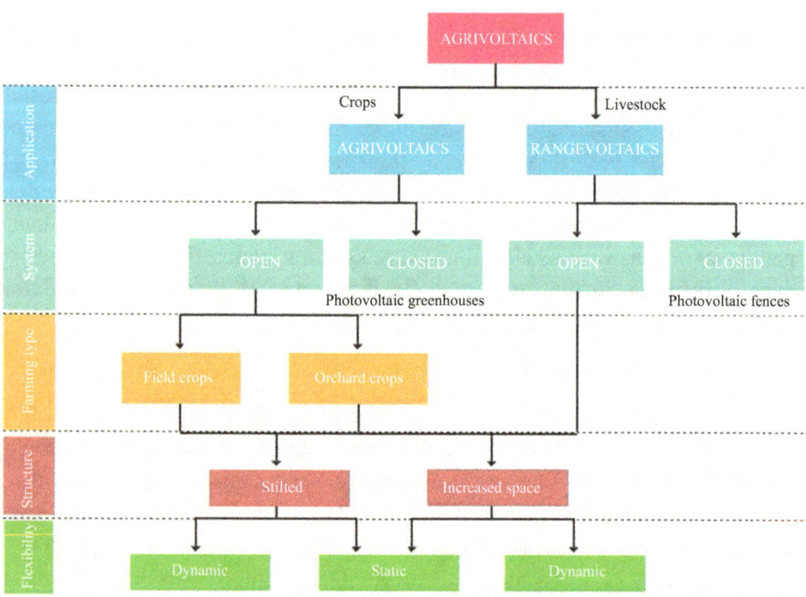

Figure 2.6 First comprehensive attempt to classify Agri-PV applications by Willockx et al. in 2020

generation, it also seems reasonable that Brecht *et al.* distinguish which type of agricultural activity takes place on the project site, e.g. arable farming, fruit growing or livestock farming. The flexibility indicator refers to the fact that certain Agri-PV applications deviate from the optimal solar power production, e.g. when a tracker system is moving the PV modules away from the ideal power generation so that more photosynthetic active irradiation reaches the crop at a required moment in the vegetation period. What Brecht *et al.* did not consider were translucent modules, which also deviate from the ideal solar power generation, as well as agricultural activities in the area of fish and seafood production on land (not in the sea), as occurs especially in the Asian region. In the second version of the Fraunhofer ISE Agri-PV-Guide, which was published in April 2022, these characteristics were considered and the approach of Willockx *et al.* was supplemented and re-structured [27], see Figure 2.7.

In May 2021, three publications with impact on the definition of Agri-PV were published: (a) the German Agri-PV pre-standard DIN SPEC 91434 [9], (b) associated background information to DIN SPEC 91434 by Schindele [13], and (c) the Agrisolar Best Practices Guidelines Version 1.0 by SolarPowerEurope [28]. SolarPowerEurope (SPE) started a workstream on Agrisolar in April 2020 and the Best Practices Guidelines (BPG) was an output of this workstream. The report reflects the experiences and views of a considerable share of solar companies active in the field of Agrisolar in Europe. The term "Agrisolar" was established as a broader umbrella to include PV technologies that are of value to the farmer, but that are not necessarily Agri-PV technologies. Non-Agri-PV solutions that are used by the farmer are e.g. Building integrated PV (BiPV) on farm buildings like stables and barns, Vehicle integrated PV (ViPV) on agricultural robotics, or off-grid PV system operating water pumping and irrigation systems. The definition of Agrisolar can even include solar thermal applications that are used for heating [29], see Figure 2.8.

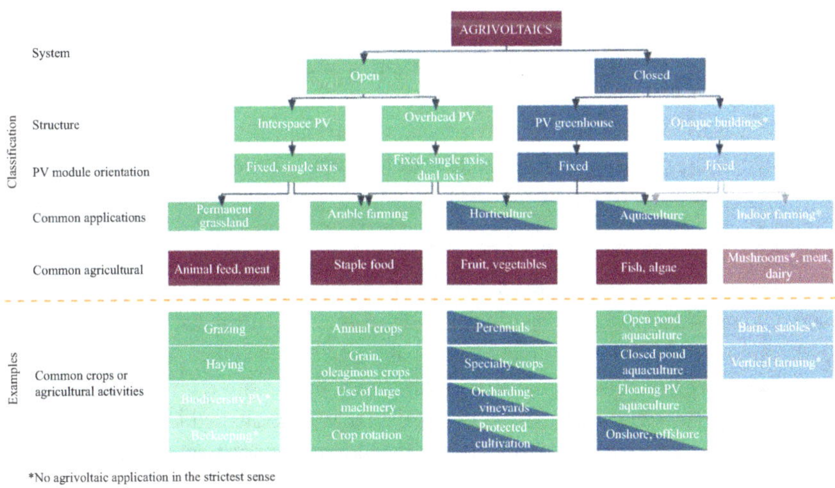

Figure 2.7 Fraunhofer ISE derives the Agri-PV classification from Willockx et al. and supplements it in 2022

Figure 2.8 Examples of solar thermal and PV installations on farm buildings (top row) and agricultural robotics (both pictures in lower row) representing Agrisolar but not Agri-PV applications as there is no techno-ecological synergies between farming and solar energy production

In contrast, in the multifunctional land use configuration Agri-PV, the agricultural production is only considering power generation using the photo-electric effect. Another important distinguishing feature of Agri-PV compared to the overarching term Agrisolar is that Agri-PV always makes use of the interaction and reciprocal effect between the agricultural activity and solar power generation, e.g. by using PV modules as a crop protection, or by providing shade and lowering the direct solar radiation on the crops, or reducing water evaporation in the soil and transpiration in the crops. Most Agri-PV projects are open-space on permanent grassland combined with grazing activities, e.g. sheep, chicken, and in future also more often with cattle grazing, or on arable land, where the agriculture activity takes place between the PV rows (Inter-row). The BPG of SPE also included SolarGreenhouses as Agri-PV, as also in this technology the sharing of sunlight between photosynthesis and photovoltaics is a characteristic of the system. The "Controlled Environment Agriculture (CEA)" that is achieved by SolarGreenhouses also provides protection for the agricultural activity and with climate change impacts such as extreme weather events becoming increasingly common, also Agri-PV projects with CEA record a growing demand. In June 2023, Version 2 of SPE's BPG on Agrisolar was published [30].

As described above, the DIN SPEC 91434 pre-standard serves to define and differentiate Agri-PV from conventional GM-PV projects. The demarcation was essentially achieved by setting minimum requirements for agricultural activity within the Agri-PV project area, e.g. only a certain proportion of the project area may be used for the installation of PV technology or a minimum agricultural yield of 66% compared to a reference must be achieved. In the publication of background information on the DIN SPEC 91434 process, Schindele was able to present a scientific position that was not distorted by compromises and decisions by absolute majority votes as in the DIN SPEC process. Schindele also looked at the key performance indicators (KPIs) of the German Sustainability Strategy to differentiate Agri-PV from other PV technologies and considered what relative advantage Agri-PV offers and which KPIs Agri-PV contributes to increasing the level of sustainability that cannot be achieved by other PV technologies. Schindele identified three relative advantages and distinguishing criteria of Agri-PV against GM-PV:

1. **No land use change from agricultural land to industrial land due to main agricultural activity on project area**: to maintain the status of agricultural land for the Agri-PV project and in order for there to be no land use change to industrial land in regional spatial planning, the primary purpose of the Agri-PV land use must be the agricultural activity for grass, crop cultivation or animal husbandry. In a conventional GM-PV project, there is no EU CAP eligible farming activity.
2. **Increase of land use efficiency from an agricultural yield perspective:** in order to ensure a minimum agricultural yield in the Agri-PV project, the land equivalent ratio (LER) value of the agricultural yield in a multifunctional Agri-PV land use must be higher than in the case of separate monofunctional land use on the very same area, which keeps food production and solar power generation separate. Only if the LER value for agriculture (LER agri) is higher in a multifunctional Agri-PV project compared to a monofunctional land use configuration with GM-PV and agricultural production separately on the same area, the Agri-PV is showing higher LER (Agri) and thus makes environmentally sense. Figure 2.9 shows the LER (Agri) calculation.

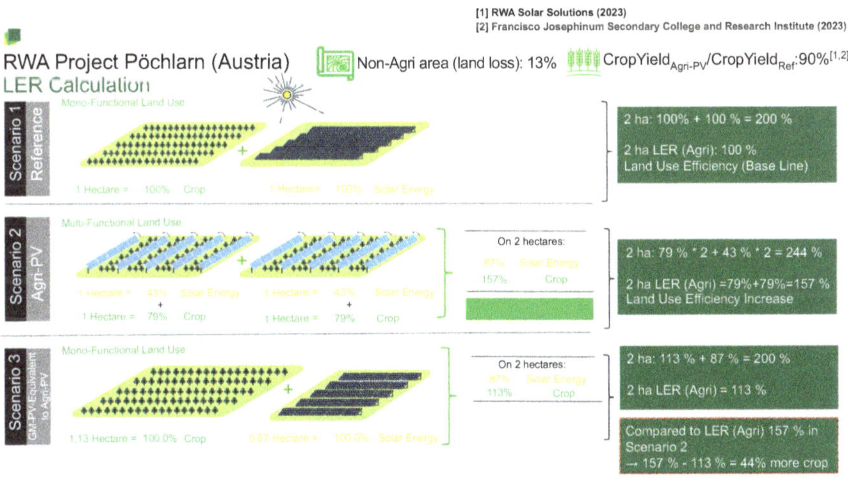

Figure 2.9 Land Equivalent Ratio (LER) focusing on agricultural yield measurement in project Pöchlarn in Austria, RWA Solar Solution & BayWa r.e.

3. **The PV generator must be adapted to the needs of the agricultural activity on the project area to enable techno-ecological synergies**: To integrate PV technology into agricultural processes, the requirements of the agricultural concept must be considered in the PV system design. PV use in the Agri-PV project is thus secondary and must enable TES-outputs as described by Hernandez *et al.* for both production systems equally. Considering these three must-have criteria for Agri-PV, the following basic definition for Agri-PV can be derived: "Agricultural-photovoltaics (Agri-PV or APV) is a multifunctional land use configuration on one and the same agricultural land area where solar power generation is integrated into an agricultural activity thereby leveraging techno-ecological synergies and increasing agricultural land use efficiency." Schindele suggested using these three must-have criteria (farmland, increase land use efficiency, and TES) as basic definition for an international valid minimum standard for Agri-PV-projects. According to this definition two Agri-PV configurations are enabled: (1) Overhead and (2) Inter-row; further, two classifications are feasible: (A) productive agricultural activity and (B) non-productive agricultural activity; and finally, two land use categories are applicable: (I) cropland, which is subdivided into (I.I) arable land, and (I.II) land under permanent crops, or (II) land under permanent pastures, which is subdivided into (II.I) cultivated and (II.II) naturally growing. He stressed that the basic definition of Agri-PV with the three must-have criteria can be complemented at national level by integrating further optional indicators to reflect the regional diversity of agricultural structures. As a result of the Agri-PV impact on the German Sustainability Strategy, the following four additional should-have indicators are proposed as extended requirements to the basis definition for Agri-PV applications in Germany.
4. **PV generator should support the prime agricultural activity to adapt to climate change**: In the upcoming decades with high probability the risk of climate change to agricultural output will further increase. To diminish the risk climate change to the farming process, e.g. risk of draughts, hail, strong wind and rain events, the Agri-PV application must increase the ecological resilience of the project area and the economic resilience of the farm. To enable Agri-PV applications, such as Fruit-PV, to take on protective functions for different agricultural land uses with high economic value, e.g. berries, pome and stone fruit, these applications must be highly elevated. It is therefore recommended that overhead Agri-PV applications should be primarily promoted.
5. **Small-scale and large-scale Agri-PV implementation should be enabled**: For self-generation of solar power the Agricultural Ministry of Germany supports farmers with its directive on energy efficiency and CO_2-reduction in agriculture. In this funding scheme only farmers are eligible to receive funding for the implementation of a small-scale Agri-PV electrifying the farm. Yet, to attract investment into rural areas and to enable large-scale Agri-PV the investment strength of institutional investors is demanded. For this reason, a support scheme attracting investment for Agri-PV applications that

feed into the grid only, e.g. the auctions within the renewable energy act (EEG), are necessary. To avoid conflicts of interest within the solar industry, both sub-sectors should have access to Agri-PV market development. On the one hand, the solar trade and local solar installers who can offer small to medium-sized Agri-PV systems for the farmer's own electricity needs and, on the other hand, project developers who focus on the implementation of larger Agri-PV systems with the participation of institutional investors.

6. **Agricultural activity in the Agri-PV project should be eligible for the EU CAP subsidy**: To support a permanent agricultural activity within the Agri-PV project, the agricultural activity should be eligible, evaluated, and controlled by the EU CAP subsidy mechanism.
7. **External expert must check the criteria 1–3 and should check criteria 4–6 as part of the building permit procedure**: To confirm that the agricultural concept and a sustainable, economically viable agricultural activity is feasible within the Agri-PV project, an external expert must approve the consistency of criteria 1–3 above and should also check criteria 4–6 to assess if the project is aligned with the German National Sustainability Strategy. The project developer that is participating in a financial mechanism supporting the implementation of the Agri-PV project must provide evidence to the administrator of the support scheme, e.g. the BNetzA in Germany, that criteria 1–3 are confirmed and 4–6 were considered. Only if this proof of feasibility by an external expert is provided with the request of funding or bid, the project becomes eligible to the support scheme.

Few months after the DIN SPEC 91434 was published in Germany, the French norming organization AFNOR published the first version of the French Agri-PV pre-standard called "Label Projet Agrivoltaïque" in December 2021 [11].

In August 2022, the National Renewable Energy Laboratory (NREL) published "The 5 Cs of Agrivoltaic Success Factors in the United States: Lessons From the InSPIRE Research Study" [31]. The 5C concept aims to maintain a high quality within Agri-PV project development, reducing the risk for the farmer and PV developer alike. These aspects include:

- **C1: Conditions, Climate, Soil, and Environmental:** This involves understanding and working with the specific ambient conditions and factors of the location where Agri-PV systems are being implemented. Factors such as temperature, rainfall and soil quality can influence the success of both solar and agricultural activities.
- **C2: Configurations, Solar Technologies, and Designs:** The choice of solar technology and the design of the Agri-PV system are crucial for maximizing light availability and solar generation. Proper site layout and infrastructure planning can also affect the overall efficiency of the system.
- **C3: Crop Selection and Cultivation Methods, Seed and Vegetation Designs, and Management Approaches:** The selection of crops, vegetation designs, and agricultural practices play a significant role in Agri-PV projects.

The choice of crops should be compatible with the solar infrastructure and should not hinder solar generation.

- **C4: Compatibility and Flexibility:** The design and configuration of the solar technology should be compatible with the needs of both solar owners/operators and agricultural practitioners. Flexibility is essential to ensure that the system can adapt to changing requirements and advancements in technology.
- **C5: Collaboration and Partnerships:** Collaboration and partnerships among stakeholders and sectors are crucial for the success of Agri-PV projects. This includes community engagement, obtaining necessary permits, and establishing legal agreements to ensure smooth operations.

Page 15 of the NREL report also includes the following definition of Agri-PV: "Agrivoltaics is defined as a land use configuration where solar energy generation and sunlight dependent agricultural activities are directly integrated and there is a layer of agricultural productivity within the boundaries of the solar infrastructure. The hallmark characteristic of agrivoltaics is thus the sharing of sunlight between the two energy conversion systems: photovoltaics and photosynthesis. Agricultural activities include practices that satisfy human food, fiber, and fuel needs as well as activities that enhance environmental quality and the natural resource base upon which the agricultural economy depends (adapted from the U.S. Department of Agriculture (USDA) 2007)." The basis of the report was the InSPIRE project, which was funded since 2015 by the U.S. Department of Energy (DOE). Over 25 Agri-PV sites across the USA that span crop production, pollinator habitat, ecosystem services, and livestock production were examined. NREL publishes the Agrivoltaics Map that contains over 360 Agri-PV sites in the USA [32]. By the end of 2022, approximately 40 GWp of utility-scale PV projects will have been installed in the USA [33]. Most utility-scale PV installation took place as conventional GM-PV in the southwest and south of USA, where GM-PV projects are installed on desert areas and non-agricultural land. In contrast, a cluster of Agri-PV projects is recognized in the north and north-east of the USA, mainly solar pollinator habitats and solar grazing with sheep, because fertile agricultural land is available in these regions and agricultural activity can therefore be carried out on the project area. According to data from the NREL Agri-PV map, just over 1.2 GWp of Agri-PV is installed in the USA, which represents a market share of around 3% in the utility-scale PV market segment (more information see Chapters 1 and 11 in this book).

In February 2023, the Italian norming organization published a pre-standard on Agri-PV called "Comitato Elettrotecnico Italiano (CEI); PAS 82-93" [10]. In the same month, the German norming organization DIN announced the start of DIN SPEC 91492, which defines quality criteria for animal husbandry within Agri-PV projects. DIN SPEC 91492 serves as an extension to DIN SPEC 91434, which enabled animal husbandry, but did not define quality parameters on animal keeping within Agri-PV projects.

In July 2023, Christian Dupraz published his ideas on how the Ground-Cover-Ration (GCR) of Agri-PV systems can act as a proxy for the potential crop productivity [34]. The logic behind the concept is simple: the lower the area covered

Regulatory requirements in agrivoltaic policies of various countries (including maximum cropped area loss and minimum crop relative yields).

Country	Maximum "dead areas"	Minimum relative crop yield	Minimum Vertical Clearance	Ground Coverage ratio limit	Minimum relative electricity yield	Rate of Agricultural subsidies	Reference
France	Not considered	100% (no reduction allowed)	Not considered Kim et al. (2022)	Not considered, but 50% indicated in the AFNOR label	Not considered	100% if GCR <= 30% No grants if GCR > 30%	French Law (République Francaise 2023) AFNOR label (AFNOR 2021)
Germany	Category 1 (Overhead AV): 10% Category 2 (Interspace AV): 15%	66%	Category 1: 2.1 m Category 2 no clearance requested	Not considered	Not considered	85%	DIN Standard 91434 (Deutsches Institut für Normung 2021)
Italy	30%	Not defined	1.3 m with animals 2.1 m with crops	40%	60%	Undecided	Ministerio de la transizione ecologica (2022)
Japan	Not considered	80%	Not considered	Not considered	Not considered	Unknown	Cited in Tajima and lida (2021)
South Korea	Under discussion	Under discussion. 80% considered	Under discussion	Not considered	Not considered	Unknown	Kim er al. (2022)

Figure 2.10 Regulatory requirements in agrivoltaic policies of various countries, published by Dupraz in 2023

by PV infrastructure, the higher the crop productivity and vice versa. In a meta-analysis of selected published Agri-PV research projects, Dupraz finds that a GCR smaller than 25% is required to ensure that most crops show a relative yield above 80%. The article also includes an overview of regulatory requirements in Agri-PV policies of various countries, see Figure 2.10.

There are some points of criticism regarding the use of GCR as a differentiator criterion from GM-PV and its suitability for estimating the agricultural yield in Agri-PV projects. For example, in the case of Agri-PV application in combination with permanent crops, the GCR is higher than 80% and the agricultural yield is almost constant compared to the reference under plastic covers [35] or in the case of solar grazing with sheep and cows, where the GCR is between 55% and 60%, and the agricultural yield (wool, meat, milk, grass) can be higher than without Agri-PV application due to less heat stress in the summer months and protection of the animals in case of strong rains or hail. In Chapter 11 of this book, there is a debate between Dupraz and Klyk/Schindele whether the GRC is a "good" Agri-PV quality criteria and if the Agri-PV definition should be more liberal (position of Klyk and Schindele) or more restrictive (position of Dupraz).

In a nutshell: Agricultural activity must take place in an Agri-PV project. No agricultural activity takes place in a conventional GM-PV project. To avoid alibi Agri-PV projects and to clarify the distinction between GM-PV and Agri-PV on agricultural land, clear quality criteria are required for the agricultural activity in the Agri-PV project. To ensure that agricultural activity can take place without restrictions and that synergy effects between the two sectors can be exploited through the dual use of agricultural land, the PV generator must be technically adapted and differentiated from conventional ground-mounted PV that is purely focusing on maximizing solar power yield per area, e.g. in an Agri-PV project there must be a minimum distances between the rows of PV modules (inter-row) allowing enough light for the vegetation to develop and allowing agricultural machinery to pass through, or raising the PV modules higher (overhead) or integrating the PV generator into the greenhouse shell without significantly reducing agricultural production. Several norming organizations in the EU have developed and published so-called pre-standards together with Agri-PV experts. They are built on scientifically developed and published knowledge. PV applications that provide a net benefit for agricultural operations, e.g. building- or vehicle-integrated PV on stables or in agricultural robotics, can be classified under the umbrella term Agrisolar, but are not Agri-PV technologies because there are no direct synergy effects between the agricultural output and the solar power yield. The main unique selling points of Agri-PV technology compared to GM-PV are as follows: First, the land status of agricultural land is maintained throughout the project life cycle; second, the land use efficiency must be higher in relation to the agricultural yield in a mixed land use than in a monofunctional land use; third, techno-ecological synergies arise between the sectors. In particular, the requirements for agricultural activity in the Agri-PV project can be recorded and documented in a standardized way and shared with interested stakeholders, e.g. authorities, landowners, farmers, funding bodies, the public, investors and managers of the project. Agri-PV definitions can be very narrow and restrictive or more liberal and open. The pre-standardization process is also an industrial policy process in which Agri-PV actors try to strengthen their Agri-PV offerings in the technology competition by restricting the quality criteria to gain a competitive advantage. A debate on the advantages and disadvantages of a restrictive vs. a liberal Agri-PV definition is included in Chapter 11.

2.1.4 Proposal for a globally and universally valid definition of agrivoltaics

The term "agricultural photovoltaics": This book honors the fact that the term "agrivoltaics" is already scientifically established worldwide. From a purely technical point of view, this neologism is made up of two words: "agriculture" and "photovoltaics." An **agricultural activity** is understood (i) to be the biomass production of food, feed, fuel, fiber, (ii) measures to promote biodiversity and the environment, as well as (iii) livestock keeping. **Photovoltaics** uses solar energy by converting the sun's radiation directly into electrical energy. Since "agrivoltaics" is always about a fair ratio of photons per agricultural area and thus the optimization of photosynthesis and photoelectric effect, it is unfortunate that the term "agrivoltaics" has eliminated the "photon." As a proposal for a globally valid definition of the technology, this deficiency should be corrected or at least, as suggested in this book, the abbreviation of agrivoltaics should contain the photons. Therefore, the term "agrivoltaics" has been used in the title and headings of the book and the abbreviation "Agri-PV" or "APV" has been used predominantly in the body text. Regardless of which term will ultimately prevail globally on the long-run—agricultural photovoltaics, agrivoltaics, Agri-PV, or APV—the principle of **Agrivoltaics land utilization can be defined as follows:**

"Agricultural Photovoltaics (APV) is a multifunctional land use configuration on one and the same agricultural land area where solar power generation is integrated into an agricultural activity thereby leveraging techno-ecological synergies and increasing the land use efficiency regards agricultural yield compared to the two monofunctional land uses on the same area."

The proposed definition and understanding of APV are therefore similar to what NREL has proposed in its 5C-publication [31].

2.1.5 Varieties of agrivoltaics: terminology according to agricultural concept

Agri-PV can be classified according to agricultural activity and the level of technology adaptation. For example, the Agri-PV application in combination with cows could be called "Cow-PV," or that with fruit "Fruit-PV." This would immediately clarify the agricultural use within the terminology. When illustrating the different Agri-PV classifications, it becomes clear that certain agricultural activities in the Agri-PV project require a special adaptation of the PV generator. For example, in the context of special fruit crops, the PV modules must be elevated (overhead) or to produce vegetables and cereals, the PV module rows must be moved further apart so that enough solar radiation is received for the plant vegetation and that agricultural machinery can work in between (inter-row). Based on a conventional GM-PV, which has the lowest LCOE due to the intensive solar power generation per area, the costs for inter-row Agri-PV applications increase between 20% and 40% and for overhead between 40% and 60%, depending on how much wider the row-pitch has been increased, the height of the PV modules installation, translucent PV modules have been utilized, and whether a PV-tracking software has been altered to allow dynamic shading during the vegetation period promoting agricultural yields. Figure 2.11 illustrates the different types of Agri-PV applications, the target, type of standard, status of land, and LCOE changes compared to GM-PV.

In general, it can be observed that overhead Agri-PV projects result in higher LCOE than inter-row projects. Whether the higher implementation costs can also be justified by higher techno-economic synergies and more pronounced increases in land use efficiency in a cost-benefit analysis cannot be assessed across the board but depends on the land use and agricultural concept in connection with the Agri-PV project. When comparing the different Agri-PV land-use configurations (inter-row and overhead) and assuming that all configurations within a certain Agri-PV application lead to the same quality and quantity of agricultural yield, the Agri-PV configuration that offers the highest cost-benefit should prevail in the long term.

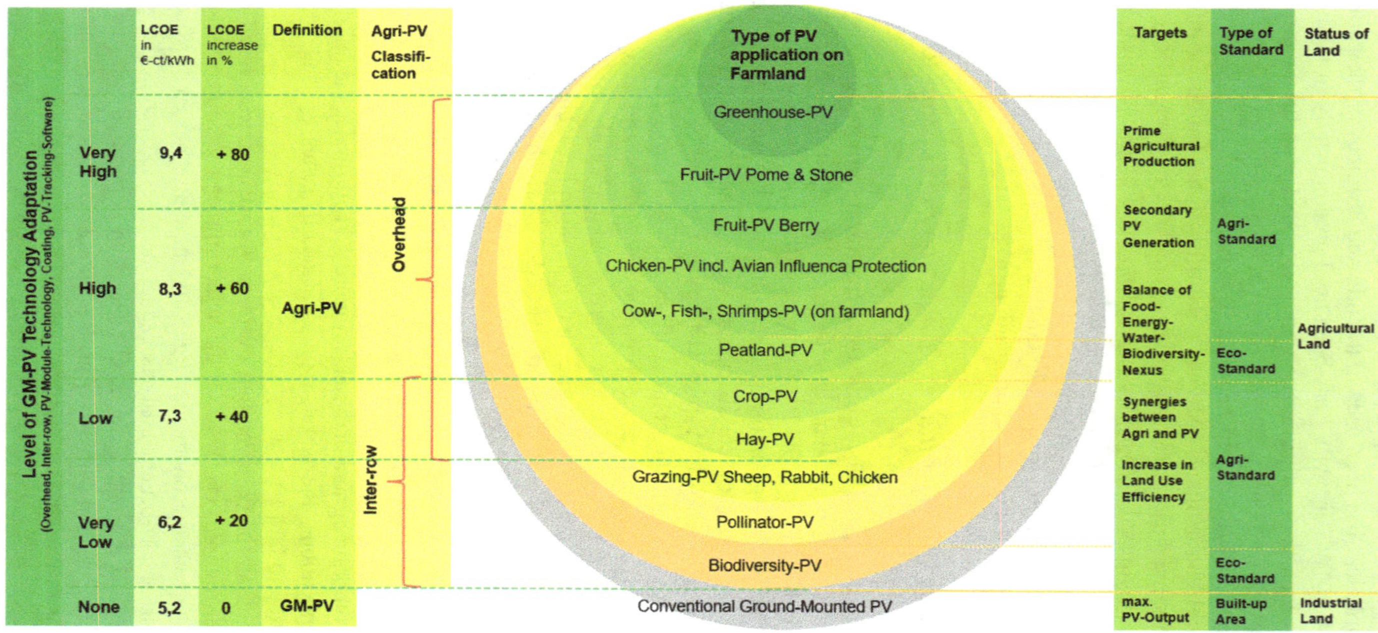

Figure 2.11 Agri-PV classification according to agricultural activity, overhead and inter-row, and level of technology adaptation and its Impact on LCOE, BayWa r.e.

2.2 Global area and PV capacity potential of agrivoltaics

In 1990, the Intergovernmental Panel on Climate Change (IPCC) completed its first comprehensive assessment of the climate change issue. Since five other reports have followed, all highlighting that we are hitting the ceiling of the planet's capacity to support humanity. Following those reports, in 2015 a binding agreement was signed with the Paris Agreement: for the first time 195 states were brought together to fight climate change. Through its adoption, parties agreed on pursuing their efforts "to limit the temperature increase to 1.5 °C above pre-industrial level" [36].

By signing the Paris Agreement, every stakeholder recognized that society urgently needs a transformation toward a sustainable future. According to the United Nations (UN) estimations, only one generation is left to achieve the 1.5 °C goal. Delivering this transition on time will require drastic transformation, from how we are producing food and feeding the population, to the production of electricity and land management. In a period where land use conflicts are increasing and future farmland must be protected, there is a high probability that multifunctional land use options are politically stimulated, and an increasing amount of solar power will be provided by Agri-PV. By harvesting on the same area sun for agricultural and energy purposes, Agri-PV has a significant asset: it maximizes cross-sectoral synergies among the water, energy, food, and biodiversity sectors. Therefore, it is of importance to analyze and assess its potential worldwide, to leverage its large-scale deployment. The following inquiries are essential: How much agricultural land is necessary for PV deployment to follow the 1.5 °C target? What is the feasible PV capacity of Agri-PV at both the worldwide level and across individual continents?

2.2.1 Methodology and approach to estimate global area potential

The estimation of global area potential is grounded in the idea that only 1% of the world's agricultural land used for Agri-PV project implementation can satisfy our global solar PV production objectives. This concept finds support in the 14th Five-Year-Plan for Promoting Agricultural and Rural Modernization by the Central Committee of the Communist Party of China and the State Council [7] as well as by research conducted by the European Commission's Joint Research Centre (JRC), as detailed in their report titled "Overview of the Potential and Challenges for Agri-Photovoltaics in the European Union, 2023" [37]. According to the JRC's assessment, allocating 1% of the EU's land area to Agri-PV could result in solar PV production ranging from 315 to 1.415 GWp, depending on power density. Meeting the EU Solar Energy Strategy's goal of 750 GWp (DC) by 2030 would then be attainable. Oregon State University similarly operates on the assumption that converting just 1% of American farmland into Agri-PV systems would be sufficient to meet their national renewable energy targets [38]. Under this assumption, calculations are made for 1% of the utilized agricultural area (UAA), on a global scale for each continent and individual country. These calculations rely on the latest FAOSTAT data from 2021 [39]. In each country, the UAA is further categorized into three primary groups: (i) Land under permanent cultivated pastures, (ii) Land under temporary crops and vegetables and (iii) Land under permanent crops, as described in sub-chapter 2.2.1.1. Countries are organized into sub-regions, following the UNs' classification of countries [40]. This list was cross-referenced with the available FAOSTAT data for agricultural land use in each country from 2021. These datasets are then integrated with the corresponding installed PV capacity per hectare (agricultural land use) for the specific Agri-PV application. The PV capacity installed per hectare is calculated using hypotheses detailed in Section 2.2.1.2.

2.2.1.1 Respective categories of utilized agricultural area

The FAOSTAT (Food and Agriculture Statistic) domains land use is available for 198 countries and 43 territories over 1961–2021 and is used as a standard by the UN System of Environmental and Economic Accounting (SEEA); the UN Framework for the Development of Environmental Statistics (FDES), and the World Census of Agriculture. According to the FAO, agriculture area, which is the land used for agricultural purposes, can be categorized into two different categories:

Land under protective cover: Land used for agriculture occupied by dwellings, operating buildings (hangars, barns, cellars, greenhouses, silos), buildings for animal production (stables, cowsheds, pig sheds, sheep pens, poultry yards), family gardens and farmyards. Within this category, the potential for Agri-PV development is either nonexistent or extremely limited. PV integration into greenhouses, barns, and other controlled environmental atmospheres are considered building integrated PV (BiPV) Agrisolar solutions, but do not account for Agri-PV.

Agricultural land: Land used for cultivation of crops and animal husbandry. This category can be divided in "cropland," which combined "arable land" and "permanent crops," and "permanent meadows and pastures," which combined cultivated meadows and pastures and natural growing. Agri-PV solutions are primarily designed for lands falling within this category and its sub-categories. The FAO land-use matrix is shown in Figure 2.12.

According to the FAO land use matrix (see below), agricultural land is subsequently classified into the following four categories:

Arable land: Land used for cultivation of crops in rotation with fallow, meadows and pastures within cycles of up to five years.

Land under permanent crops: Land cultivated with long-term crops which do not have to be replanted for several years (such as cocoa and coffee), land under trees and shrubs producing flowers (such as roses and jasmine), and nurseries (except those for forest trees, which should be classified under "Forestry"). It refers to fruits and berries, nuts, oil-bearing crops (permanent only), spices, condiments and aromatic herbs and other permanent crops.

Land under cultivated permanent meadows and pastures: Land used permanently (five years or more) to grow herbaceous forage crops through cultivation.

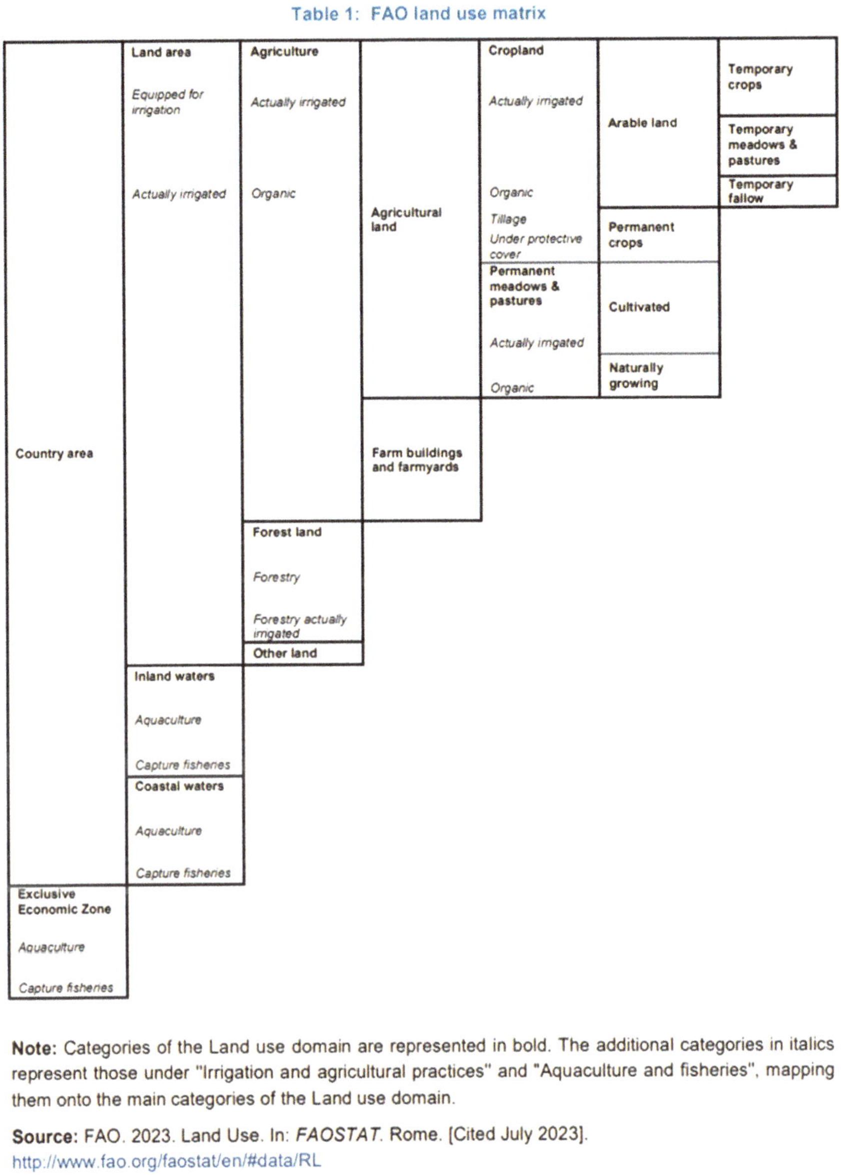

Figure 2.12 FAO Land Use Matrix, 2021

Land under naturally growing permanent meadows and pastures: Land used permanently (five years or more) to grow herbaceous forage crops naturally (wild prairie or grazing land). In many countries, data regarding naturally and cultivated growing permanent meadows and pastures are not available. Therefore, in the following these two categories are combined into a single category referred to as "**permanent meadows and pastures.**"

In the forthcoming study, three FAO categories (arable crops, permanent crops, and permanent meadows and pastures) are considered on a global and continental scale. The estimation of the Global Area Potential for Agri-PV involves allocating 1% of the area within each category.

2.2.1.2 Estimation of the PV capacity per hectare

Estimations of Agri-PV systems' PV capacity per hectare exhibit substantial variation based on their specific design. Numerous factors influence the ultimate PV capacity per hectare in Agri-PV projects, such as:

- Localization (longitude and latitude)
- Modules transparency
- Pitch between PV module rows
- System type: tracking or fixed-tilt
- Table setup
- PV modules tilt (Portrait (P) or Horizontal (H)
- PV modules orientation (east-west or south/north)
- PV modules electrical output
- Area topography

Agri-PV system designs are as diverse as farms themselves. They vary significantly depending on the practices of the farmer, the type of production, the shape of the land. According to JCR (2023), the range of installed PV power obtained can vary from 0.2 to 0.9 MW/ha, depending on the specific design of the Agri-PV system.

In the following sections, different Agri-PV capacities per hectare are assumed, categorized by land-use type. These assumptions are based on existing projects or internal calculations from BayWa r.e., utilizing 575 Wp modules.

Estimation of Agri-PV capacity per hectare on *arable land*

Agri-PV on arable lands refers to a wide range of designs, which can be either overhead or inter-row configuration. In the subsequent discussion, it is assumed that inter-row solutions will be prioritized against overhead designs in this land use category due to their better price-performance ratio (PPR) resulting from lower steel requirements for the substructure, leading to lower expected Levelized Cost Of Electricity (LCOE) while at the same time harvesting the same agricultural quality and quantity.

Arable lands are used for cultivation of crops in rotation with fallow, meadows and pastures within cycles of up to five years. It implies the use of a diverse array of machinery. Only projects that accommodate machinery with working widths ranging from 9 m to 18 m are considered.

Table 2.1 provides details on the capacity per hectare for three systems, assuming the use of 575 Wp modules

- Tracking 1P: tracking system with one row of portrait modules
- Tracking 2P: tracking system with two row of portrait modules
- Fixed-tilt 2P: fix system with two rows of portrait modules.

Given these assumptions, a capacity of **0.4** MWp/ha is applied to arable lands. The project Pöchlarn of RWA in Austria, an inter-row Agri-PV installation on cropland constructed in 2022, aligns with this assumption, boasting a PV capacity of 0.4 MWp/ha.

Estimation of Agri-PV capacity/ha on *permanent crops*

Permanent crops refer to all fruit trees, all citrus fruit trees, all nut trees, all berry plantations, all vineyards, all olive trees and all other permanent crops used for human consumption or other purposes. Overhead designs are predominant on permanent crops. Higher panel transparency is required from one culture to another, together with a lower pitch to protect the culture above from natural hazards.

Table 2.2 provides details on the capacity per hectares of overhead systems from 2 m to 5 m pitch, assuming 300 Wp modules (high transparency).

Considering these assumptions, we assume a capacity of **0.8** MWp/ha for permanent crop areas. This assumption is consistent with the observed PV capacity per hectare in existing projects on permanent crops (Table 2.3).

Estimation of Agri-PV capacity per hectare on *cultivated permanent meadows and pastures*

Agri-PV installations on cultivated meadows and pastures refer to a wide range of designs, as these lands serve multiple purposes, including animal breeding and hay production. In the subsequent discussion, priority is given to inter-row

Table 2.1 Estimation of crop-PV, inter-row, capacity per hectare on arable land

Max machinery Working Width (m)		**9**	**12**	**15**	**18**
Installed capacity (kWp/ha)	Tracking 1P	335	261	214	181
	Tracking 2P	645	507	418	355
	Fixed-tilt 2P	497	411	350	305
Average of installed capacity per working width		492	393	327	280
Total average (kWp/ha)					373

Table 2.2 Estimation of fruit-PV, overhead, capacity per hectare on permanent crops

Pitch (m)	2	3	4	5
Estimated installed capacity (kWp/ha)	1190.48	793.65	595.24	476.19
Average				763.89

Table 2.3 Observed fruit-PV, overhead, capacity per hectare on permanent crops

Project Name	Developer(s)	Area (ha)	Energy production (MWp)	Crop(s)	GM-PV, overhead, Interspace	PV capacity (MWp/ha)
Enspijk	BayWa r.e. GroenLeven	0.12	0.11	Cherries	Overhead	0.88
Wadenoijen	BayWa r.e. GroenLeven	1.50	1.18	Red currants	Overhead	0.79
Randwijk	Baywa r.e. GroenLeven	0.12	0.13	Pears	Overhead	1.08
Gelsdorf	Fraunhoffer Institute/ BayWa r.e.	0.35	0.26	Apples	Overhead	0.74
SEFRA	Sun'Agri	2.7	1.90	Vineyard	Overhead	0.70
Tressere	Sun'Agri	4.50	2.10	Vineyard	Overhead	0.47
Babberich	BayWa r.e.	3.00	2.67	Raspberry	Overhead	0.89
					Average	0.79

Table 2.4 Estimation of Hay- and Cow-PV, inter-row, capacity per hectare on cultivated permanent pastures

Max machinery Working Width (m)		**6**	**9**	**12**	**15**
Installed Capacity (kWp/ha)	Tracking 1P	467	335	261	214
	Tracking 2P	887	645	507	418
	Fixed tilt 2P	629	497	411	350
Average of installed capacity per working width		661	492	393	327
Total average					468

solutions, as they require less steel for the substructure, resulting in a lower expected LCOE compared to other Agri-PV applications.

Cultivated meadows and pastures are utilized for cultivation of crops in rotation with fallow, meadows and pastures are land used permanently (five years or more) to grow herbaceous forage crops through cultivation (FAO definition). Machinery with smaller widths is employed compared to arable land. Only projects permitting machinery with working widths ranging from 6 m to 15 m are considered.

Table 2.4 provides details on the capacity per hectare for three systems, assuming the use of 575 Wp modules:

- Tracking 1P: tracking system with single row of portrait modules
- Tracking 2P: tracking system with two rows of portrait modules
- Fixed-tilt 2P: fix system with two rows of portrait modules.

Based on those assumptions, a capacity/ha of 0.5 MWp/ha is assumed for cultivated permanent meadows and pastures. Next2Sun declares that the installed capacity of their Agri-PV systems (vertical fixed-tilt alignment of the bifacial solar modules, with 2 modules in landscape position) on grasslands is 0.4 MW/ha, which is in accordance with the previous results.

Estimation of Agri-PV capacity/ha on *natural growing permanent meadows and pastures*

Naturally growing permanent meadows and pastures refers to land used permanently (five years or more) to grow herbaceous forage crops naturally (wild prairie or grazing land). On these lands, sheep breeding and ecovoltaics are assumed to be the dominant activities. Therefore, the same PV capacity as ground-mounted systems, which is **0.9 MWp**/ha according to [33], is considered. The estimation of Agri-PV capacity on permanent meadows and pastures is derived by averaging the Agri-PV capacity per hectare for cultivated and naturally growing permanent meadows and pastures, resulting in an estimate of **0.7 MWp/ha**.

2.2.2 Global *area and PV capacity potential of agrivoltaics*

In 2021, the Worldwide UAA covered 5.3 billion ha (FAOSTAT Land Use domain). This area includes around 3.6 billion ha of permanent meadows and pastures, 1.5 billion ha of arable land and 0.2 billion ha of permanent crops. Permanent meadows and pastures accounted for the majority of the UAA, making up 68% of it. Arable lands followed, comprising 28% of the UAA, while permanent crops represented only 4%.

Table 2.5 presents the potential capacities achievable by covering 1% of the UAA with Agri-PV systems, assuming various installed PV power capacities per Agri-PV specification land area category. With 1% of the worldwide UAA covered with Agri-PV solutions, the installed capacity worldwide would be 32.847 GWp

Table 2.5 Worldwide utilized agricultural area (UAA) and PV capacity potential per land use category, if 1% of UAA is implemented with Agri-PV

	UAA worldwide per category (1000 ha)	**UAA worldwide per category (%)**	**1% of the area (ha)**	**PV capacity per category (MW/ha)**	**Potential (GW) for 1% coverage**	**Potential (%) for 1% coverage**
Arable lands	1.506.134	28	15.061.342	0.4	6.025	18
Permanent crops	202.394	4	2.023.945	0.8	1.619	5
Permanent meadows and pastures	3.600.507	68	36.005.070	0.7	25.204	77
TOTAL	5.309.036	100	530.90.357		32.847	100

i.e. 32 TWp. Permanent meadows have the greatest potential, contributing a total of 25.2 TWp, which accounts for 77% of the total capacity. Arable lands come in second at 18%, while permanent crops are behind at 5%, despite assuming a higher PV capacity per hectare. Globally, the total installed capacity of solar PV reached 1.2 TWp by the end of 2022, according to Solar Power Europe [41]. By covering only 1% coverage of its UAA, the already installed capacity could be reached more than 26 times.

According to the International Renewable Energy Agency (IRENA), 14 TWp of PV capacity must be installed worldwide in 2050 to stay on track with the 1.5 °C target [42]. Achieving this goal, with 50% of the capacity deployed on farmland, could involve covering approximately 0.2% of the world's agricultural land with Agri-PV by 2050. However, considering the expected increase in the global population, we might require additional land to continue following the 1.5 °C target after 2050. Moreover, this estimate considers various countries and continents where technical and political limitations are severely restricting solar development.

Scientists at Fraunhofer ISE and the Potsdam Institute for Climate Impact Research (PIK) assume higher numbers with between 20 and 80 TWp of PV capacity to be installed in 2050 and between 80 and 170 TWp in 2100 [43]. As approximately 50% of this capacity is expected to be deployed on farmland (see Chapter 1), achieving this goal involves covering between 0.3% and 1.2% of the world's agricultural area by 2050 and between 1.2% and 4.1% in 2100.

2.2.3 European Union, United Kingdom, and Switzerland *area and PV capacity potential of agrivoltaics*

The 27 countries of the European Union (EU) as well as the United Kingdom and Switzerland, due to their close situation, are considered in the following. Overseas territories and departments are neglected.

In 2021, the collective UAA for these nations amounted to nearly 182 million ha, based on data from FAOSTAT Land Use domain. It included 64 million ha of permanent meadows and pastures, 105 million ha of arable land and 12 million ha of permanent crops. The majority consisted of arable lands, making up 58% of the total UAA, while permanent meadows and pastures accounted for 28%, and permanent crops represented 7%.

Table 2.6 presents the potential capacities attainable by covering 1% of the UAA in the European Union, the United Kingdom, and Switzerland with Agri-PV systems, considering various assumed installed power capacity per land categories.

With 1% of the UAA covered with an Agri-PV system, the installed capacity would be 968 GW, i.e. 1 TW. Even if permanent meadows and pastures rank second regarding the total UAA, it is first when it comes to Agri-PV potential, with a total amount of 451 GW, i.e. almost 47% of the total amount. Arable comes second with 44% as its PV capacity installed per area is assumed to be smaller than for permanent meadows and pastures. Permanent crops are last with 10%.

The total capacity and production (CAP) of solar PV in the European Union, the United Kingdom and Switzerland reached 217 GW in 2022 (Irena, 2022). By

Table 2.6 European Union, United Kingdom, and Switzerland utilized agricultural area (UAA) and PV capacity potential per land use category, if 1% of UAA is implemented with Agri-PV

	EU27 + UK + CH: 1% agricultural area (1000 ha)	**PV capacity per category (MW/ha)**	**Potential (GW) for 1% coverage**	**Potential (%) for 1% coverage**
Arable land	1.052	0.4	421	44
Permanent crops	120	0.8	96	10
Permanent meadows and pastures	644	0.7	451	47
TOTAL	1.816		968	100

covering only 1% coverage of the European Union UAA (without considering UK and CH), the CAP could be reached 4.5 times.

According to SolarPower Europe, the European Union needs to deploy 870 GWp of solar by 2030 to set the most cost-efficient trajectory towards climate neutrality in 2050. This goal of 870 GWp of installed PV capacity in 2030, which implies 435 GWp produced on farmland (see Chapter 1) is widely achieved with the 1% coverage of Agri-PV on agricultural land in EU, which represents 968 GW of installed Agri-PV capacity.

2.2.4 Asia and Oceania *area and PV capacity potential of agrivoltaics*

2.2.4.1 Asia

The assessment of Asian Agri-PV potential includes the following countries:

Asian countries and areas	Afghanistan; Bangladesh; Bhutan; Brunei; Darussalam; Cambodia; China; Democratic People's Republic of Korea; India; Indonesia; Japan; Kazakhstan; Kyrgyzstan; Lao People's Democratic Republic; Maldives; Malaysia; Mongolia; Myanmar; Nepal; Pakistan; Philippines; Republic of Korea; Singapore; Sri Lanka; Tajikistan; Thailand; Timor-Leste; Turkmenistan; Uzbekistan; Viet Nam

In 2021, the combined UAA of Asia amounted to almost 1.345 billion ha, as recorded in the FAOSTAT Land Use domain). This area includes 820 million ha of permanent meadows and pastures, 434 million ha of arable land and 91 million ha of permanent crops. Permanent meadows and pastures represent the larger part of the UAA (60%). This category is followed by arable lands with 32% of the UAA, and permanent crops with only 7% of the UAA.

Table 2.7 Asia utilized agricultural area (UAA) and PV capacity potential per land use category, if 1% of UAA is implemented with Agri-PV

	1% of UAA of Asia (1000 ha)	**PV capacity per category (MW/ha)**	**Potential (GW) for 1% coverage**	**Potential (%) for 1% coverage**
Arable land	4.339	0.4	1.735	32
Permanent crops	910	0.8	728	7
Permanent meadows and pastures	8.205	0.7	5744	61
TOTAL	13.454		8.207	100

Table 2.7 presents the potential capacities attainable by covering 1% of the UAA in Asia with Agri-PV systems, considering various assumed installed power capacity per land categories. When 1% of the UAA is equipped with Agri-PV systems, the installed capacity would reach 8.207 GWp, equivalent to 8.2 TWp. Permanent meadows and pastures top the list in terms of Agri-PV potential, contributing a total of 5.7 TWp, which makes up 61% of the overall capacity. Arable land ranks second, but with a significantly lower share of just 32%. Permanent crops occupy the last position with 7% of the total capacity.

In 2022, the cumulative capacity and production (CAP) of solar PV in Asia reached 604.1 GWp, with China contributing 65% of this capacity. Remarkably, by installing solar panels on just 1% of the agricultural land in Asia, the CAP could potentially increase by a factor of 14.

2.2.4.2 Oceania

The assessment of Oceanian Agri-PV potential includes the following countries:

Oceanian countries and areas	American Samoa, Australia; Cook Islands; Fiji; French Polynesia; Guam; Kiribati; Malaysia; Marshall Islands; Micronesia (Federated States of); Nauru; New Caledonia; New Zealand; Niue; Norfolk Island; Northern Mariana Islands; Palau; Papua; New Guinea; Samoa; Solomon Island; Tokelau; Tonga Tuvalu; Vanuatu; Wallis and Futuna Islands

The total UAA in Oceania stands at 384 million ha, according to data from the FAOSTAT Land Use domain for the year 2021. This area is comprised of 342 million ha of permanent meadows and pastures, 33 million ha of arable land, and 9 million ha of permanent crops. Permanent meadows and pastures make up the majority of the UAA at 89%, followed by arable lands at 9%, and permanent crops at just 2%.

Table 2.8 Oceania utilized agricultural area (UAA) and PV capacity potential per land use category, if 1% of UAA is implemented with Agri-PV

	1% of UAA of Oceania (1000 ha)	**PV capacity per category (MW/ha)**	**Potential (GW) for 1% coverage**	**Potential (%) for 1% coverage**
Arable land	332	0.4	133	5
Permanent crops	92	0.8	74	3
Permanent meadows and pastures	3423	0.7	2396	92
TOTAL	3847		2603	100

Table 2.8 provides insights into the potential capacities that could be attained by covering 1% of Oceanian UAA with Agri-PV systems, assuming various installed power capacity per land area category.

By covering 1% of the UAA with Agri-PV systems, the total installed capacity would amount to 2.603 GWp, equivalent to 2.6 TWp. Notably, permanent meadows and pastures exhibit the highest potential for Agri-PV, contributing a total of 2.4 TWp, which represents 92% of the overall capacity. Arable land lags significantly with just 5%, while permanent crops occupy the smallest share at 3%.

The total capacity and production (CAP) of solar PV in Oceania reached 30.3 GWp in 2022 [42], with Australia providing 98% of it. By covering only 1% coverage of the Oceanian's agricultural lands, the CAP could be reached 86 times.

2.2.5 America area and PV capacity potential of Agrivoltaics

2.2.5.1 North America

The assessment of North America Agri-PV potential includes the following countries:

North American countries and areas	Bermuda; Greenland; Mexico; Saint Pierre and Miquelon; Canada; USA

The total UAA in North America is recorded at 560 million ha, as documented in the FAOSTAT Land Use domain for the year 2021. Within this expanse, there are 338 million ha of permanent meadows and pastures, 216 million ha of arable land, and 6 million ha of permanent crops. Permanent meadows and pastures account for the majority of the UAA, constituting 60%. Arable lands follow with 39% of the UAA, while permanent crops represent only 1% of the total UAA.

Table 2.9 North America utilized agricultural area (UAA) and PV capacity potential per land use category, if 1% of UAA is implemented with Agri-PV

	1% of UAA of North America (1000 ha)	**PV capacity per category (MW/ha)**	**Potential (GW) for 1% coverage**	**Potential (%) for 1% coverage**
Arable land	2161	0.4	864	27
Permanent crops	57	0.8	45	1
Permanent meadows and pastures	3384	0.7	2369	72
TOTAL	5602		3278	100

Table 2.9 presents insights into the potential capacities that could be realized by covering 1% of North America's UAA with Agri-PV systems, assuming various installed power capacity per land area categories.

By implementing Agri-PV on 1% of the UAA, the total installed capacity would reach 3.278 GWp, equivalent to 3.3 TWp. Notably, permanent meadows and pastures exhibit the highest potential for Agri-PV, contributing a total of 2.4 TW, which accounts for 72% of the overall capacity. Arable land follows with a distant 27%, while permanent crops represent just 1%. These ratios are notably similar to those observed in Asian countries.

The total capacity and production (CAP) of solar PV in North America reached 127.7 GW in 2022 (Irena, 2022), with USA providing 89% of it. By covering only 1% coverage of the North American agricultural lands, the CAP could be reached 26 times.

2.2.5.2 South America

The assessment of South America Agri-PV potential includes the following countries:

South American countries and areas	Argentina; Bolivia; Brazil; Chile; Colombia; Ecuador; Falkland Islands (Malvinas); French; Guiana; Guyana; Paraguay; Peru; Suriname; Uruguay; Venezuela

The total UAA of South America is around 535 million ha (FAOSTAT Land Use domain, 2021). This area includes 394 million ha of permanent meadows and pastures, 124 million ha of arable land and 17 million ha of permanent crops. Permanent meadows and pastures represent the larger part of the UAA (74%). This category is followed by arable lands with 23% of the UAA, and permanent crops with only 3% of the UAA.

Table 2.10 shows the capacities that could be achieved if 1% of the UAA of South America was covered with Agri-PV, with different assumed installed power capacity per land area categories.

Table 2.10 South America utilized agricultural area (UAA) and PV capacity potential per land use category, if 1% of UAA is implemented with Agri-PV

	1% of UAA of South America (1.000 ha)	PV capacity per category (MW/ha)	Potential (GW) for 1% coverage	Potential (%) for 1% coverage
Arable land	1.238	0.4	495	15
Permanent crops	168	0.8	134	4
Permanent meadows and pastures	3.938	0.7	2.757	81
TOTAL	5.344		3.386	100

With 1% of the UAA covered with Agri-PV system, the installed capacity would be 3.386 GWp, i.e. 3.4 TWp. Permanent meadows and pastures rank first regarding Agri-PV potential, with a total amount of 2.8 TWp, i.e. 81% of the total amount. Arable comes second by far with 15%. Permanent crops are last with only 4%.

The total capacity and production (CAP) of solar PV in South America reached 32.8 GWp in 2022 (Irena, 2022), with Brazil providing 73% of it. By covering only 1% coverage of the South America's agricultural lands, the CAP could be reached 100 times. South America holds significant potential for Agri-PV, but there has been relatively limited solar installation in the region.

2.2.5.3 Central America and Caribbean

The assessment of Central America Agri-PV potential includes the following countries:

Central America and Caribbean countries and areas	Anguilla; Antigua and Barbuda; Aruba; Bahamas; Barbados; Belize; British Virgin Islands; Cayman Islands; Costa Rica; Cuba; Curacao; Dominica; Dominican Republic; El Salvador; Grenada; Guadeloupe; Guatemala; Guyana; Haiti; Honduras; Jamaica; Martinique; Nicaragua; Panama; Puerto Rico; Saint Barthélémy; Saint Kitts and Nevis; Saint Lucia; Sint Maarten; St Martin; Saint Vincent and the Grenadines; Trinidad and Tobago; Turks and Caicos Islands; United States Virgins Islands

The total UAA in Central America and the Caribbean stands at approximately 31 million ha, according to data from the FAOSTAT Land Use domain for the year 2021. This region comprises 16 million ha of permanent meadows and pastures, 11 million ha of arable land, and 4 million ha of permanent crops. Permanent meadows and pastures make up the majority of the UAA at 50%. Arable lands follow with 36% of the UAA, while permanent crops account for a significant 14% of the total UAA.

Table 2.11 Central America and Caribbean utilized agricultural area (UAA) and PV capacity potential per land use category, if 1% of UAA is implemented with Agri-PV

	1% of UAA of Central America and Caribbean (1000 ha)	PV capacity per category (MW/ha)	Potential (GW) for 1% coverage	Potential (%) for 1% coverage
Arable land	112	0.4	45	24
Permanent crops	42	0.8	34	18
Permanent meadows and pastures	157	0.7	110	58
TOTAL	311		189	100

Table 2.11 provides insights into the potential capacities that could be achieved by covering 1% of Central America and the Caribbean's UAA with Agri-PV systems, assuming various installed power capacity per land category.

With Agri-PV systems applied to 1% of the UAA, the total installed capacity would reach 189 GWp, equivalent to around 0.2 TWp. Notably, permanent meadows and pastures exhibit the highest potential for Agri-PV, contributing a total of 110 GWp, which constitutes 58% of the overall capacity. Arable land comes in second place, but with a substantial gap at 24%. Permanent crops fall last with a share of 18%. It's worth mentioning that in Central America and the Caribbean, the prevalence of permanent crops is significantly higher compared to other regions of the world.

The total capacity and production (CAP) of solar PV in Central America and Caribbean reached 3.9 GWp MW in 2022 [42]. By covering only 1% coverage of the Asian's agricultural lands, the CAP could be reached more than 45 times.

2.2.6 African continent area and PV capacity potential of agrivoltaics

The assessment of African Agri-PV potential includes the following countries:

Countries/ islands located in Africa	Algeria; Angola; Benin; Botswana; Burkina Faso; Burundi; Cameroon; Cape Verde; Central African Republic; Chad; Comoros; Congo; Côte d'Ivoire; Democratic Republic of the Congo; Djibouti; Egypt; Equatorial Guinea; Eritrea; Eswatini; Ethiopia; Gabon; Gambia; Ghana; Guinea; Guinea-Bissau; Kenya; Lesotho; Liberia; Libya; Madagascar; Malawi; Mali; Mauritania; Mauritius; Mayotte; Morocco; Mozambique; Namibia; Niger; Nigeria; Réunion; Rwanda; Saint Helena; Sao Tome and Principe; Senegal; Seychelles; Sierra Leone; Somalia; South Africa; South Sudan; Sudan; Togo; Tunisia; Uganda; United Republic of Tanzania; Western Sahara; Zambia; Zimbabwe

Table 2.12 Africa utilized agricultural area (UAA) and PV capacity potential per land use category, if 1% of UAA is implemented with Agri-PV

	1% of UAA of Africa (1.000 ha)	**PV capacity per category (MW/ha)**	**Potential (GW) for 1% coverage**	**Potential (%) for 1% coverage**
Arable land	2.534	0.4	1.014	14
Permanent crops	398	0.8	319	4
Permanent meadows and pastures	8.685	0.7	6.079	82
TOTAL	11.617		7.412	100

The total UAA in Africa is the largest, covering approximately 1.2 billion ha, as reported in the FAOSTAT Land Use domain for the year 2021. This extensive area comprises 869 million ha of permanent meadows and pastures, 253 million ha of arable land, and 40 million ha of permanent crops. Permanent meadows and pastures make up the majority of the UAA at 75%. Arable lands follow with 22% of the UAA, while permanent crops represent only 3% of the UAA.

Table 2.12 presents data on the potential capacities that could be achieved by covering 1% of Africa's UAA with Agri-PV systems, assuming various installed power capacity categories per land area.

With 1% of the UAA covered by an Agri-PV system, the installed capacity would reach 7.412 GWp, equivalent to 7.4 TWp. Notably, permanent meadows and pastures exhibit the highest potential for Agri-PV, contributing a total of 6.1 TWp, which accounts for 82% of the overall capacity. Arable land follows at a considerable distance with 14%, while permanent crops fall last with only 4%.

The total capacity and production (CAP) of solar PV in Africa reached 12.5 GWp in 2022 (Irena, 2022). This number is significantly low regarding the total area of the continent. This can be explained by difficulties to have access to a reliable grid as well as the economic and political instability in some countries. By covering only 1% coverage of the Africans agricultural lands, the CAP could be reached 593 times. This ratio is the highest of every region considered.

In a broader context, with a 1% coverage ratio, we could reach 140 times the total CAP of renewable energy of the continent (211 times the CAP of hydropower, 1030 times the wind CAP).

2.2.7 Conclusion

Covering only 1% of the UAA offers a glimmer of hope in reaching global solar energy targets while maintaining our course to limit global warming to 1.5 °C. Agri-PV presents an effective opportunity to address both energy and climate goals. Nevertheless, swift political action is needed to protect future farmland and to integrate Agri-PV applications into the farming process.

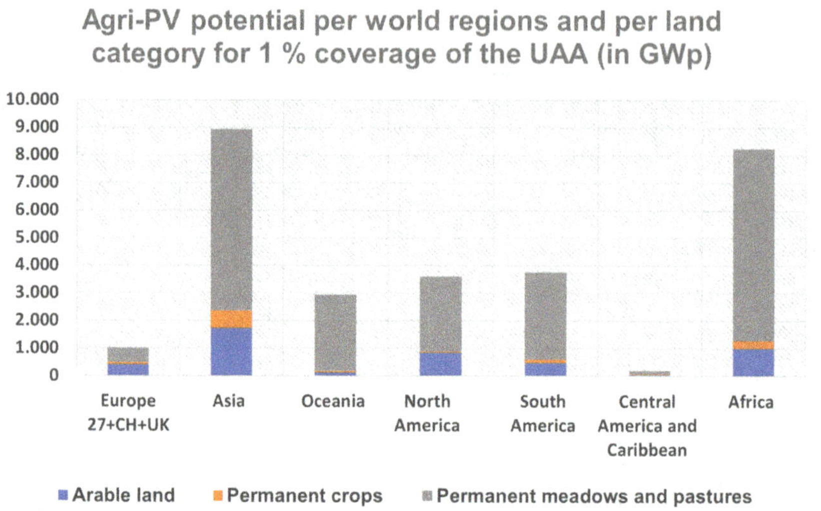

Figure 2.13 Agri-PV potential in GWp per world region and per land use category for 1% coverage of the utilized agricultural area (UAA)

Permanent pastures consistently emerge as the most promising land category for Agri-PV development across various regions. This pattern holds true for all the regions examined in this document, highlighting the vast potential in this land category, see Figure 2.13.

The analysis also emphasizes the prominent role of Asia in the global potential for Agri-PV solutions, with this region ranking first (8.207 GWp). Following closely behind, Africa exhibits substantial potential as well.

But it is of importance to acknowledge the diverse challenges that face regions considered in the study. Due to a lack of grid access and stability, political and economic frameworks, as well as topography, some regions might not qualify as best-suited for the development of Agri-PV despite their large area potential. E.g., even though the African continent shows vast area potential for the deployment of Agri-PV it might partially not support the security for investments to be attracted. Therefore, it is crucial to consider region-specific nuances.

In areas like Europe, Oceania, Asia, and North America, where per capita CO_2 emissions are high and vast agricultural land with low technical constraints against the deployment of Agri-PV exists, the question arises whether aiming for 1% UAA coverage is sufficient, and if a uniform covering ratio/goal is appropriate. These regions offer a favorable landscape for the installation of Agri-PV solutions, presenting an opportunity to address both energy and environmental concerns effectively. Adjusting the coverage ratio would then be a way for each country to shoulder their role in the matter of CO_2 emissions over the last centuries.

In navigating the transition to sustainable energy and combating climate change, recognizing the unique circumstances of each region will be essential for tailoring strategies that maximizes the potential of Agri-PV, while addressing local challenges and opportunities.

References

[1] Renn O. *Methoden und Verfahren der Technikfolgenabschätzung und der Technologiebewertung*. Universität Stuttgart; 1982. Available from https://elib.uni-stuttgart.de/bitstream/11682/7250/1/ren2.pdf.

[2] Goetzberger A and Zastrow A. On the coexistence of solar-energy conversion and plant cultivation. *International Journal of Solar Energy*. 1982 [cited 2022 Apr 25];1982(1):55–69. Available from DOI:10.1080/01425918208909875.

[3] Goetzberger A, Swanson R, Werner T and Yamaguchi M. Energy Farming: Für einen intelligenteren Gebrauch von Agrarsubventionen. *Solarzeitalter*. 2006; 2006 (4):44–48.

[4] Beck M, Bopp G, Goetzberger A, Obergfell T, Reise C and Schindele S. Combining PV and Food Crops to Agrophotovoltaic – Optimization of Orientation and Harvest [5 pages / *27th European Photovoltaic Solar Energy Conference and Exhibition*; 4096-4100] 2012.

[5] Dupraz C, Marrou H, Talbot G, Dufour L, Nogier A and Ferard Y. Combining solar photovoltaic panels and food crops for optimising land use: Towards new agrivoltaic schemes. *Renewable Energy*. 2011 [cited 2022 Apr 27];36(10):2725–32. Available from https://www.sciencedirect.com/science/article/abs/pii/S096014811100l194#!

[6] PRI. *Japanese Farmers Producing Crops and Solar Energy Simultaneously: Revitalizing the Farming Sector for the Next-Generation and Could Provide More Than Enough Electricity for the Entire Country*. [Restriction to solar PV installation on farmlands removed.]. Permaculture Research Institute; 2013 [cited 2022 May 30]. Available from https://permaculturenews.org/2013/10/16/japanese-farmers-producing-crops-solar-energy simultaneously/.

[7] GIZ. *An Overview of the Current Situation of Agrivoltaic Development in China and Some Recommendations: Sino-German Energy Partnership*. The Sino-German Energy Partnership; 2022.

[8] REMTEC. *Borgo Virgilio: Agrovoltaico Plant Mantova - Italy*. REM TEC S. r.l.; 2020 [cited 2022 Apr 27]. Available from https://remtec.energy/en/agrovoltaico/installations/29-borgo-virgilio.

[9] DIN. *DIN SPEC 91434:2021-05: Agri-Photovoltaik-Anlagen - Anforderungen an die landwirtschaftliche Hauptnutzung*. DIN; 2022 [cited 2022 Apr 26]. Available from https://www.beuth.de/de/technische-regel/din-spec-91434/337886742.

[10] Comitato Elettrotecnico Italiano. PAS 82-93 Impianti Agrivoltaics; 2023 2023 Apr 29 [cited 2023 Dec 19]. Available from https://www.certifico.com/normazione/210-documenti-riservati-normazione/organismi-normazione/documenti-cei/19523-cei-pas-82-93.

[11] AFNOR. Label Projet Agrivoltaïque. [cited 2023 Dec 19]. Available from https://certification.afnor.org/energie/label-agrivoltaique-positif

[12] Ketzer D, Weinberger N, Rösch C and Seitz SB. Land use conflicts between biomass and power production – citizens' participation in the technology development of Agrophotovoltaics. *Journal of Responsible Innovation.* 2019;15(2):1–24.

[13] Schindele S. Feldfrüchte und Strom von Agrarflächen: Was ist Agri-Photovoltaik und was kann sie leisten? *GAIA - Ecological Perspectives for Science and Society.* 2021 [cited 2022 Apr 26];30(2):87–95. Available from https://www.ingentaconnect.com/content/oekom/gaia/2021/00000030/00000002/art00007.

[14] Schindele S. Nachhaltige Landnutzung mit Agri-Photovoltaik: Photovoltaikausbau im Einklang mit der Lebensmittelproduktion: Szenarioanalyse zur Inanspruchnahme landwirtschaftlicher Nutzflächen durch Photovoltaik in Deutschland bis 2050. *GAIA - Ecological Perspectives for Science and Society.* 2021 [cited 2022 Apr 27];30(2):96–105. Available from https://www.oekom.de/_files_media/zeitschriften/artikel/GAIA_2021_02_96.pdf.

[15] Schindele S. *Nachhaltige Landnutzung Dank Doppelernte: Eine Mehrdimensionale Politikanalyse der Agri-Photovoltaik-Diffusion in Deutschland [Dissertation]*. Tübingen, Germany: Universität Tübingen; 2023.

[16] Hill J L. Publicly available specification. Standards; 1995.

[17] British Standards Institutions. *A Standard for Standards — Principles of Standardization.* BSi Standards Publication; 2012 2012. Available from https://www.bsigroup.com/globalassets/localfiles/en-gb/standards/bs-0-2021—a-standard-for-standards.-principles-of-standardization.pdf [cited 2023 Apr 1].

[18] de-academic.com. Normierungsorganisation: Academic [cited 2023 Dec 19]. Available from https://de-academic.com/dic.nsf/dewiki/1029747#Einzelnachweise.

[19] Matland R E. Synthesizing the implementation literature: The ambiguity-conflict model of policy implementation. *Journal of Public Administration Research and Theory: J-PART.* 1995;5(2):145–74. Available from http://www.jstor.org/stable/1181674.

[20] DIN. *Geschäftsplan für ein DIN SPEC-Projekt nach dem PAS-Verfahren zum Thema Agri-Photovoltaik-Systeme – tierhaltungsspezifische Anforderungen.* DIN; 2023 May 16 [cited 2023 Dec 19]. Available from https://www.din.de/de/wdc-beuth:din21:364494320/pdf-3414657.

[21] Scharpf F W, Reissert B and Schnabel F. *Politikverflechtung. Theorie und Empirie der kooperativen Föderalismus in der Bundesrepublik.* Kronberg/Ts.: *Scriptor-Vrlag*; 1976. (Monographien Ergebnisse der Sozialwissenschaften; vol 1).

[22] Benz A. Konstruktive Vetospieler in Mehrebenensystemen. In: Mayntz R, ed. *Die Reformierbarkeit der Demokratie: Innovationen und Blockaden; Festschrift für Fritz W. Scharpf. Frankfurt/Main*, New York: Campus-Verl.; 2003. pp. 205–36 (Schriften aus dem Max-Planck-Institut für Gesellschaftsforschung; Bd. 45) [cited 2022 May 18]. Available from https://books.google.de/books?hl=de&lr=&id=uGRKNR7mLBIC&oi=fnd&pg=PA205&dq=Benz,+Arthur,+2003:+Konstruktive+Vetospieler+in+Mehrebenensystemen,+in:+Mayntz,+Renate+/+Streeck,+Wolfgang+(Hrsg.),+2003:+Die

+Reformierbarkeit+der+Demokratie,+Innovationen+und+Blockaden,+Frankfurt+/+Main,+205+%E2%80%93+238.&ots=xSl3fr12AS&sig=P9DFKP37eh5OyPxBEHiaG4focpA&redir_esc=y#v=onepage&q&f=false.

[23] Finanzbehörden der Länder. Zurechnung und Bewertung von Agri-Fotovoltaik-Anlagen: Bewertungsgesetz; BStBl I 1226. 2022 Jul 15 15.

[24] Dinesh H and Pearce J M. The potential of agrivoltaic systems. *Renewable and Sustainable Energy Reviews.* 2016 [cited 2022 May 18];54:299–308.

[25] Hernandez R R, Armstrong A, Burney J, *et al.* Techno–ecological synergies of solar energy for global sustainability. *Nature Sustainability.* 2019 [cited 2022 May 18];2(7):560–8. Available from https://www.nature.com/articles/s41893-019-0309-z.

[26] Willockx B, Uytterhaegen B, Ronsijn B, Herteleer B and Cappelle J. A Standardized Classification and Performance Indicators of Agrivoltaic Systems [4 pages / *37th European Photovoltaic Solar Energy Conference and Exhibition*; 1995–1998] 2020.

[27] Fraunhofer ISE. *Agri-Photovoltaik: Chance für Landwirtschaft und Energiewende: Ein Leitfaden für Deutschland / Stand April 2022.* Freiburg: Fraunhofer Institute for Solar Energy Systems ISE; 2022 [cited 2022 Aug 11]. Available from https://www.ise.fraunhofer.de/content/dam/ise/de/documents/publications/studies/APV-Leitfaden.pdf.

[28] SolarPower Europe. Agrisolar Best Practice Guidelines: Version 1.0; May 2021 [cited 2022 May 16]. Available from https://www.solarpowereurope.org/insights/thematic-reports/agrisolar-best-practice-guidelines.

[29] Brumme D. Was bringt Solarthermie im Winter?; 2017 [cited 2023 Dec 19]. Available from https://blog.paradigma.de/was-bringt-solarthermie-im-winter/.

[30] SolarPower Europe. *Agrisolar Best Practice Guidelines: Version 2.* SolarPowerEurope; 2023 [cited 2023 Dec 19]. Available from https://www.solarpowereurope.org/insights/thematic-reports/agrisolar-best-practice-guidelines-version-2-2.

[31] Macknick J and Hartmann H. *The 5 Cs of Agrivoltaic Success Factors in the United States: Lessons From the InSPIRE Research Study.* NREL; August 2022 NREL/TP-6A20-83566 [cited 2023 Dec 19]. Available from https://www.nrel.gov/docs/fy22osti/83566.pdf.

[32] NREL. Agrivoltaics Map: InSPIRE Project; 2023 [cited 2023 Nov 19]. Available from https://openei.org/wiki/InSPIRE.

[33] Bolinger M and Seel J. *Utility-Scale Solar, 2023 Edition: Empirical Trends in Deployment, Technology, Cost, Performance, PPA Pricing, and Value in the United States*. October 2023 [cited 2023 Dec 19]. Available from https://emp.lbl.gov/sites/default/files/utility_scale_solar_2023_edition_slides.pdf.

[34] Dupraz C. Assessment of the Ground Coverage Ratio of AgriVoltaic systems as a proxy for potential crop productivity. 2023.

[35] Sunbiose. Sunbiose is a 4 year project for the development of Agri-PV solutions. 2023 [cited 2023 Nov 7]. Available from https://sunbiose.nl/.

[36] Paris Agreement. 2015 [cited 2023 Dec 19]. Available from https://unfccc.int/sites/default/files/english_paris_agreement.pdf.

[37] Chatzipanagi A, Taylor N, Thiel C, Jaeger-Waldau A, Dunlop E and Kenny R. Agri-photovoltaics (Agri-PV): How multi-land use can help deliver sustainable energy and food; JRC129225; 2022 May 5. Available from https://publications.jrc.ec.europa.eu/repository/handle/JRC129225.

[38] Hassanpour Adeh E, Selker J S and Higgins C W. Remarkable agrivoltaic influence on soil moisture, micrometeorology and water-use efficiency. *PLoS ONE*. 2018;13(11):e0203256.

[39] FAO. *FAOSTAT: Land Use*. Food and Agriculture Organization of the United Nations; 2022 [cited 2022 May 17]. Available from https://www.fao.org/faostat/en/#data/RL.

[40] Series M. No. 49. Methodology: Standard country or area codes for statistical use (M49): United Nations. Available from https://unstats.un.org/unsd/methodology/m49/ [cited 2023 Nov 8].

[41] Hemetsberger W, Schmela M and Cruz-Capellan T. *Global Market Outlook For Solar Power 2023–2027*. SolarPower Europe; June 2023 ISBN: 9789464669046. [cited 2023 Dec 20]. Available from https://www.solarpowereurope.org/insights/outlooks/global-market-outlook-for-solar-power-2023-2027/detail#global-solar-market-update-2000-2022.

[42] IRENA. *World Energy Transitions Outlook 2023*. International Renewable Energy Agency; 2022 [cited 2023 Nov 8]. Available from https://www.irena.org/Digital-Report/World-Energy-Transitions-Outlook-2023.

[43] Goldschmidt J C, Wagner L, Pietzacker R and Friedrich L. Technological learning for resource efficient terawatt scale photovoltaics. *Energy and Environmental Science* 2021 [cited 2023 Nov 7];14:5147–60. Available from https://pubs.rsc.org/en/content/articlelanding/2021/ee/d1ee02497c.

Chapter 3

Powering growth: technological insights into photovoltaics in agrivoltaic systems

Constantin Klyk[1] *and Peter Bendix*[2]

The integration of agriculture and photovoltaics, commonly known as agrivoltaics, is a rapidly emerging field that has the potential to play an important role in a fully renewable energy system that is compatible with ecosystem restoration, nature conservation and agriculture. This chapter provides a comprehensive overview of the technological aspects of agrivoltaic systems, with a specific focus on the design and implementation of solar panels and sub-structures, as they are the main components differing from a standard PV installation. Further than this, special needs and developments in inverter design are showcased and challenges in the power-plant planning are discussed.

Agrivoltaic systems are designed to utilise the same land for both food production and energy generation, thereby increasing land use efficiency and reducing the environmental impact of both industries. Solar panels are a crucial component of these systems, as they convert sunlight into electricity.

The design and implementation of solar panels for agrivoltaics systems requires careful consideration of factors such as shading, orientation and tilt angle to ensure optimal performance. Emerging module technologies, such as transparent solar panels, are also being explored in agrivoltaics systems as they have the potential to further increase the potential symbiosis in agrivoltaics.

The sub-structure is a key component of agrivoltaics systems, as it provides the necessary support for the solar panels while also ensuring enough space for the growth of crops beneath. The design of the sub-structure is mainly driven by the type of agrivoltaic system and can be divided into the three categories named overhead agrivoltaic, interrow agrivoltaic or solar greenhouses. Factors such as wind loads, snow loads and the integration of agricultural subsystems as growing wires in fruitvoltaics, water irrigation systems or rainwater harvesting gutters, must be take into account in the design and construction of these sub-structures. Of course, the system layout of the Agri-PV park is also of importance, e.g. the pitch between PV rows allowing agricultural machinery to pass through according to the

[1]Agri-PV Consulting, Jörg-Zürn-Str. 22 , 88339 Bad Waldsee, Germany
[2]Next2Sun Technology GmbH, Franz-Meguin-Str. 10a, Germany

specific crop being grown, placement of transformer station, inverters, fencing, entrance and exit gates in the fence, or mobile roads during construction avoiding soil compaction. In this chapter, the technical details of these components and designs are presented, discussing the latest research and developments in the field. Also, the challenges and opportunities associated with the integration of agrivoltaics systems into existing agricultural operations will be examine. In addition, we will present case studies of successful agrivoltaics projects, providing a practical perspective on the implementation of this technology. Overall, this chapter aims to provide a comprehensive understanding of the technological aspects of photovoltaics in agrivoltaics systems, highlighting the potential of this innovative approach to addressing the challenges of food security and energy production. As the demand for sustainable land use and energy solutions continues to grow, the development and implementation of agrivoltaics technology will play an increasingly important role in shaping our future.

3.1 Reaching new heights: exploring overhead agrivoltaic systems

This chapter explores the diverse approaches to overhead agrivoltaic systems, specifically focusing on fixed tilt systems, trackers and rope suspended systems. These configurations offer high protective effects for crops while enabling dual land use, combining renewable energy production with productive agriculture. The technical aspects, benefits and considerations of these systems are described, providing insights for implementing effective overhead agrivoltaics.

3.1.1 Influence of module placement: height and row spacing

An overhead agrivoltaic system is a unique way to combine solar energy production and agricultural production in a single space. The system consists of a solar panel array installed above a crop canopy, allowing for the simultaneous growth of crops and generation of electricity. The design of the system, including the height and row spacing of the solar panels, plays a crucial role in determining the light availability and an equal light distribution for the crop cultivation beneath [1].

One of the main factors that affects the light available for the crops is the height of the solar panels. The higher the panels are installed, the more even will the light be distributed in the crop canopy [2]. At the same time the height of the system will determine the possibilities of access for workers and machinery to the project area. A decreased height might lead to an increase of labour cost in the agrivoltaic setting. However, there is a trade-off between the height of the panels and the cost of the system. Higher panels require more structural support and are more expensive to install [3]. Additionally, higher installed solar panels may result in increased wind loads, which can further increase the cost and complexity of the system design [2,3].

Module spacing is another important factor to consider in the design of an overhead agrivoltaic system. The closer the PV-Modules are installed to each other, the less light will reach the crops. Module Spacing can be considered in both direction while having different influence on the crop and soil below. Increasing the Module spacing will increase the possibilities of Agriculture while closer spacing of the modules can increase economic efficiency of the PV-array by increasing the installed capacity per area and decreasing the cost of the structure per module [3]. A dense installation can also result in insufficient light availability throughout the agricultural area. This can be a particular problem for crops that are shade intolerant [4–6]. Further the module spacing will influence the distribution of water, especially in case of usage of gutters at the module edges.

The ideal row spacing and height of the solar panels will depend on the specific crop being grown, as well as the local climate and weather conditions. For example, shade intolerant crops will require wider row spacing to ensure sufficient light reaches the crop canopy and possibly higher solar panels to provide a more even light distribution [7]. However, shade loving crops can benefit from a low and dense installation by the protective effect of the PV array and the higher changes in Microclimate [4,5]. In addition to the height and row spacing of the solar panels, other factors that influence the light availability for the agricultural use include the angle and orientation of the panels, as well as the type and transparency of the materials used for the solar panel and support structure. The angle and orientation of the panels can affect the amount and distribution of light reaching the crop canopy. Also the transparency of the PV-module and the design of the support structure can impact the homogenous light distribution [5].

The pattern of the PV-array influences how uniform the light is spread under the modules. Changing the pattern can adjust the distribution of light, with chessboard patterns resulting in more uniform light distribution compared to a row patterns [1]. At the same time a chessboard pattern will increase the cost of the system, as the fixation will get more complicated. Figure 3.1 shows a comparison between a line configuration and a chessboard configuration in the agrivoltaic system.

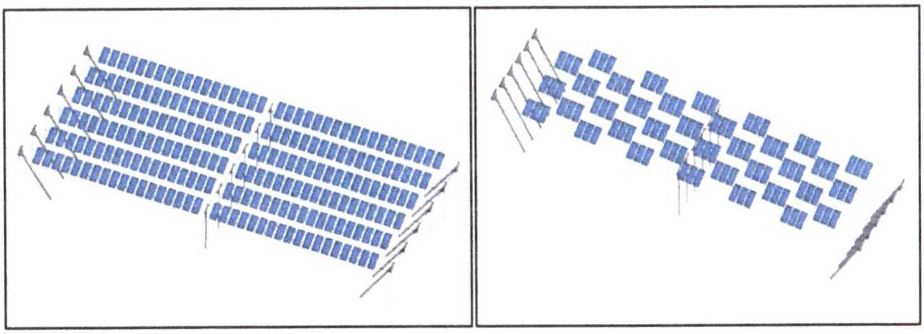

Figure 3.1 A line configuration of a high elevated agrivoltaic system at the example of a rope suspended system (left) and a chessboard configuration in a suspended system (right) © REM Tec

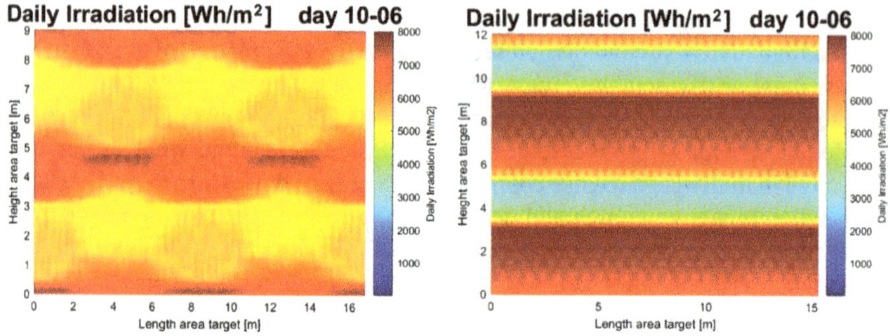

Figure 3.2 Shadow frequence in comparison between a chessboard (left) and a line (right) configuration. The line configuration shows much sharper lines and has areas with very little direct sunlight and other areas with a lot of direct sunlight. © REM Tec.

The orientation of the modules can also have a significant influence on the distribution of light under the PV-array. In a south-facing system, there may be areas that are strongly shadowed throughout the year, while other areas receive full sunlight. In east or west-facing systems, the course of the pattern is more agile due to the movement of the sun and the light is spread more evenly throughout the year. According to a study by Toledo and Scognamiglio [1], a proper orientation of the solar panel can help to achieve a better balance between energy production and crop growth in an agrivoltaic system [1,6].

Figure 3.1 shows a comparison between a row configuration of PV modules compared to a chessboard configuration of the modules [8]. Figure 3.2 shows the associated shadow frequencies. It is evident that in the chessboard layout, the distribution of solar light is considerably more homogeneously, whereas in the row configuration, certain portions of the agricultural plot receive very little irradiation under the modules while others receive significantly more irradiation. These issues may be especially problematic for crops. The crops that receive a lot of sunlight will almost certainly mature differently from the crops that receive a lot of shadow. This will end up in a more complex harvest strategy.

In conclusion, adjusting module placement strategies is a key factor in optimising the performance of an agrivoltaic system. By adjusting the density and pattern of the modules, as well as the orientation of the panels, it is possible to influence the amount of light available for the crops and the distribution of shadows. Future research should focus on developing optimal module placement strategies that balance energy production with crop growth.

3.1.2 Tracking systems

To optimise the angle of the modules toward the sun, there are a range of mechanisms for overhead agrivoltaic systems. Single-axis tracking and dual-axis tracking are two categories into which these methods can be divided.

Figure 3.3 Single-axis tracker with a height of 5 m and a pole distance of 9 m © Zimmermann PV-Stahlbau

Single-axis tracking systems [9] are usually set up in a North–South orientation, allowing the modules to follow the sun from east to west throughout the day. These systems consist of a long structure with one or more motors that control the inclination of the modules. Single-axis tracking systems require fewer moving parts, hence they are simpler to design and build, but they may not achieve the highest possible energy yield per installed kilowatt peak. Figure 3.3 shows a highly elevated single-axis tracker that has been built with Zimmermann PV-Tracker technology in southern France.

Dual-axis tracking systems [8], on the other hand, are able to exactly follow the position of the sun. Having the modules always in the best possible direction towards the sun promises the highest possible yield per module, but it also requires a more complex system to be build. Dual-axis tracking systems typically use one engine to control the east-west inclination of the modules and another engine to control the north-south inclination. This mix allows an higher energy production than for the single-axis tracker. The less cost-effectiveness per installed kWp compared to the single-axis tracking can be partly compensated by the higher lifetime yields. The projected shadow on the ground has the highest precision during the complete day tracking for a dual-axis tracker. Figure 3.4 shows a high elevated dual-axis tracker. The primary tube, as well as the secondary axis allows the module to follow the sun like a sunflower. Complementary to advantages related to tracking systems, the shadow control allows to optimise light irradiation of the crops in order to improve crop production. This system is designed and built by REM Tec. It is in operation in Italy over 45 ha since 2011 and in many other countries in the world (Figure 3.4).

Tracking systems have several advantages over fixed tilt systems, such as a higher energy yield per installed kilowatt peak [10]. The capacity of agrivoltaic systems to completely shade the crops below or provide them with all of the sun's radiation is quite intriguing and creates new opportunities for optimising the light available to the plants [8,10]. This can be combined with the possibility to control the tracker according to power prices. At low feed-in prices the tracker can pass all

Figure 3.4 High elevated dual-axis tracker in France. Differences in shading between full light tracking (left) and full shadow tracking © REM Tec

the available irradiance to the plants while producing power at the times of higher prices. Additionally tilting the modules in case of snowfall can help to reduce snow loads [9]. By tilting the modules in the best angle towards the wind, tracking systems also help to reduce loads on the structure. These factors allow tracking systems to be built with higher spans between the poles compared to fixed tilt installations [9]. Another major advance of tracker systems is that they can be tracked according to demand of the crop, e.g. in very hot days the crop can benefit from full shade or in days with little direct irradiation the crop can benefit from full sunlight through a anti tracking algorithm.

Case study: high elevated dual-axis tracking system Castelvetro

The high elevated dual-axis agrivoltaic tracker built by REM Tec in 2011 in Castelvetro, Italy is a 1.293 MW solar system installed on a 7-ha site. The project is a unique combination of solar energy and agriculture, with the solar panels mounted on a dual-axis tracker system that moves to follow the sun's trajectory throughout the day. The influence on the crops below has been minimised while the energy generation from the solar panels has been optimised thanks to this technology.

Figure 3.5 shows the dual-axis tracker system consisting of individual tracker rows, interconnected with wire ropes to take vertical loads. It was built with a post distance of 12 m allowing for agricultural machinery to pass. The panels are mounted on a structure that allows them to rotate on two axes, following the sun's movement throughout the day to maximise energy generation as well as crop production and optimise shadow precision.

The agrivoltaic tracker was installed with the aim of investigating the potential for solar energy systems to coexist with agricultural crops, as well as the environmental impacts of such systems. The project is mainly focused on large cultures like maize, sunflowers and many other crops, with a specific emphasis on analysing the impact of shading on crop yields and water consumption.

The results of the study were positive, indicating that the agrivoltaic system was able to produce a significant amount of energy while also maintaining crop yields. The dual-axis tracker system was found to provide more uniform shading

Figure 3.5 High elevated dual-axis tracker in Castelvetro (Italy) © REM Tec

than fixed-angle systems, resulting in minimal impacts on crop growth and development. Additionally, the shade provided by the solar panels reduced evapotranspiration rates, resulting in decreased water consumption by the crops.

The project also evaluated the environmental impacts of the agrivoltaic system. The dual-axis tracker system was found to have a smaller footprint than other solar energy systems, which allowed for more land to be used for agricultural purposes. Additionally, the project found that the agrivoltaic system had a positive impact on soil moisture, which lead to improve soil biodiversity and health as well as higher crop yields in the long term

Case study: high elevated single-axis tracker St. Etienne

The high elevated single-axis tracker built in St. Etienne, France by the Zimmermann Company in 2019 is a 3.2 MW solar energy project that spans over 8 m and stands at a height of 5 m. The innovative design of the system allows for regular farming below, making it an ideal solution for areas with limited land availability.

Figure 3.6 shows the system consisting of a rigid steel frame with a single-axis tracker on top that moves to follow the sun's trajectory throughout the day, optimising energy production from the solar panels. The tracker is mounted at a height of 5 m, which allows for regular farming activities to take place below the solar panels. The system is designed to accommodate changing crops, with enough space between rows to allow the passage of tractors and other farm equipment.

The project's research focused on analysing the benefits of using a high elevated single-axis tracker in combination with regular farming activities. The system was found to have several positive impacts on the crops and soil. The shading provided by the solar panels reduced the intensity of the sun's rays, which led to lower temperatures and reduced water consumption by the crops. The tracker system was also found to create a microclimate under the panels, which could potentially lead to improved soil health and higher crop yields in the long term.

Figure 3.6 Highly elevated single-axis tracker being built in St. Etienne, France © Zimmermann PV-Stahlbau

The high elevated single-axis tracker in St. Etienne, France is an example of how solar energy and agriculture can coexist in a mutually beneficial manner. While minimising the negative effects on crop yields and the environment, the system enables the effective use of scarce land resources. Additionally, the project's research has provided valuable insights into the potential benefits of combining solar energy with regular farming activities.

3.1.3 Fixed tilt overhead systems

Fixed tilt agrivoltaic systems are characterised by the installation of PV panels at a fixed angle, which is typically between 10 and 30 degrees, and do not move throughout the day to follow the sun's movement [3]. The height and spans of fixed tilt systems are determined by the crops below and the machinery that will be used in the agricultural operation [3].

According to the German standard DIN SPEC 91434, an overhead system must be at least 2.1 m high [11]. In practice, installations can go up to 6 m [3,12]. These systems have Spans between 2 m and 19 m [12]. The use of more vertical supports, or stand tubes, reduces the overall amount of material needed, which is an important consideration when designing an agrivoltaic system, as materials can represent a significant cost [3]. Usually, this criterion is in conflict with the farmers perspective, who needs the maximum spans possible to have less restrictions in his farming strategies.

In fixed tilt agrivoltaic systems, most systems up to date use Standard modules installed in a pattern. The scattered pattern ensures an adequate amount of light to reaches the crops below, allowing for the simultaneous production of energy and crops. In more recent systems, an increased use of semi-transparent modules (compare Section 3.4.3.) is seen that allow for a denser placement of modules. This is advantageous for plants with a high need of protection from weather [13]. The specific degree of transparency required may vary depending on the crop type, location and other environmental factors, and should be carefully evaluated in site-specific studies to determine the optimal configuration [14].

Fixed tilt systems are well suited for orchards, providing a high level of protection for the plants, and the option to integrate hailnets or foils for weather protection can provide additional benefits [3,15]. Further integration of rainwater harvesting is a possibility with high benefits in fixed tilt agrivoltaic systems. This type of system allows the integration of PV panels in a way that does not disrupt the crops below and enables simultaneous growth of crops and generation of energy [3,15].

While the energy yield per installed kilowatt peak may be lower compared to single or dual-axis tracking systems, fixed tilt systems are simpler to build and maintain, making them a cost-effective option for many agricultural operations [3,9]. It is important to consider the crop and location when choosing the right system, and the best way to determine the optimal configuration is to conduct a study of the specific site and its conditions [13].

Different overhead systems have very different amounts of steel consumption. While low canopy systems with small pillar distances use only slightly more Steel than a standard GM-PV system overhead systems with great spans and big heights can use up to four times the amount of this or even go higher. These amounts are not only affected by the system design but also highly affected by the loads at the project size. Table 3.1 shows a comparison between different overhead agrivoltaics systems and their steel consumption per kWp.

Case study: high elevated fixed tilt Agri-PV system with high spans Heggelbach

In Heggelbach, Southern Germany, an agrivoltaic system was implemented with a unique design in 2016. The system has pillar distances of 13 m in the

Table 3.1 Investigation on steel consumption in different fixed tilt systems

Crop	Type of system	Hight of lower module edge	Machinery working width	Steel consumption (kg/kWp)
Raspberry	E/W Canopy System	2.5 m	2.8 m	40.04
Apple	E/W Canopy	3.5 m	3.2 m	46.16
Wine	E/W Top Row	2.5 m	8 m	130.3
Arabel	Rope suspended System	5.5 m	25 m	98

1The values of systems 1–3 are based on standardised calculations by Zimmermann PV-Stahlbau based on real systems.
The value of the arable land system is based on REMTecs Rope suspended system.

Figure 3.7 Agrivoltaic demo plant in Heggelbach. The great pillar distances and big clearance height allow for the use of big agricultural machinery like full harvesters. © Hofgemeinschaft Heggelbach.

east–west direction and 19 m in the north–south direction. The system has a clear working height of 5.5 m and a pitch of 9.5 m with a configuration of 2 modules in portrait, producing a power output of 194 kWp. The system utilised bifacial Solar World modules, which are designed to capture sunlight from both sides of the panel [12] (Figure 3.7).

The agrivoltaic system was installed on a farm with a variety of crops, including wheat, celery, clover grass and potatoes. It was designed to provide shade for the crops while still allowing enough sunlight to reach the plants for photosynthesis. A monitoring system was implemented to investigate the impact on crop growth, yield and quality [12].

The study found that the combined output of the system depended strongly on the crop and the weather. While recording significant decreases between 5% and 19% in 2017, in the following year a change in harvest could be seen between a decrease of 8% to an increase to 12%. These values still exclude the land loss of the system due to structural elements which accounted for an additional 8%. In total an overall increase in the LER between 1.57 and 1.87 could be observed. Additionally, the shade provided by the system helped to reduce water loss through evaporation, resulting in improved water use efficiency. The system also helped to reduce temperature fluctuations, which can be beneficial for some crops [12].

The unique design of the system, with its large pillar distances, allowed for a more efficient use of land while still providing shade for the crops. At the same time, it is a system with a massive amount of steel used kWp installed coming along with high cost. The system was also able to generate electricity, reducing the farm's reliance on non-renewable energy sources [12].

Case study: high elevated east-west Agri-PV system Babberich

The agrivoltaic plant in Babberich, Netherlands, built in 2020 by BayWa r.e., is a remarkable example of a solar farm that also functions as a food-producing system. The plant has a power capacity of 2.6 MW and utilises 10.500 semi-transparent photovoltaic modules that protect 31.000 raspberry plants from weather influences. BayWa r.e. has integrated an irrigation system, which ensures the plants receive sufficient water, resulting in high-quality crops.

The system shown in Figure 3.8 utilises a light steel framing by company Zimmermann with pole distances of 2.8 m. The photovoltaic modules are semi-transparent and are installed over each row of plants, protecting the crops from rain and hail. Vertical poles are set up in the rows of plants to support the modules. Raspberry plants are grown under the photovoltaic modules, and no limitations are imposed on the usual farming practices of raspberries. Especially the east–west orientation of the modules give high amount of weather protection to the plants. At the same time, the East/west configuration on the modules create problems in the distribution of the rainwater between the rows. While every second row under the ridge has very little water availability, every other row has a lot of water, creating problems in wet seasons.

Further an irrigation system was implemented into the agrivoltaic system, ensuring the crops receive the necessary amount of water and the quality of the fruit is of high standard.

Babberich is a great example to show how photovoltaic systems can be integrated into agriculture and replace the former agricultural infrastructure. In other farms, the raspberries are still protected by artificial roofs made mostly of plastic that serve only one purpose: to protect the raspberries. These are being replaced with photovoltaic modules, which serve a dual purpose and promote symbiosis between agricultural and the solar system.

Figure 3.8 Agrivoltaic system in Babberich. One row of semi-transparent PV modules over each row of raspberries [3]; and a comparison of raspberry production under PV modules (B: left corner) and raspberry production under plastic tunnels (B: right corner) © BayWa r.e.

3.2 Finding balance: interrow fixed and tracker agrivoltaic systems for dual harvesting

Interrow agrivoltaic systems utilise the space between photovoltaic panels for agricultural purposes. This approach to integrating agriculture and renewable energy generation is a different concept that aims to utilise agricultural spaces for multiple purposes. In comparison to overhead agrivoltaic systems, Interrow agrivoltaic systems offer different techno-ecological synergies, but they still allow the growth of crops and the generation of energy in the same area. The amount of space available for crops to grow can vary, depending on the type of inter-space agrivoltaic system chosen.

Inter-space agrivoltaic systems can be classified into three main types: fixed tilt, tracking and vertical PV. Fixed tilt inter-space agrivoltaic systems are the simplest to install and maintain, and they are typically set at a fixed angle between 10 and 30 degrees. Tracking inter-space agrivoltaic systems, on the other hand, are designed to follow the sun's movement, which increases the energy yield per installed kilowatt peak. Vertical PV inter-space agrivoltaic systems, as the name suggests, have the PV modules mounted vertically, which maximises the available space for crops to grow and give a unique yield curve with two peaks.

3.2.1 Interrow fixed tilt systems

Interrow fixed tilt systems are a type of agrivoltaic systems that require relatively low adaptation to agriculture. They are often elevated from the ground and mounted on a single stand tube. The biggest difference from a normal ground-mounted fixed tilt system is the inter-row distance, which is increased to accommodate the width of the agricultural machinery used in the field. This distance is dependent on the type of crop and the size of the agricultural machinery.

Interrow fixed tilt systems are usually built with opaque standard modules and a standard configuration of the sub-structure. By increasing the height of the system, these agrivoltaic systems can also provide more space for the agricultural machinery to work, further reducing the impact on agriculture and minimising the area that cannot be cultivated. This makes interrow fixed tilt systems an attractive solution for farmers who want to incorporate renewable energy into their operations without disrupting their daily activities.

Interrow Systems faces a unique challenge that is often encountered in their operations – dealing with diverse climate zones within a single row. A visual representation, depicted in Figure 3.9 demonstrates the fluctuation of both irradiance and humidity levels between two rows of photovoltaic (PV) panels, depending on their location. While the area just south of each PV row exhibits the highest levels of irradiance and the lowest humidity levels, the opposite is observed north of each row. This significant contrast in environmental factors can lead to varying growth conditions for the plants, making it particularly challenging to achieve uniformity in crop cultivation. To address this issue, one possible approach is to introduce a range of different crops in the distinct climate zones between the

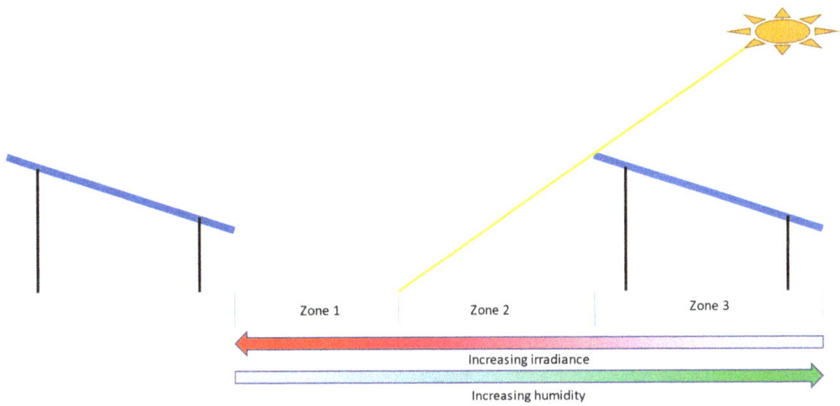

Figure 3.9 Different climate zones in interrow fixed tilt systems and the distribution of irradiance and humidity on the ground

PV rows. By strategically selecting crops that are better suited to the specific conditions of each zone, the adverse effects of the differing microclimates can be mitigated. However, it is important to note that implementing this strategy would require additional efforts and resources in terms of farming activities. Fixed-tilt interrow solutions offer a great benefit with the possibility of collecting water as it is already shown in existing powerplants. This can be especially crucial in arid or semi-arid areas where water is a limited resource. Water is already needed for the operation of a powerplant and will become more crucial when hydrogen production will be a common way to utilise the power at times in which the electricity price is low.

Case study: interrow Agri fixed tilt system Alhendin

The Alhendin Agrivoltaic plant is an innovative project built by BayWa r.e. in Spain. The plant is part of a larger 50 MW photovoltaic plant and has a capacity of 5 MW. The plant utilises a 2-portrait system with interrow cropping, allowing for the passage of big agricultural machinery. Rainwater is collected and utilised for cleaning the PV modules. The plant primarily farms wheat, which is a critical staple crop in Spain [16].

The plant has a pitch of 10 m ensuring that the machinery can pass through without interference of the PV system.

BayWa r.e. has implemented a system to collect rainwater, which is used for cleaning the PV modules. This way is chosen to avoid the conflict for water between PV interest and farming. As the cleaning process is essential to maintain the efficiency of the modules, rainwater is stored in a reservoir to be used for cleaning. Water conservation is promoted by the use of rainwater, a vital factor in arid regions such as Spain.

The plant has a positive impact on the environment, as it generates renewable energy and promotes sustainable agriculture practices. The plant's capacity of 5 MW and its location within a 50 MW photovoltaic plant underline the potential of

Figure 3.10 Alhendin interrow system with rainwater collection system © BayWa r.e.

agrivoltaic systems to provide large-scale renewable energy. The Alhendin Agrivoltaic plant is a critical project that has potential to inspire other organisations to invest in similar projects, leading to a shift towards sustainable agriculture practices globally (Figure 3.10).

3.2.2 Vertical PV systems

The concept of Vertical PV systems consists of rows of vertically mounted PV Modules with interspacing in between the rows. With this kind of installation, the system is minimising ground usage for the PV installation and providing the availability to maintain the agricultural usage of the field. It is an optimised racking system for bifacial PV modules as the direct sunlight is used at front and backside of the PV modules to produce power.

When installed in north–south oriented rows, this leads to an energy production curve with two peaks, one in the morning and one in the evening and enables to efficiently realise the energy transition with the existing grid infrastructure (Figure 3.11).

The idea of vertical installation had been invented in 2015 by Next2Sun and together with the reinvention of bifacial Modules, which came into the PV Market slowly from 2017 on and grew to an immense market share these days and the exploding request for Agri-PV Systems from 2020 on, the vertical racking systems offers a new system setup for the Agri-PV usage.

The various existing systems of Agri-PV provide benefits based on the needed aspects for different cultivation habits. The vertical racking system is mainly designed for usage on large scale agricultural surfaces with the availability to do machinery-based agriculture with working widths of 5–20 m. The vertical bifacial racking system does need more ground area for installing the requested power compared to other interrow Agri-PV systems. About 450 kWp/ha of power density is achieved with vertical bifacial Agri PV systems with a common interrow of 10 m and an active height of 2 m, if installed with three modules in height as upcoming within 2024 the achievable power will be about 675 kWp/ha. The row interspacing

Figure 3.11 Oat harvest within vertical bifacial racking © Next2Sun

can be easily adjusted based on the plant physiology requests and the available ground surface to increase the power density for this racking system.

Regarding the shading of plants, the vertical bifacial system is not designed for intentional shading of crops but providing high and evenly distributed sunlight availability on the agricultural field of about 85% of the total irradiance. In comparison to other interrow Agri-PV Systems the vertical-bifacial system provides much more sunlight availability for the maturation process of the crops.

The vertical bifacial racking system is achieving the benefit of a uniform maturity process within the cultivation in between the rows and additionally the minimised shading leads to reduced drought stress in plants and increase water use efficiency.

The distribution of sunlight within the agricultural field is presented in Figure 3.12 and shows that the inhomogeneity with the vertical racking system is about 20% in comparison to other Agri-PV Plants where inhomogeneity is about higher. Figure 3.12 shows this impact for system with an active height of 2 m with a pitch of 10 m.

The shading produced by this system setup is mainly used for minimising the evaporation of groundwater in morning hours. Here the vertical bifacial system is providing longer availability dew water in the morning hours due to the inter row shading but evaporation during midday is not affected.

Compared to a tracker interrow Agri-PV system which is in principle able to adjust tracking angle advantageous or disadvantageous to the needed plant physiological requests, the vertical bifacial racking system is providing a fixed system with the main objective to provide a simple and fixed racking structure with low maintenance costs.

Compared to PV Trackers which need to move into horizontal position in case of high windspeeds to minimise the windloads on the racking system, the vertical

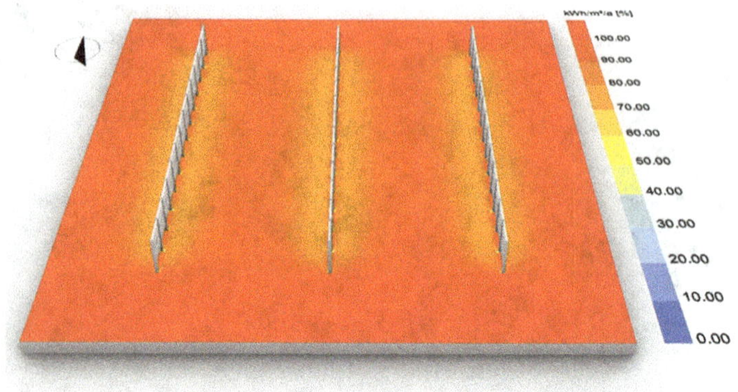

Figure 3.12 Irradiation availability within vertical bifacial racking [Next2Sun]

bifacial system is designed to withstand the appearing windloads. This leads to the disadvantage, that the needed amount of steel is significantly higher due to the deeper grounding of this system. But on the other hand, it offers the agricultural advantage to use the system as windbreaker and protect the crops inside the field.

Due to its separation into elements of only one or two modules in length the installation inside hilly terrain is easier for the vertical bifacial, than for systems with table lengths of many meters. The impact of water distribution inside the field is not changed by the vertical PV system. This provides the disadvantage of not being able to install water collecting systems and water storage in dry regions but also provides the advantage of not affecting the rain distribution inside the field to maintain an evenly rainwater distribution and maturity.

The benefits in concern of costs, additional agricultural yield and water distribution during rain events need to be considered to find the perfect system setup for every requested plant site.

3.2.3 Interrow trackers

Interrow agrivoltaic trackers are a novel approach to combining solar power with agriculture. The system involves using standard solar trackers with minor modifications to allow for the installation of crops in the gaps between the panels. The trackers' modifications enable greater height than it would be used in a sole PV powerplant allowing sufficient space for crop growth.

It is critical for Interrow systems to concentrate as much power as possible on one axis. This reduces the PV system's influence on agriculture. As a result, in an agrivoltaic arrangement a tracker system with two modules in portrait (2P) is an intriguing option to consider, as it allows for greater row spacings while maintaining the same land-use ratio [17]. At the same time a 2P tracker has a much greater shading rate compared to a 1P tracker [17] and more diverse climate zones in between two rows of trackers, increasing the influence on crops and making the farming activities more complicated.

Further, building 2P trackers can be challenging due to the static problems caused by wind loads. As the trackers are designed to move, wind loads can cause significant strain on the system. Therefore, it is crucial to ensure that the trackers are built with the necessary structural reinforcements to withstand wind loads.

Interrow agrivoltaic trackers offer several advantages over fixed tilt systems. One of the main advantages is the ability to have a closer row spacing with the same installed power. The possibility of turning the modules to 90 degrees means that there is very little coverage of the ground, allowing for optimal crop growth and maximum space for farming. Additionally, by moving the solar panels to their optimal position in each situation, the system can increase electric yields when the sun is shining or reduce loads on the trackers in case of high winds, allowing the use of less steel and therefore be more cost efficiently. Similar to overhead trackers or standard PV trackers also interrow trackers offer greater energy yields per installed peak power [17]. Another major advantage of tracker systems is that they can be tracked according to demand of the crop, e.g. in agrivoltaics on grassland, when grass/hay is cut and needs fast drying, the tracker can move the PV modules in anti-Tracking and enable morning sun hitting the ground and grass to have fast drying and hay-harvest compared to a fixed-tilt system. Also a tracker can be tilted to the horizontal position to lower risk from stone flicking during agricultural processes.

The combination of solar power and agriculture offers numerous benefits, including water conservation and efficient land use. As such, interrow agrivoltaic trackers offer a promising future for sustainable agriculture practices and renewable energy production.

Case study: interrow EWS Sonnenfeld – Bruck/Leitha

The EWS Sonnenfeld initiative is a pioneering project focused on identifying the best plant cultures and management techniques for simultaneous solar energy production in an interrow configuration. The project employs the innovative use of 2P tracker solution in combination with agriculture. This solution has a profound impact on both agricultural and energy yields. The arrangement of table rows with different spacings between 6 m and 12 m is another critical element under study. The Sonnenfeld and its two benchmark areas are equipped with advanced sensors and tools to assess the influence of various factors on farming and crop productivity. The research area is divided into eight unique segments, each dedicated to different facets of the project. This initiative marks a significant step forward in the optimal use of land for both farming and energy production (Figure 3.13).

3.2.4 Layout adaption in interrow systems

The technology used in interrow agrivoltaic systems for tracking and mounting solar modules is similar to that of standard PV plants, with only minor differences. However, the layout of agrivoltaic systems must be carefully planned and adapted to suit the needs of agricultural machinery. The spacing between rows must be adjusted to accommodate the machinery, and the layout should allow turning. Additionally, transformer or inverter stations placed within the field must be

Figure 3.13 2P Agri tracker with a pitch of 12 m. Due to the 2P configuration, a high GCR is possible even with a high pitch [18].

considered, especially when project size is increasing. A transformer or a central inverter will interfere with the driveway of machinery when not carefully placed. Especially for this reason in an agrivoltaic context string inverters are to be preferred, as they have less interference in the layout and are easier to integrate on limited surfaces.

A typical transformer is using mineral oils for the cooling of its components. Using transformers with biodegradable oils reduces the risk of contamination of the adjutancy Agriculture. Further than this, transformers with Mineral oil have to cope with higher construction constraints [19].

Another important consideration is the fencing of agrivoltaic systems. Traditional photovoltaic plans have fencing to protect against theft and hazards associated with electric power plants. However, these fencing strategies must be reevaluated to accommodate farming strategies. It is important to consider that farmers may require more than one access point to their fields. In big-sized agrivoltaics farms more than one access point to the agricultural area should be considered. This will easy the agricultural activity and decrease the interference with typical farming practices. Next2Sun has shown that agrivoltaics systems can also work without fences at all with erecting a 2.5 MW powerplant in Austria.

The depth of cabling is an important factor that needs to be considered while designing an agrivoltaic plant. Unlike a typical PV Plant a agrivoltaic plant likely will be ploughed very regularly. To make sure that no damages are created in this process, the farming practices and machinery to be used in the future agrivoltaic plant should be considered when determining cabling depth.

Besides this also in the construction process of the plant, the future Farming needs to be considered. Heavy machinery and unplanned driving strategy lead to soil compaction and can influence future farming for many years. To avoid this, mobile roads can be used, which spread the loads of heavy machinery and minimise soil compaction.

3.3 Solar greenhouses

Solar greenhouses are a unique and innovative solution that combines traditional greenhouse practices with renewable energy technology. These greenhouses harness the power of the sun to provide optimal growing conditions for crops, while also generating clean energy. Solar greenhouses are becoming increasingly popular as a way to enhance food security, promote sustainable agriculture and reduce the carbon footprint of traditional greenhouse operations.

This chapter provides an overview of the concept of solar greenhouses. Various components and designs categorised as solar greenhouses will be presented. Benefits of using solar greenhouses will also discuss, including improved crop yields, energy independence and cost savings. Successful solar greenhouse projects will be highlighted, and the future potential of this innovative technology will be examined.

3.3.1 Roof-integrated photovoltaics in greenhouses

In this approach of solar greenhouses, photovoltaic technology is integrated into the roofing of the greenhouse. The use of PV modules as roofing allows efficient utilisation of space, as the same area can be used for both growing crops and generating clean energy.

There are different ways to design this type of solar greenhouse. For example, very transparent PV modules can be used to maximise the amount of light that reaches the crops. Alternatively, it is possible to use modules with varying levels of transparency to adjust the intensity and evenness of the light [20–24]. Also the use of opaque standard modules in a pattern is possible and a way to reduce costs. [1,25] Another option is to retrofit light and opaque modules onto an existing greenhouse structure [26]. While the use of semi-transparent modules with an even cell spacing will offer the most even light distribution, it is the most expansive of the different solutions.

The approach to integrate PV into the roof allows the integration of common greenhouse technologies, such as shading systems and water management systems. This can result in improved crop yields, increased energy efficiency and reduced costs. Moreover, the integration of PV technology into the roofing of the greenhouse has minimal impact on the greenhouse structure, making it a cost-effective and flexible solution [27] (Figure 3.14).

3.3.2 Integrating photovoltaics inside greenhouses

Another approach to solar greenhouses involves installing PV technology under the roofing of the greenhouse. This approach requires a significant adaptation of the greenhouse structure and can impact the ability to use traditional greenhouse technologies. To support the weight of the PV modules, additional structural support may be necessary.

One example of this approach is to use very light, flexible and semi-transparent PV modules that are hung below the roof. This allows for some light to pass through the modules and reach the crops, while also generating clean energy [24].

Figure 3.14 Greenhouse with a roofing of semi-transparent glass–glass modules

While this approach may require a greater investment in terms of structural adaptations and technology, it offers the potential for increased energy yield. However, careful consideration must be given to the impact on crop growth and the ability to use other greenhouse technologies, such as shading systems and water management systems [24].

Overall, the integration of PV technology under the roofing of a greenhouse is an innovative solution that requires careful planning and consideration of the trade-offs between energy yield and the impact on traditional greenhouse practices.

3.3.3 Harnessing the power of photovoltaics outside the greenhouse

Another approach to integrating PV technology with greenhouses involves installing the solar array outside of the greenhouse. This approach has gained popularity in areas with high levels of irradiation, especially with foil tunnel greenhouses.

In this design, which is shown in Figure 3.15, PV trackers are elevated between the foil tunnels, allowing them to benefit from higher bifacial gains due to the reflection of the usually white or light-coloured foil tunnels. This results in minimal impact on the crops within the tunnels while still providing a high energy yield at a relatively low cost compared to other agricultural PV systems. SolarGik claims to be able to install trackers just on existing greenhouse structures, minimising the cost and impact of the PV array [28].

Another option is to install standard PV modules on the trackers, which can also produce a very high energy yield with a relatively low cost. This approach allows a flexible and cost-effective solution for generating clean energy in agricultural settings.

Figure 3.15 PV Trackers build in-between plastic foil tunnels © SolarGik [28]

By integrating the PV technology outside of the greenhouse, this approach provides a solution that minimises the impact on traditional greenhouse practices, while still allowing the benefits of renewable energy.

Case study: solar greenhouse Senden

The Photovoltaic Greenhouse Potthof, located in Western Germany, is a revolutionary venture that amalgamates the realms of agriculture and renewable energy. This project utilises photovoltaic technology and a specialised greenhouse structure to optimise strawberry cultivation. The project is powered by a 300 kWp system, generated from 1440 semi-transparent Brite Solar photovoltaic modules.

Designed by Zimmermann PV-Stahlbau, the greenhouse employs a substructure designed to support hanging trays for growing strawberries. The retractable walls of the greenhouse offer both high air fluctuation when necessary and adequate protection during inclement weather. A notable design feature is the roofing that consists solely of PV modules, ensuring waterproofness through overlapping. Furthermore, the project includes a gutter system for collecting water, which is then used to efficiently water the plants. Its design ensures that the greenhouse operation is self-sustaining, with the photovoltaic modules supplying the required power. The strawberries are grown in a controlled environment, which leads to higher quality produce, and the hanging tray system allows for more efficient use of space and easier harvesting.

Water usage is minimised by the in-built gutter system that collects and recycles rainwater. Moreover, the project significantly reduces carbon footprint by generating clean, renewable energy. The greenhouse's power system is capable of producing 300 kWp, which not only meets the energy needs of the greenhouse but also allows surplus power to be fed back into the grid, contributing to Germany's renewable energy targets.

The photovoltaic greenhouse project in Western Germany is a testament to the possibilities that arise when agriculture meets renewable energy technology. Not only does it provide a sustainable and efficient method for strawberry cultivation, but it also contributes to energy generation and environmental conservation. The project's success serves as an inspiring model for future innovation in sustainable agriculture (Figure 3.16).

Figure 3.16 Photovoltaic greenhouse Potthof, a photovoltaic greenhouse for growing strawberries in hanging trays and with retractable walls © Hof Grothues-Potthof

Case study: solar greenhouse Bardzour

The Bardzour PV Greenhouse is an innovative solution that combines responsible farming with carbon-free energy production. Developed by Akuo Energy, the project aims to enhance energy autonomy in isolated areas while promoting sustainable development.

The Greenhouse Project implements a photovoltaic greenhouse system that combines ground-level and rooftop photovoltaic arrays (Agrinergie®). The system includes 6,000 m^2 of cyclone-proof photovoltaic greenhouses for market gardening, providing a dual land use solution that combines energy and food independence in isolated areas. The project includes a battery storage system with a capacity of 9 MWh.

In addition to its technological components, the Greenhouse Project addresses environmental and social concerns. A 2-ha arboretum featuring endemic plants, bee-keeping and sustainable farming education has been established, as well as funding for the renovation of the island's Insectarium. The project also provides detainee training and work, improving their living conditions and providing support to local associations.

The Bardzour project proves that agrivoltaic technology can generate a whole set of positive spin-offs for a region and the preservation of its environment, making it an excellent example of eco-innovation in action (Figure 3.17).

Figure 3.17 *The greenhouse system in the Project of Bardzour. Standard modules are facing south while full transparent modules facing North © Akuo.*

3.4 PV modules for agrivoltaics: specialised technologies for solar energy conversion

The PV module stands as a pivotal component within any solar energy system, playing a vital role in converting sunlight into electricity through the photovoltaic effect. Among the various technologies employed, crystalline silicon cells have emerged as the most prevalent and extensively optimised option over the years. While this chapter briefly acknowledges the significance of crystalline silicon technology, it primarily delves into a distinct realm of research the development of adapted modules and cell technologies specifically tailored for the unique demands of agrivoltaics.

3.4.1 Crystalline silicone cells

Crystalline silicon photovoltaic cells are the most widely used type of solar cells. These cells work by converting sunlight into electrical energy through a series of physical and chemical processes. The solar cell is made up of a thin layer of silicon that has been treated with impurities to create a p-n junction. When sunlight hits the surface of the solar cell, it excites the electrons in the silicon atoms, creating a flow of electrical current across the p-n junction. The electrical current is then collected by a conductive metal grid that is placed on top of the solar cell. The metal grid allows the current to flow out of the solar cell and into a wire, which can be connected to a battery or an electrical grid. The efficiency of crystalline silicon photovoltaic cells can be improved by using anti-reflective coatings, texturing the surface of the cell and optimising the thickness of the silicon layer [29].

3.4.2 Bifaciality

Bifaciality refers to the ability of photovoltaic cells to generate electricity from both sides of the cell. Unlike traditional PV cells that only capture light from one side, bifacial PV cells can utilise light from both the front and rear sides of the cell. This is achieved by placing the cell between two transparent layers or by using a reflective surface to bounce light back onto the rear side of the cell. Bifacial PV cells have several advantages over traditional cells, including higher energy production, improved performance in low-light conditions and increased flexibility in installation options. As the cost of bifacial PV cells continues to decrease, they are becoming an increasingly popular choice for solar panel installations [30].

Bifacial modules have gained popularity, especially in agricultural settings. The increased elevation of the modules has a positive effect on the illumination of the backside of the cell. In addition, semi-transparent modules are often utilised, allowing more light to penetrate through the module and resulting in a greater irradiation on the surface underneath. This irradiation is then reflected from the ground or plants, further illuminating the backside of the bifacial module [29]. Especially in vertical mounted agrivoltaic plants bifaciality is greatly utilised, as in different times of the day the two sides are facing the direct sunlight.

3.4.3 Semi-transparent silicon modules

Semi-transparent modules, based on the crystalline silicon technology, are commonly used in horticulture agrivoltaic installations. They make use of the very far developed stable and cheap crystalline silicon technology to convert solar, energy to electricity to ensure that sufficient light is reaching the plants, and they have a transparent backseat, or a built-in glass, glass configuration and have increased spacing between the cells. This is a very simple way to adjust the transparency of a module and gives great flexibility while doing this. These modules are easily available on the market by now the transparency of the available modules is ranging between 5% and 90%. These modules are available in different sizes that are similar to the module sizes of standard modules [31,32].

3.4.4 Glass coating technologies

LUMO technology, introduced by Soliculture in 2012, is a transparent solar panel that contains a periodic arrangement of low-density silicon PV strips on a glass panel, allowing light transmission between the strips. In addition, a thin layer of luminescent material is applied to the backside of the glass, which converts green light to red light. This optimised light spectrum enhances power production and facilitates plant growth, making LUMO solar panels ideally suited for crop production areas in greenhouses where maximum light transmission is desired.

The core LUMO technology is a light-altering dye that converts green light to red light, which has the highest efficiency for photosynthesis in plants. The absorbance of chlorophyll a and b, two pigments critical to photosynthesis, highlights that plants absorb blue and red light, but not green light. Extensive plant trials have been conducted to evaluate the efficacy of LUMO technology, which have shown that crops grown under LUMO panels exhibit positive growth responses, such as increased yield, faster time to maturation and disease resistance.

In conclusion, the LUMO technology developed by Soliculture has introduced a new approach to solar panel design that facilitates plant growth by enhancing light quality for photosynthesis. By converting green light to red light, LUMO technology optimises the light spectrum, resulting in increased power production and improved crop growth, making it an ideal solution for crop production areas in greenhouses. The positive growth responses observed in extensive plant trials demonstrate the potential for LUMO technology to revolutionise the agriculture industry [33].

3.4.5 Emerging PV technologies for agrivoltaic applications

This chapter focuses on the emergence of PV technologies specifically designed for agrivoltaic applications. As the demand for renewable energy and sustainable agricultural practices grows, there is a need for innovative PV solutions tailored to the unique requirements of agrivoltaics. By exploring novel module designs, specialised cell technologies and integration techniques, the aim is to enhance the efficiency and productivity of agrivoltaic systems. This chapter highlights the key features, performance characteristics and potential benefits of these emerging PV technologies in agricultural settings.

3.4.6 Tubesolar

Tube Solaris a new approach to photovoltaic module design that leverages the Osram glass tube technology to incorporate organic thin film PV cells within glass tubes. This unique module design offers a multitude of advantages, particularly in agrivoltaic settings. By utilising mini glass tubes, Tube Solar modules facilitate optimal water and light distribution across the agricultural surface, ensuring even coverage. Furthermore, the open design of these modules reduces wind and snow loads, minimising the overall structural needs of the installation.

One notable benefit of Tube Solar modules lies in their superior cleaning efficiency compared to conventional PV modules. The tube design allows improved self-cleaning properties, reducing the accumulation of dirt and debris. Moreover, these modules eliminate the need for angling towards the sun, as their design ensures efficient light absorption from various orientations (Figure 3.18).

3.4.7 Organic PV modules

Organic photovoltaics (OPV) is a promising technology for application in the agrivoltaic context because of its ability to split the light spectrum between plants and solar cells. By adapting organic semiconductor materials, OPV cells can transmit visible light for photosynthesis and use infrared light for electricity generation. This increases area efficiency and reduces competition for sunlight between the two uses. In addition, OPV cells can be manufactured as flexible and transparent films that can be easily integrated into greenhouses or foil tunnels. OPV cells can thus contribute to climate adaptation and mitigation by providing farmers with additional revenue from solar power while protecting crops from extreme weather conditions [23,24,34].

Figure 3.18 Prototype installation of tubesolar modules over tomato plants © TubeSolar

3.4.8 Wavelength selective technologies

Wavelength selective technologies have been gaining attraction in science due to their potential to optimise plant growth and increase the productivity of PV systems. This technology can be broadly categorised into two groups: selective cells and selective mirrors.

Selective cells are organic PV semi-transparent cells that can be fine-tuned to absorb specific wavelengths of light, while allowing other wavelengths to pass through. They are expensive and have a short lifespan of only five years, but they can be tuned to absorb only the wavelengths that are useful for the Photovoltaic effect, allowing the remaining wavelengths to pass through and ensure optimal growth conditions for the plants. However, the efficiency of these cells is relatively low when compared to crystalline silicon cells [23].

Selective mirror technology has been used in a variety of configurations, including a triangular design [35] and a concentrating design [36], to optimise the reflection of light to the PV module. The studies of selective mirror design involve always a tracker with usually two axis trackers in the concentrator design and one axis tracker in the triangular design. Both design can increase the GCR without decreasing the PAR available to the plants [35,37].

The triangular design involves two mirrors that reflect light onto a bifacial module, which generates electricity from both the front and back of the module. This design is advantageous as it ensures a more even distribution of light on the module and increases the amount of energy that can be produced [35]. The concentrating design (as shown in Figure 3.19), on the other hand, involves a selective mirror that concentrates the reflected light onto a small module. This design is more focused and can result in higher energy production in a smaller area. The materials used in the study are multiple of coated layers with different refractive indices. An

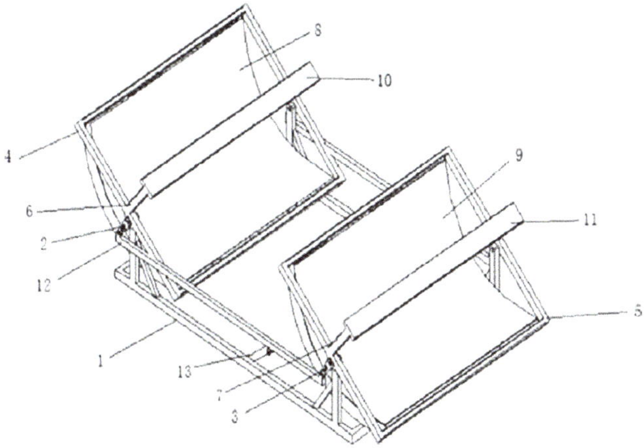

Figure 3.19 Schematic design of a spectral beam splitting tracker. The structure of the tracking system: 1 – base; 2, 3 – flip chip; 4, 5 – support frame; 6, 7 – bracket; 8, 9 – CPV groove type condenser; 10, 11 – photovoltaic chip; 12 – linkage rod;13 – one dimensional servo system [36].

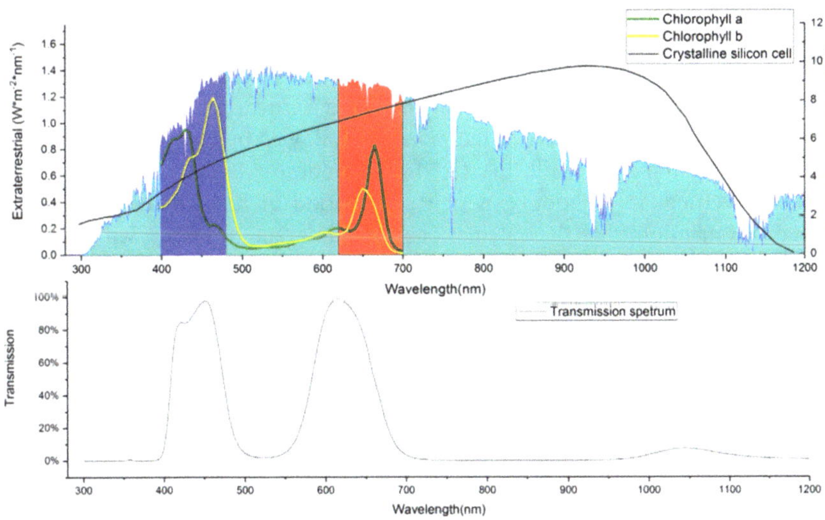

Figure 3.20 Transmission spectrum of the reflective mirror

increased growth rate was proved for different plants like lettuce, cucumber and spinach [36] (Figure 3.20).

In addition to the design of the selective mirrors, the use of trackers can enhance the effectiveness of the technology. Single-axis trackers adjust the angle of the PV system to follow the sun's movement throughout the day, while dual-axis trackers also adjust for changes in the sun's angle throughout the year. This allows the selective mirrors to capture more light and reflect it onto the PV module, resulting in increased electricity production and higher PAR on the ground.

3.5 Comparison of yields in different agrivoltaic setups

When comparing various types of agrivoltaics and focusing on their energy yield, significant differences arise. Two different methods are used to measure the energy yield: yield per installed PV capacity, measured in kWh/kWp and yield per area. A study conducted by Bousi *et al.* [17] compared five different interrow system solutions for a location in central France and Figure 3.21 shows the comparison. It found that in this location the different tracking solutions outperformed the fixed tilt solution by a significant margin in terms of specific yield. While this result was expected, the findings on the yield per area are particularly interesting. The 2P tracking system, which exhibited the highest yield per area, outperformed the vertical system with similar row spacing by almost 170%, with the same amount of area loss between these two systems. Although shading in the different fixed tilt systems, as well as the 1P tracker, was comparable, the 2P tracker resulted in almost double the shading, creating difficulties for certain crops in agriculture.

Significant intraday yield curve variations are observed among different module orientations in [38]. Figure 3.22 shows yield curves for different PV installations. The typical shape of the fixed tilt solution features a single peak in the middle of the day. However, the east-west facing vertical installation displays a very pronounced power increase in the morning, followed by two peaks throughout the day. During midday, while the fixed tilt solution exhibits maximum power output, the vertical installation experiences a period of low power output. It is important to mention, that the calculations fort his graph are done with modules with low bifacially, which is visible especially in the vertical setup, in which the two peaks a very differently high. A bifacial module with a bifaciality of up to 95%

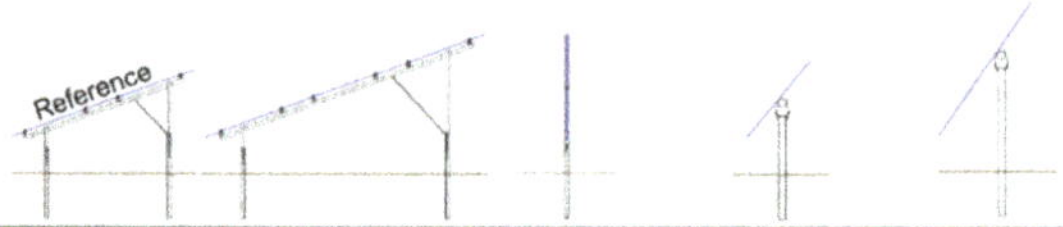

	Fixed-Tilt 2P	Fixed-Tilt 3P	Fixed-Tilt Vertical	Tracking 1P	Tracking 2P	
Pitch (m)	14,3	16,4	10,3	10,6	10,6	PV
[2] Specific Yield (kWh/kWp/a)	1297	1285	1075	1599	1479	
Specific Installed Capacity (kWp/ha)	484	631	334	326	652	
PV Yield (MWh/ha/a)	18836	24344	10771	15644	28939	
Relative change to reference	0%	29%	-43%	-17%	54%	
Land losses (%)	24%	29%	9%	10%	10%	Agri
Shading rate* (%)	19%	27%	22%	22%	46%	

Figure 3.21 Comparative analysis of PV configurations for agrivoltaic systems in France © BawWa R.E [17]

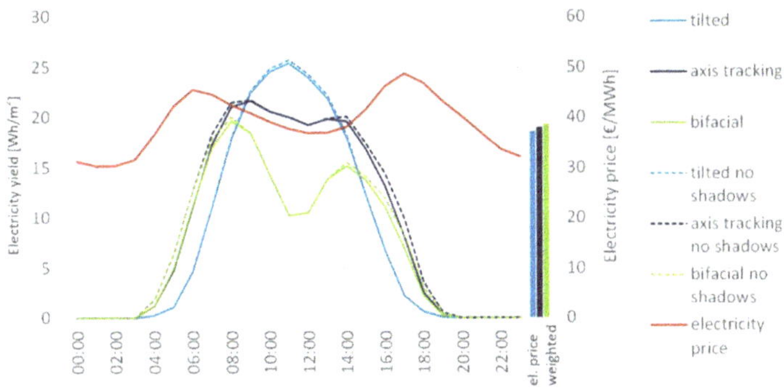

Figure 3.22 Comparison between the power generation in different configuration of agrivoltaics and the electricity price [38]

would increase the second peak and could boost overall yield by up to 12% in this graph are The characteristic curve of a tracker, including both single- and dual-axis trackers, features a rapid power increase in the morning, followed by a plateau at a higher level. The main difference between one and two axis trackers is primarily the amplitude of the plateau.

3.6 Inverters, batteries and transformer stations in agrivoltaic context

Inverters are electronic devices that are used to convert the direct current (DC) electricity generated by photovoltaic panels into alternating current (AC) electricity that can be utilised by appliances and the electrical grid. In photovoltaic systems, inverters are of paramount importance for managing the power output of the solar panels. Inverters regulate the voltage, frequency and phase of the electricity to ensure it complies with the requirements of the grid and the end-user. Additionally, inverters optimise power generation by adjusting the voltage to the maximum power point, ensuring that the system is efficient and cost-effective. It is worth noting that the use of standard inverters in agrivoltaic systems is viable. In the common agrivoltaic setup they need a higher amount of protection than in a Standard PV system. This is due the use of pesticides and fungicides to protect the crops or the possibility of animals grazing under the PV array damaging the hardware. Usually the inverters are highly mounted to avoid these kind of problems (Figure 3.23).

Battery systems represent a valuable add on to any photovoltaic system. This is also the case for agrivoltaic systems as they store excess energy generated by the solar panels for future use. Battery systems offer a reliable and steady energy supply for both agricultural and non-agricultural activities, ensuring uninterrupted power supply. Furthermore, the stored energy in battery systems can be utilised during peak demand periods, reducing energy costs.

Figure 3.23 An inverter mounted at a height of 3 m directly under the high elevated modules. The inverter is better protected from weather but also influence of fungicides and pesticides are smaller

The integration of battery systems in agrivoltaic systems yields multiple benefits. One of the key benefits is an increased level of energy independence. By incorporating inverters and battery systems, agrivoltaic systems provide a dependable and stable energy supply, reducing reliance on the electrical grid. This promotes energy independence, lowers energy costs and provides a consistent and reliable power source for farmers.

3.7 Designed for harmony: addressing special considerations in the construction of agrivoltaic systems

Adapting the technology and layout planning is not the only consideration when constructing an agrivoltaic system. The construction strategy also needs to be tailored to minimise negative impacts on agriculture, such as soil compression. The use of heavy machinery during construction can have detrimental effects on soil fertility, especially for agrivoltaic plants where typical mounting strategies are used. To avoid this, solid steel plates can be used to construct roads and distribute loads over larger surfaces, reducing soil compression. However, this strategy can be costly and time-consuming. Another approach is to use mobile cranes to lift heavy materials and place them without requiring significant road construction. However, this also comes at an increased cost and may not be practical for large areas far from existing roads.

During construction, safety considerations must be taken into account, especially for high elevated agrivoltaic systems where workers will be operating at heights of 5 m or more. Workers need access to these heights while also handling heavy steel parts and modules on uneven ground. Farming equipment which is often available especially in horticulture farms can be repurposed for these tasks.

3.8 Symbiotic solutions: harnessing crop adaptation in agrivoltaics for enhanced PV yields

The integration of photovoltaic arrays over agricultural surfaces and plants provides numerous benefits for the plants, making it an advantageous approach. While the advantages and disadvantages to the plant will be discussed in Chapter 5, here we speak about the advantages that a PV array and the respective yield can gain in reverse.

The PV arrays help in maintaining lower temperatures under the modules due to the water evaporating, providing a cooling effect. This cooling effect reduces the PV temperature and enhances its efficiency, thereby improving its energy yield. Also, the integration of PV arrays over agricultural surfaces results in lower land costs since the land is shared, and parts of the land cost can be supported by agriculture. This shared land-use approach reduces the pressure on farmland and decreases the costs associated with obtaining suitable land for PV installation. Additionally, the integration of PV over agricultural surfaces is more acceptable than other approaches, as no valuable land area is lost exclusively for energy generation. This approach is more sustainable, environmentally friendly and less intrusive, making it a more socially acceptable alternative. The higher elevation required for plants and farming machinery, results in increased backside irradiation for bifacial modules. This additional energy yield increases the overall efficiency and performance of the PV array.

In conclusion, the integration of PV arrays over agricultural surfaces and plants has several advantages, including cooling effects, reduced land costs, increased social acceptability and increased backside irradiation for bifacial modules. This approach presents an opportunity for sustainable, low-impact and high-efficiency energy generation in the agriculture sector.

References

[1] Toledo C and Scognamiglio A. Agrivoltaic systems design and assessment: A critical review, and a descriptive model towards a sustainable landscape vision (three-dimensional agrivoltaic patterns). *Sustainability* 2021;13(12):6871.

[2] Macknick J, Hartmann H, Barron-Gafford G, *et al.* The 5 Cs of agrivoltaic success factors in the United States: Lessons from the InSPIRE Research Study. 2022.

[3] Zimmermann PV-Stahlbau GmbH & Co. Kg, ed. *Agri-PV Systems: Agri-PV Solar Mounting Frames* [Technical Datasheet]. 2023; 2023 Jun 7.

[4] Marrou H, Guilioni L, Dufour L, Dupraz C and Wery J. Microclimate under agrivoltaic systems: Is crop growth rate affected in the partial shade of solar panels? *Agricultural and Forest Meteorology* 2013;177:117–32.

[5] Reasoner M and Ghosh A. Agrivoltaic engineering and layout optimization approaches in the transition to renewable energy technologies: A review. *Challenges* 2022;13(2):43.

[6] Kwon O-H and Lee K-S. Agrophotovoltaic designs: Irradiation analysis on and under PV modules. *KSES* 2021;41(2):9–23.

[7] Valle B, Simonneau T, Sourd F, *et al.* Increasing the total productivity of a land by combining mobile photovoltaic panels and food crops. *Applied Energy* 2017;206:1495–507.

[8] REM TEC, ed. *Agrivoltaico REM Tec: Green Power to the People*. 2022.

[9] Zimmermann PV-Tracker GmbH & Co. Kg, ed. *Zim-Track: Horizontal Single-Axis Tracker*. 2023 2023 Jun 7.

[10] Niazi KAK and Victoria M. Comparative analysis of PV configurations for agrivoltaic systems in Europe. 2022 Nov 1. Available from http://arxiv.org/pdf/2211.00331v1.

[11] DIN Deutsches Institut für Normung e. V. *Agri-Photovoltaic Systems – Requirements for Primary Agricultural Use*. English translation of DIN SPEC 91434:2021-05. Beuth; 2021, May 2021.

[12] Trommsdorff M, Kang J, Reise C, *et al.* Combining food and energy production: Design of an agrivoltaic system applied in arable and vegetable farming in Germany. *Renewable and Sustainable Energy Reviews* 2021;140:110694.

[13] Max Trommsdorff, and apl. and Prof. Dr. Petra Högy. *Agri-Photovoltaik: Chance für Landwirtschaft und Energiewende: Ein Leitfaden für Deutschland*. April 2022; 2022.

[14] Zainol Abidin MA, Mahyuddin MN and Mohd Zainuri MAA. Solar photovoltaic architecture and agronomic management in agrivoltaic system: A review. *Sustainability* 2021;13(14):7846.

[15] Erik van Nieuwenhuijzen. *Green Meteor: Power in Simplicity*. 2023 [cited 2023 Aug 22]. Available from https://greenmeteor.nl/.

[16] Diermann, R. Baywa r.e. baut in Spanien Agri-Photovoltaik-Anlage für PPA-Vertrag mit Velux. 2022 [cited 2023 Aug 22]. Available from https://www.pv-magazine.de/2022/11/15/baywa-re-baut-in-spanien-agri-photovoltaik-anlage-fuer-ppa-vertrag-mit-velux/.

[17] Bousi E. *Land Equivalent Ratio as a Key Assessment Indicator for Utility Scale Interrow Agri-PV*. Daegu, Korea; 2023.

[18] EWS. EWS Sonnenfeld - Bruck/Leitha (NÖ), Agri-PV Referenzanlage. 2023.

[19] Bioöl-Transformator: Fortschritt in der Höchstspannungsklasse; 2023 [cited 2023 Aug 28]. Available from https://www.transnetbw.de/de/unternehmen/portraet/innovationen/biooel-transformator.

[20] Bambara J and Athienitis AK. Energy and economic analysis for the design of greenhouses with semi-transparent photovoltaic cladding. *Renewable Energy* 2019;131:1274–87.

[21] Brite solar. *Solar Glass: Ultra Thin Solar Glasses [Agrivoltaics – A New Path to Sustainable Agriculture.]*. 2023 [cited 2023 Aug 28]. Available from https://www.britesolar.com/products/solar-glass/.

[22] Cossu M, Yano A, Li Z, *et al.* Advances on the semi-transparent modules based on micro solar cells: First integration in a greenhouse system. *Applied Energy* 2016;162:1042–51.
[23] Di Wang, Liu H, Li Y, *et al.* High-performance and eco-friendly semi-transparent organic solar cells for greenhouse applications. *Joule* 2021; 5(4):945–57.
[24] La Notte L, Giordano L, Calabrò E, *et al.* Hybrid and organic photovoltaics for greenhouse applications. *Applied Energy* 2020;278:115582.
[25] Tritec. 2.5 MW Photovoltaikanlage auf Gewächshäusern in Montélimar installiert. 2023 [cited 2023 Aug 28]. Available from https://www.tritec-energy.com/news/photovoltaikanlage-gewaechshaeusern-montelimar/.
[26] Sunman. Integration of Lightweight modules ontop of Greenhouse structure. 2023 [cited 2023 Aug 28]. Available from https://www.sunman-energy.com/lightweight-solar-applications/details/212/.
[27] Akuo Energy. Club Ademe International's Members: Their Solutions to Climate Change. Bardzour.
[28] Solargik. Greenhouse solar trackers. 2023 [cited 2023 Aug 28]. Available from https://www.solargik.com/greenhouse-solar-panel-solution/.
[29] Stein, JS., Reise, C, Bonilla Castro, J, *et al.* IEA PVPS Task 13 performance, operation and reliability of photovoltaic systems - bifacial photovoltaic modules and systems: Experience and results from international research and pilot applications. 2021 [cited 2023 Aug 28].
[30] Katsikogiannis OA, Ziar H and Isabella O. Integration of bifacial photovoltaics in agrivoltaic systems: A synergistic design approach. *Applied Energy* 2022;309:118475.
[31] Shanghai Aerospace Automobile Electromechanical Co., Ltd. HT40-18X (PD)-F Bifacial Module: Module Datasheet [cited 2023 Aug 28].
[32] Brite Solar. Greenhouse solar glass | GreenGlass TM: Module Datasheet BSG-115/77.2023 [cited 2023 Aug 28]. Available from https://www.britesolar.com/products/solar-glass/.
[33] Soliculture. LUMO: The first commercially available, mass produced luminescent solar collector.
[34] Magadley E, Teitel M, Peretz MF, Kacira M and Yehia I. Outdoor behaviour of organic photovoltaics on a greenhouse roof. *Sustainable Energy Technologies and Assessments* 2020;37:100641.
[35] Shalom BA, Mittelman G, Kribus A and Vitoshkin H. Optical and electrical performance of an agrivoltaic field with spectral beam splitting. *Renewable Energy* 2023;219.
[36] Liu W, Liu L, Guan C, *et al.* A novel agricultural photovoltaic system based on solar spectrum separation. *Solar Energy* 2018;162:84–94.
[37] Ali Abaker Omer A, Li M, Liu W, *et al.* Water evaporation reduction using sunlight splitting technology. *Agronomy* 2022;12(5):1067.
[38] Ali Khan Niazi K and Victoria M. Comparative analysis of photovoltaic configurations for agrivoltaic systems in Europe. *Progress in Photovoltaics: Research and Applications* 2023;31.

Chapter 4

Adoption of agricultural processes with agrivoltaic applications

Shiva Gorjian[1,2], Erion Bousi[3], Omid Fakhraei[1] and Manijeh Khanian[1]

The adoption of agricultural processes enhanced by agrivoltaic applications represents a transformative approach to addressing pressing food and energy security challenges in an era of population growth and climate change. Agrivoltaic systems that combine agriculture with solar energy generation contribute to both food and energy security and are seen as promising solutions to food and energy security. Mechanization is emerging as a key driver for the development of efficient agricultural systems that enable the transition from subsistence to market economies. While mechanization is critical to efficient agriculture, its integration with agrivoltaics must overcome technical and operational challenges to ensure a seamless transition, given the role of farm machinery in the operational phase of agrivoltaic systems. If these elements are carefully considered in the project development phase, multiple solutions can be found for an agrivoltaic system where agronomic and PV operations are optimally coupled. In this regard, this chapter explores the potential synergies between agrivoltaics and agricultural machinery, equipment, and smart farming practices, focusing on improving crop productivity and increasing energy production. In addition, the replacement of traditional agricultural nets with PV panels in agrivoltaic systems is discussed. The potential benefits and challenges of this transition are explored, offering insights into more efficient and environmentally friendly agrivoltaic designs.

4.1 Introduction

The rapidly growing global population and the mounting concerns over climate change have put immense pressure on agricultural practices to evolve and become

[1]Biosystems Engineering Department, Faculty of Agriculture, Tarbiat Modares University (TMU), Iran
[2]Fraunhofer-Institut für Solare Energiesysteme ISE, Germany
[3]BayWa r.e. Solar Projects, Kaiser-Joseph-Str. 263, 79098 Freiburg, Germany

more sustainable and efficient. Agriculture, which can continuously provide food and other resources for the world's growing population, is vital to human existence and therefore to any human activity. However, many problems threaten the ability of agriculture to meet current and future human needs [1], including climate change; high rates of biodiversity loss; land degradation through soil erosion, compaction, salinization, and pollution; reduction and pollution of water resources; increase in production costs; The ever-increasing number of farms and related to it, the poverty and the reduction of the agricultural rural population not only have to face these problems, but the way it has been used in recent decades is the main cause of all these issues [2]. With its 17 Sustainable Development Goals (SDGs), the 2030 Agenda for Sustainable Development seeks to balance the needs of people with those of the environment. Human resilience, livelihoods, and overall well-being depend on the functioning of the biosphere. Therefore, with a biosphere-centric approach, the SDGs contribute significantly to our society, economy, and overall quality of life [3].

Mechanization is the key driver of efficient agricultural systems. This possibility provides the transition from subsistence agriculture to market-oriented agriculture, makes off-farm jobs attractive for women and youth, and accelerates rural development. Mechanization options include agricultural tools, equipment, and machinery for land preparation, crop management, harvesting, post-harvest activities, processing, and all actions in the agro-food value chain [4]. Agricultural mechanization provides the necessary power and equipment for soil preparation and establishment, maintenance, and processing of agricultural products in the field. One important type of agricultural automation that can transform agriculture and rural living conditions is motorized mechanization. Faster and more effective farming operations can lead to increased agricultural production and flexibility, higher agricultural yields, labor and cost savings, and other benefits for producers [5]. Sustainable mechanization is a means to an end. The use of new technologies that are compatible with the environment enables farmers to produce crops more efficiently using less energy. Mechanization of sustainable agriculture can also contribute to the development of value chains and food systems, as it has the potential to make post-harvest, processing, and marketing activities and practices more efficient, effective, and environmentally friendly [6]. Usually, the total amount of greenhouse gases (GHGs) released by agricultural activities is expressed as carbon equivalent ($CO_{2\text{-eq}}$) [7]. According to a study by Yousefi *et al.* [8], of the total 2,042.091 kg $CO_{2\text{-eq}}$/ha emissions in sunflower production, agricultural machinery alone contributed 67.176 kg $CO_{2\text{-eq}}$/ha (3.29%). While the figures for wheat and corn cultivation were reported as 25% and 20%, respectively, due to fossil fuel consumption, the total carbon footprint of agricultural machinery in China for rice cultivation was reported as 13% [9]. The carbon footprint of cotton production in Australia was estimated to be 7% of the total carbon footprint, with emissions from machinery accounting for about 16% of agricultural activities. It should be emphasized that cultivation techniques, the type of machinery used, the degree of mechanization, and the volume of production all influence the emissions caused by fossil fuel combustion in various agricultural activities [10]. The major issues facing the agricultural sector that also make it unsustainable include rising energy costs, faster depletion of fossil fuels, diminishing water supplies, resistance to agrochemicals, and environmental protection [11]. By increasing productivity and income while increasing the energy required for agricultural activities, optimization of agricultural practices, automation, mechanization, and better use of equipment and resources can stabilize the agricultural industry. In order to meet the increasing global demand for food, it is also important to increase the agricultural area and the number of workers [12]. In a long-term vision and planning, targeted national strategies to financially support farmers and contractors to invest in infrastructure, ensure optimized use of the fleet, and access to or adapt to the best available CO_2-compatible technologies to produce energy efficient crops is needed. Therefore, in this regard, there is no single solution, but a comprehensive strategic plan is needed that is compatible with the progress of other sectors, the availability of energy carriers and technologies, and the specific needs of work and machines [13]. The European Union's Common Agricultural Policy (CAP) created a system of agricultural subsidies and other programs. An intelligent, sustainable, competitive, resilient, and diversified agricultural industry is promoted through CAP strategic plans that also ensure long-term food security. They also support efforts to combat climate change, protect natural resources, conserve and enhance biodiversity, and strengthen the socio-economic base of rural communities [14]. Agricultural tools are used to maximize the full potential profit for each sector and final market [12]. Although GHG emissions from agriculture, forestry and land use (AFOLU) have roughly stabilized over the past 25 years, the agricultural sector still produces nearly 24% of total global GHG emissions [13].

In response to these challenges, the concept of agrivoltaic systems has emerged as a promising solution that combines agriculture and solar energy generation in a mutually beneficial manner. Agrivoltaics involves the integration of solar panels with traditional agricultural practices, creating a symbiotic relationship that enhances land productivity and optimizes energy production [15]. While the concept of agrivoltaics has garnered attention for its potential to address food and energy security, its widespread adoption has faced challenges due to factors such as the compatibility of solar structures with mechanized agriculture and the need for advanced monitoring and control systems [16]. Therefore, the adoption of an approach to bridge the gap between agrivoltaics and modern farming techniques by leveraging mechanized equipment and smart farming technologies to enhance the integration, efficiency, and scalability of agrivoltaic systems is crucial. By integrating cutting-edge automation, data analytics, and artificial intelligence into agrivoltaic setups, farmers can optimize crop growth, water usage, and energy generation, ultimately leading to increased productivity and reduced environmental impact. In this context, this chapter aims to explore the challenges and opportunities in adapting agrivoltaic systems to mechanized agriculture and smart farming. It will delve into the various technologies, methodologies, and best practices that can be employed to maximize the potential of agrivoltaics while ensuring sustainable agricultural practices.

4.2 Different farming processes in different agrivoltaic systems

Till now, various classifications of agrivoltaic systems have been developed, each tailored to suit specific crop types, climates, and farming objectives. Crop and food production, livestock production, ecosystem services through crop management, and solar greenhouses are just a few of the agrivoltaic applications (see Figure 4.1). Because these applications are not mutually exclusive, multiple activities can occur simultaneously at a given site, even on the same land at different times of the year. While some projects use different zones for different objectives, such as targeted grazing in pollinator habitat zones to strategically manage vegetation at certain times of the year, other initiatives use the same zones for multiple objectives. It should be mentioned that agricultural and ecovoltaic projects are not incompatible with each other [17]. Systems for growing seasonal and perennial food, fiber, or specialty crops can be deployed under and around solar infrastructure. Partial shading of the solar infrastructure affects the cropping systems; plants can be planted directly under the panels and between the rows of panels. Crops can be grown by hand or with machinery [18]. Infrastructure can be modified in terms of panel height, spacing between panels, spacing between rows, or other design features to allow for variations in sunlight/shade patterns and adherence to agricultural activities. Crops can be grown within a conventional ground-mounted solar array. Various types of crops and horticulture are supported by novel technologies and systems, including systems for orchards, viticulture, field crops, and other fruits [15].

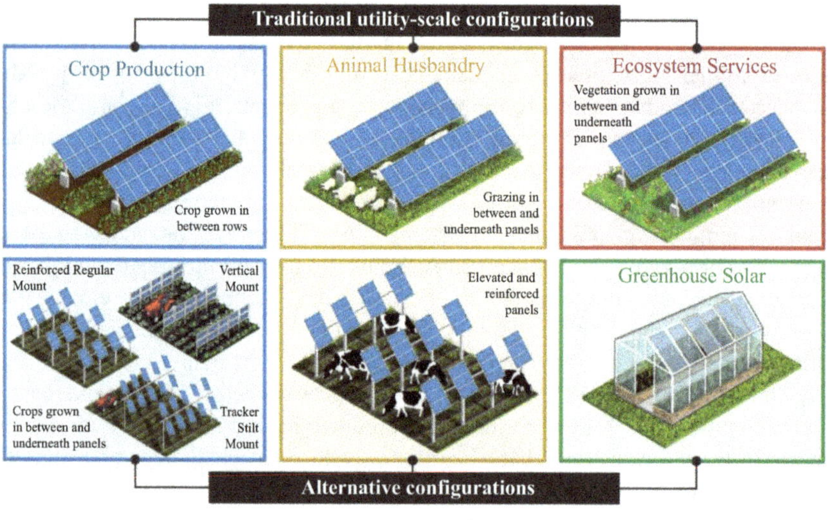

Figure 4.1 Types of agrivoltaics systems that have been deployed commercially [17]

Agrivoltaic systems for livestock and animal husbandry production and management include livestock grazing and control in close proximity to solar infrastructure [19]. According to the literature, although other animal husbandry systems are conceivable, agrivoltaic systems are typically implemented with sheep, cattle, poultry, honeybees, and rabbits. Depending on the requirements of the solar site and the needs of the livestock operation, animals may be on-site year-round, sometimes, or as needed. Animals may not obtain all of their food from on-site resources and may occasionally be transferred or receive additional food from off-site sources [17]. Although not all countries classify this as an agrivoltaics use, the integration of solar infrastructure into aquaculture operations is an emerging application [20] (see Figure 4.2(a)). In many cases, conventional solar infrastructure does not need to be significantly modified to support grazing and livestock operations. However, other changes can be made to the site, such as vegetation planting and management, fencing, water supply, and livestock access, to make the system more compatible with livestock management practices.

Ecosystem support Agrivoltaic projects include structures built to improve soil, provide other ecosystem services, and create or restore habitat. In addition to beneficial insects (e.g., pest controllers and pollinators), the habitat may also support species of interest such as birds, reptiles, mammals, and other wildlife. The potential for these areas to provide relevant services is largely determined by raising the elevation of the slabs to a sufficient level to allow vegetation to become established with carefully selected beneficial seed mixes. These sites typically require only minor modifications to conventional solar designs [17]. Solar greenhouse agrivoltaic projects are greenhouse installations where PV technologies are placed over or integrated with greenhouse infrastructure to provide partial shade and light regulation. To vary the availability and quality of light during the day, solar technologies and configurations can include transparent, semi-transparent, or opaque modules that rest on or are integrated with the building structure (Figure 4.2(b)). Some countries do not recognize solar greenhouses as agrivoltaic projects, like the way they do not recognize ecosystem services projects [22]. Therefore, during the design phase of the agrivoltaic system,

Figure 4.2 (a) Aquavoltaic system installed in Les Cedres, La Réunion [20], and (b) a glass greenhouse with an opening on the roof to prevent shading from PV panels [21]

developers should consider the types of agricultural practices that are or could be used at each site based on anticipated agricultural activity and whether those practices are compatible with solar infrastructure. The size of the agrivoltaic project can have an important influence on the types of agricultural activities that can be undertaken on-site and their economic viability. If agricultural or pastoral activities require equipment, machinery, or fencing, it is important that these are compatible with the solar design and configuration and do not cause damage to the solar infrastructure. Agriculture often requires larger equipment and machinery for land preparation and maintenance that could be used for tilling, planting beds, seeding, mowing, or harvesting. For solar projects with grazing, the number of animals and the length of time they will be on the project should be considered [17]. This section addresses improving crop productivity and increasing energy production by exploring the potential synergies between agro-photovoltaics and agricultural machinery, equipment, and smart farming practices. The topic covers a range of adaptation issues in agrivoltaics, such as the use of technologies, precision agriculture methods, and real-time data analysis, all of which contribute to realizing modern, sustainable agriculture.

4.3 Farm machinery adoption with agrivoltaic plants

4.3.1 Conventional farm machinery

An agrivoltaic system combines two very different subsystems into one, an agricultural production system and a PV system. The PV system is mainly composed of PV modules, a substructure, inverters, transformers, and electrical equipment (DC wiring, AC wiring, etc.). In addition, it is important to consider the installation and O&M (operation and maintenance) costs, as well as the economic, regulatory, and social conditions that interact with the system. However, as shown in Figure 4.3, an

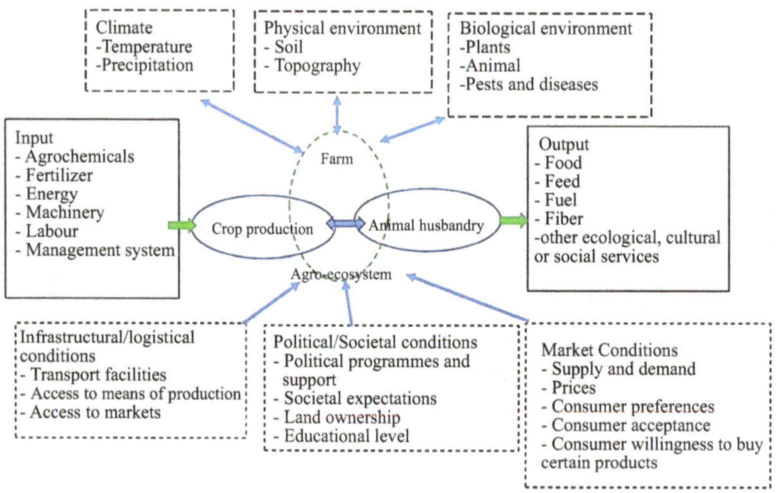

Figure 4.3 Determinants of agricultural production systems [23]

agricultural production system consists of a variety of inputs, outputs, and environmental, climatic, infrastructural, political, social, and market conditions that are either part of the system, such as the soil, and/or affect the operation [23]. The challenge is to optimally combine these subsystems. This requires a deep understanding of the agronomic aspects, the PV aspects, and especially their interaction within the agrivoltaic system.

The introduction of agrivoltaic systems on the farm has potential implications for all agricultural factors, which are discussed in detail in this chapter. This chapter focuses on farm machinery and its impact on the design and operation of agrivoltaic systems. Agricultural machinery is very diverse and serves a range of applications and sizes. To facilitate understanding of this section, agricultural machinery in agrivoltaics is divided into three levels (Figure 4.4). The first level refers to the type of cultivation: arable, grassland cultivation, and perennial crops. The second level refers to the agrivoltaic technology (interspace or overhead) and the third level refers to the project phase (design or operation). Note that the section focuses on agrivoltaic systems that require intensive use of farm machinery, so systems with livestock and ecosystem services are not included.

The degree of mechanization and the type of machinery used vary greatly according to agricultural practice. However, three distinct categories with common practices can be identified:

- **Arable farming** is the most mechanized form of cultivation, using multiple machines for tillage, seeding, fertilization, crop protection, and harvesting. Depending on the size of the farm, the crop grown and the cultivation methods, the machines have a great variety in terms of technology and size.
- In **grassland farming**, different machines are used depending on the intensity of cultivation. In semi-natural grassland with low yields, the use of machinery is usually limited and mainly focused on maintenance. However, in fertilized monocultural grassland, high grass yield is targeted, resulting in the typical cycle of tillage, seeding, and fertilization. In addition, harvesting usually involves standard tools for mowing, conditioning, tedding, and raking. After raking, the grass is usually baled into rectangular or cylindrical hay bales using automatic machines called balers. Later, they are transported and stored for later use as animal feed.

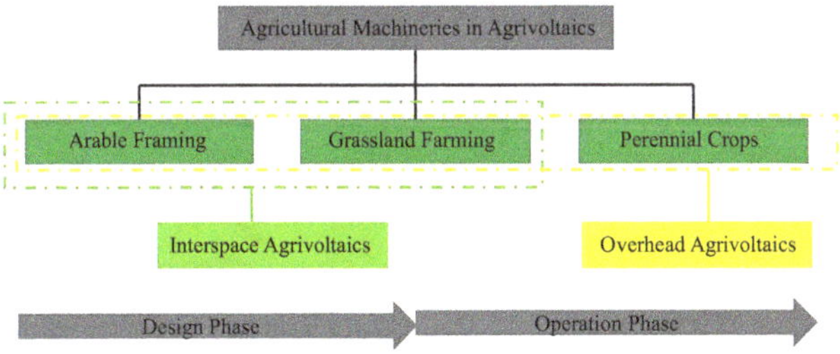

Figure 4.4 Categorization of agricultural machinery within agrivoltaic systems

- **Perennial crops** such as stone fruits, pome fruits, and berries usually use small tractors, which mainly facilitate fertilization, crop protection, and crop transportation.

Agricultural machinery has a significant impact on the design phase of agrivoltaic systems. Therefore, project-specific parameters of all agricultural machinery used by the farmer must be specified. The most important parameters needed for the agrivoltaic design are (Figures 4.5 and 4.6):

- Maximum machinery working width of all involved machineries;
- Maximum machinery height;

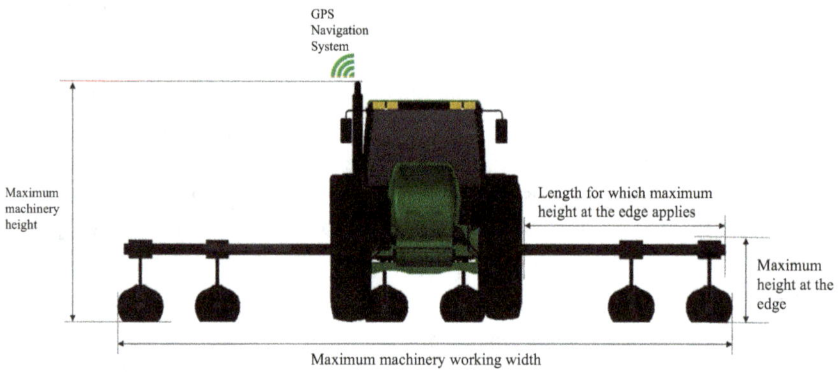

Figure 4.5 Illustration of the main machine information required for planning agrivoltaic systems © BayWa r.e.

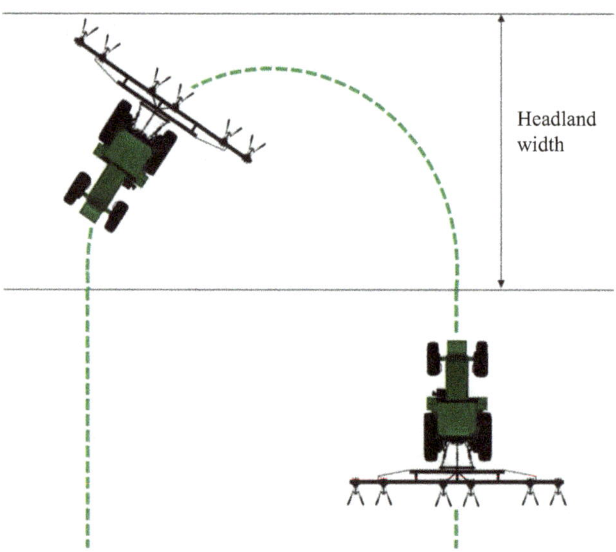

Figure 4.6 Illustration of the headland width © BayWa r.e.

- Maximum height at the edge of the tool;
- Length for which the maximum height at the tool's edge applies;
- Availability of an autonomous guidance system;
- The width of the headland, i.e., the space required for the machines to turn;
- Plowing depth.

The impact and benefit of the above machine parameters on the design differ depending on the type of agrivoltaic system. Interspace agrivoltaics can be used in both arable and grassland agriculture. Different types of systems have been developed, such as horizontal single-axis tracking systems with one module in portrait orientation (Tracking 1P) or two modules in portrait orientation (Tracking 2P), south-facing fixed-tilt systems (Fix-South), and vertically mounted east-west systems (Fix-Vertical) (Figure 4.7). Their common feature is the movement of the system between rows of modules. In the tracking systems, the modules can be rotated to a suitable position by farmer-controlled software during machine operation, vertically for Tracking 1P (Figure 4.7(c)) and to an "alpha" position for Tracking 2P (Figure 4.7(d)). As can be seen in this figure, the pitch is also affected by the width of the greenstripe. The greenstripe is an unavailable area for growing plants and is mainly used for biodiversity measurements. The size of the greenstripe largely determines land loss, which is often regulated at the country level [24]. As a result, minimization of the greenstripe is sought for both agriculture (less land loss) and PV performance (smaller pitch). The minimum greenstripe is defined based on the type of facility (fixed-south facilities have a larger greenstripe than other facilities and the minimum distance from the structure that the farmer can maintain while driving. Therefore, the use of a vehicle autonomous guidance system (VAGS) is very beneficial, allowing driving accuracy of centimeters [25]. The benefit of the autonomous guidance systems is not restricted only to the reduced greenstripe, more importantly, it can play a pivotal role in the safety of farmers' operations within the agrivoltaic system. These operations include the linear movements in between the module rows, the turnings at the headland, and the maneuvers to avoid PV components like inverters and transformers. For all types of equipment, the maximum working width of the machine defines the pitch, which is one of the most important technical parameters of the equipment. The pitch chosen can affect both PV and agronomic performance. To understand the pattern of pitch effect in the interspace, the modeling results of a study by BayWa r.e. for a tracking 1P system are discussed [26].

As can be seen in Figure 4.8(a), increasing the width of the machine improves the specific electricity yield in kilowatt-hours (kWh) per kilowatt-peak (kWp), but significantly reduces the specific installed capacity in kWp per hectare (ha), resulting in an overall lower annual yield in megawatt-hours (MWh) per ha. In addition, increasing the pitch leads to higher PV system costs due to higher cable consumption and substructure costs. Overall, increasing the width of the machine (i.e., increasing the pitch) is expected to worsen the PV performance. However, the agronomic performance of the Agri-PV system improves when the working width of the machines is increased due to the decrease in area losses and shading rates

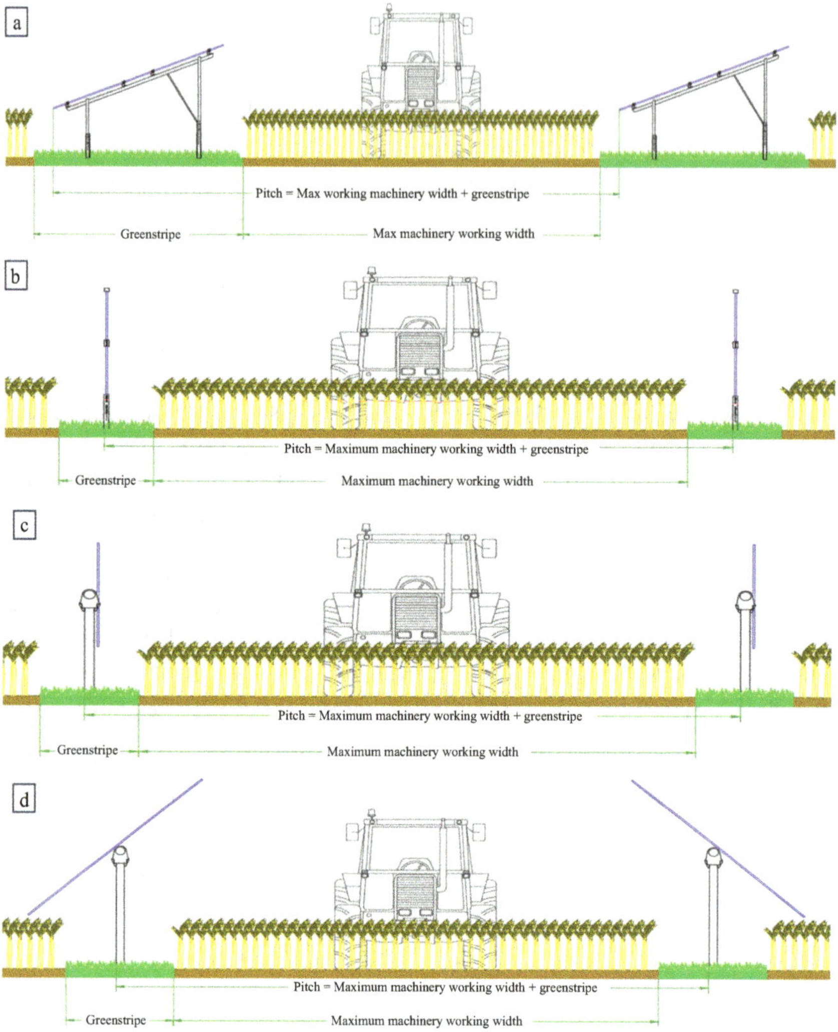

Figure 4.7 Illustration of interspace agrivoltaic system types during machinery operation for (a) fixed-south, (b) fixed-vertical, (c) tracking 1P, and (d) tracking 2P (d). © *BayWa r.e.*

(Figure 4.8(b)). A performance indicator that combines agronomic and PV performance is the land equivalent ratio (LER), which expresses land use efficiency. The study found that LER is reduced for three crop categories with different shade tolerance when the working width is increased (Figure 4.8(c)).

The above effects of larger machine working widths have also been reported in other studies for different interspace agrivoltaic system topologies [27–29], showing the need for smaller machine widths in agrivoltaic applications. In grassland

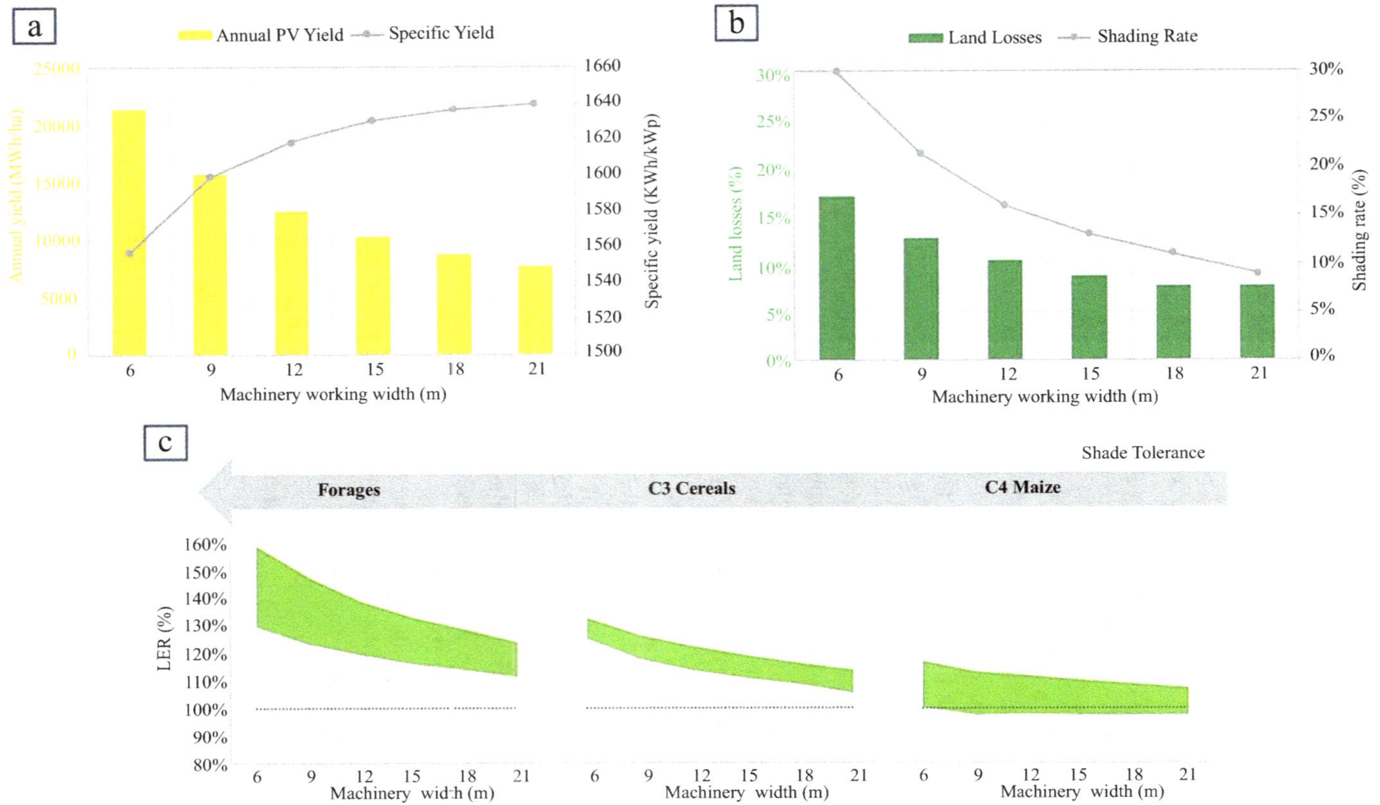

Figure 4.8 *Effect of the machinery width on the electricity yield: (a) land losses and shading rate, (b) land equivalent ratio, and (c) tracking 1P interspace system. Based on modeling results of © BayWa r.e.*

agriculture, the maximum machine width is mainly determined by the hay rake, which can have a working width of up to 14 m [30], but smaller working widths in the range of 6–9 m are also common. The main problem arises in arable farming, where there is a historical trend towards larger machines, which have significantly improved the performance of large farms in terms of yield and efficiency [25]. In arable farming, the maximum working width is mainly determined by the spreaders (either for fertilization or crop protection), which can be up to 39 m [31]. The development of agrivoltaics for such high working widths can be challenging from the point of view of economic viability and land efficiency. This results in the requirement for the use of smaller-width sprayers. The use of smaller sprayers can affect the farmer's productivity. Therefore, the impact on farmer productivity for different machine widths must be evaluated on a project-specific basis. In addition, there is the question of the availability of smaller-width sprayers and the farmer's access to these sprayers through a rental service or direct purchase [32].

The development of VAGS has been rapid in recent years, driven by the need for precision farming [25]. VAGS have been developed with various navigation sensors, such as machine vision, global navigation satellite system (GNSS), laser-based sensors, inertial sensors, geomagnetic directional sensors, and a combination of the [33]. However, the most common commercially available type is based on GNSS receivers [34], whose signal is enhanced by RTK (Real Time Kinematic) techniques, allowing centimeter accuracy in the guidance [34–36]. Many farmers have already chosen GNSS-based VAGS for the benefits of improved productivity, accuracy, and safety in operating their machinery [33]. More specifically, with a VAGS, the farmer can use complicated machinery with less effort, achieve higher productivity through increased tractor speed, and significantly reduce overlap when driving, resulting in potential savings and a reduction in soil compaction [37]. The main disadvantages are the high initial investment costs and the need for an RTK base station within 10 km [34]. However, the importance of increased safety in propulsion within systems and the agronomic advantages of VAGS make their overall application in agrivoltaics very advantageous. Two other important parameters that affect the design of the agrivoltaic are headland width and plowing depth. A high headland width means a greater distance between the periphery of the agrivoltaic system and the edge of the farm, which can significantly reduce the total installed PV power. In contrast, an insufficient headland may increase the farmer's working time or increase the risk of collisions due to the limited space at the turning areas, especially considering that the driving accuracy at the headland is reduced. Therefore, the optimal headland width must be determined based on project-specific cropping practices. In general, smaller machine widths equipped with a VAGS have advantages in minimizing headland width. However, plowing results in increased depth of the cable trench, which negatively impacts construction costs. In general, national standards regulate the cable trench depth in tillage deeper than the usual GM-PV systems. In general, intensive plowing has a negative impact on the environment, leading to a trend toward shallower plowing depths in conventional tillage. In parallel, many farms tend to have conservation and zero tillage, which account for 26% of the total cultivable land in the European Union

[38]. The size and type of agricultural machinery also play important roles in the design of overhead agrivoltaic systems. For perennial crops, the influence of agricultural machinery on the design is rather limited compared to annual crops. Specialty crops typically use small tractors that fit between rows of fruit trees (2.5 to 5 m) and allow for a limited number of operations. At the same time, the system is higher than the maximum harvest height, between 2.5 and 5 m, so the tractor has enough space to drive under the modules. In addition, in conventional fruit growing, farmers are used to driving in confined spaces that are limited in width (rows of trees) and height (shade netting/foil). Overall, the use of machinery in agrivoltaic systems with perennial crops can usually be accomplished without challenges or adjustments. A very important cultivation technique for these systems is the shading elements applied, which are discussed in detail in Section 4.4. The exception to this is in viticulture systems, where there is a trend toward the use of heavy machinery for pruning, canopy maintenance, and harvest [39]. As can be seen in Figure 4.9, these machines enclose the row of vines from above, which makes it difficult to operate within an agrivoltaic system because of the columns present. In addition, the height is much higher than that of the vineyards, which leads to a greater height of the substructure.

For overhead systems in arable or grassland farming, many aspects related to the required driving precision and the effect of tillage depth and headland width remain the same as discussed above for interspace systems. However, the movement of the machines under the modules results in differences in the interaction between the machine dimensions and the design of the agrivoltaic system. First, for overhead systems, the maximum machine height defines the clear height of the system. Consequently, higher machines lead to higher system costs. In arable farming, a maximum machine height of 4.0 m can be expected, which is mainly determined by the harvester. As an example, the height in the agrivoltaics pilot project in Heggelbach, Germany, was set at 5.0 m (Figure 4.10) [40]. A second major difference is that the pitch is decoupled by the maximum working width of the machine or the applied greenstripe. In overhead systems, the greenstripe is

Figure 4.9 Example of machineries used in vineyard farms: (a) a hedger bards removal [39], and (b) a mechanical grape harvester. © Lakeview Vineyard Equipment Inc.

Figure 4.10 View of the agrivoltaic facility in Heggelbach and illustration of the main dimensions driven by the applied agricultural machinery © BayWa r.e.

minimal to begin with because the modules are overhead, and the only possible interference point is the vertical columns. Trommsdorff *et al.* [41] report about 8% area loss due to the piers for the Heggelbach pilot. In addition, the available machine width may be higher than the pitch of the system because vertical piers are installed in every second row of modules. In the Heggelbach example, the available machine width is 19 m, compared to a much smaller pitch of 9.5 m (Figure 4.10) [40]. Thus, the pitch can be selected based on the light availability requirements of the plants and the span of the system based on the maximum working width of the machines. However, it is important to note that the greater the system's span, the greater the technical challenges related to static stability. This requires measurements to be taken for reinforced horizontal and vertical beams, which significantly increase the steel consumption and the cost of the system's substructure compared to an interspace agrivoltaic system [42]. In summary, the maximum width of machines in overhead systems does not affect the total installed capacity. However, the overall dimensions of the machines are the main reason for the very large dimensions of the systems (height and span), which consequently lead to high steel consumption and high costs for the substructure. Measures to reduce steel consumption are taken by introducing substructures based on [43].

Agricultural machinery plays an essential role in the operational phase of the agrivoltaic system. On the one hand, the presence of PV components affects the agricultural practices of the machines, and on the other hand, the operation of the machines brings new risks and challenges to the PV operation. PV components affect machinery operation in arable and grassland farming more than interspace or overhead systems. With the introduction of agrivoltaics, an open field is transformed into several sub-fields with limited width and/or height. This creates new boundary conditions for the farmer within the agrivoltaic systems: along the rows of modules (interspace system) or the substructure pillars (open space systems) and near other PV system components (e.g., inverters, transformers, fence). This means that areas with headland-like conditions multiply in agrivoltaics systems. According to Karl Wild of the Dresden University of Applied Sciences [32], this phenomenon of increased boundary effects can hurt soil productivity by reducing the efficiency of fertilizer application at the edges. In addition, the agrivoltaic system makes the farmer's driving much more complicated, which can lead to reduced working speed and increased driver fatigue. Finally, the greatest risk is the possibility of a collision with the system, which could result in injury. The above challenges can be overcome to some degree by applying VAGS to agricultural machinery. Precise driving can make boundary deposition more effective, and at the same time, driving is accomplished with less effort and greater efficiency. More importantly, safety is greatly enhanced by reducing the risk of collision. Additional measures need to be taken for collision risk, such as sufficient safety distances, boundary labeling, and preventive fencing of high-risk components such as the transformer stations. Movement with heavy agricultural machinery within PV components introduces some new risk factors for the PV side of the system:

- The risk of collision as described above.
- The possibility of damage caused by stone-throwing is possible in machines with rotating elements, especially in grassland farming. This risk can be minimized by carefully cleaning the field of stones before use. In the case of the interspace tracking system, it may be advantageous to rotate the modules horizontally during the use of machines with a high risk of stone-throwing.
- In all categories of cultivation, various chemical elements are used to fertilize and protect the plants, which can be attached to the PV components. The exact effect of these chemicals in each component is not known, and further research in this direction is needed. This risk is generally higher in interspaced and overhead systems over specialty crops, where the chemicals are deposited very close to the PV components. One precaution in arable and grassland cropping is to apply the chemicals as close to the soil as possible [32]. Another solution for interspace systems is to rotate the trackers to a horizontal position during spraying. Finally, cleaning of the modules and visual inspections are recommended.
- Fire caused by a PV component and spread by dry plants is also an important risk factor. Risk mitigation measures could include careful cleaning of the environment surrounding high-risk components such as inverters and transformers and increased preventive fire protection measures.
- Another aspect that affects PV performance is the dust generated during the operation of the machines, which can lead to higher power losses due to pollution compared to GM-PV system [44]. On this topic, more frequent cleaning cycles and/or adaptation of cleaning cycles to farms may be needed.

As qualitatively discussed, there are several risks for both agriculture and photovoltaics. However, it must be emphasized that there is very little research on these issues, making quantification difficult. More research in the future could help quantify the magnitude of the impact of the above risks at the technical level and examine the frequency of occurrence. In addition, it is important to translate the impact of the above risks to the economic performance of the agronomic and PV sides. Finally, the impact on the existing insurance contracts can be analyzed. Many large-scale agrivoltaic systems have been installed in recent years, which will allow more knowledge to be gained on these important issues.

4.3.2 Electric farm machinery

The use of agricultural machinery such as tractors, combines, wagons, loaders, pickups, and other vehicles in small to large farms around the world gives them a key position in the agricultural industry [45]. By limiting human intervention and overcoming labor shortages, automation of various agricultural processes such as planting, spraying, weeding, and harvesting can also help achieve sustainability goals. A variety of farm machinery is used on farms, but tractors remain the primary source of power for many agricultural operations. Through various attachments, tractors are used to mechanize a range of agricultural activities. In addition to traditional grass cutting, landscape maintenance, fertilizer application, and pesticide application, modern tractors are now used for plowing, planting, harvesting, and mowing [46]. Most agricultural vehicles on the road today use internal combustion engines (ICEs), that burn fossil fuels and emit a lot of pollutants into the environment. Due to the time and fuel required for tillage, emissions from agricultural machinery are significantly higher [7]. The fuel consumption of agricultural equipment used for plowing depends on several factors, including soil texture, tractor size, equipment used, and depth of tillage. In general, fuel consumption ranges from 18.0 to 46.0 L/ha, 12.4 to 20.2 L/ha, and 5.6 to 11.2 L/ha, respectively, for share plows, chisel plows, and tandem disks. Instead of tractors powered by fossil fuels, electric motors powered by renewable energy sources such as solar, wind, and hydroelectric power can be used in agricultural equipment because they have higher conversion efficiency, more flexible torque and speed control, and lower emissions, and can also be easily adapted to mechanized agricultural tasks [47]. Automated motorized mechanization automates agricultural tasks such as plowing and milking. Diagnostic and decision-making processes can be automated using digital automation technology. These technologies can increase environmental sustainability and resilience, increase production, and improve accuracy. The transition from manual tools to animal traction, motorized mechanization, computerized equipment, and finally robotics with artificial intelligence can be considered the technological evolution in agriculture (AI) [4] (see Figure 4.11).

The integration of agrivoltaics with farm machinery and robots holds immense potential for revolutionizing modern agriculture. This synergy not only addresses the need for sustainable energy sources but also enhances the efficiency and environmental sustainability of farming practices. Agrivoltaic systems can power electric farm machinery, irrigation systems, and other agricultural equipment, making farming practices more sustainable and energy efficient. This integration reduces the environmental impact of agriculture and improves operational efficiency and resource management. Smart tractors equipped with GPS and precision seeding technologies can efficiently plant crops in the shaded regions between solar panels, ensuring precise spacing and proper light distribution. Moreover,

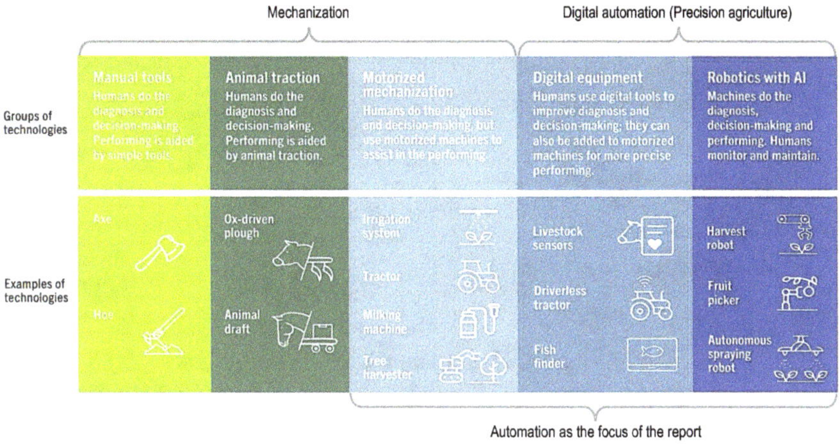

Figure 4.11 Evolution of agricultural automation [4]

sensor-equipped robots can continuously monitor soil moisture, nutrient levels, and crop health, providing real-time data to farmers and enabling them to make informed decisions regarding irrigation, fertilization, and pest control. The combination of precision farming with agrivoltaics enhances crop yields while promoting sustainable resource management. Incorporating farm machinery and robots into agrivoltaic systems allows for the collection of vast amounts of data. Machine learning algorithms and artificial intelligence can analyze this data to optimize various aspects of the system. For instance, the data can be used to predict energy production, determine the most suitable crops for specific shading patterns, and develop adaptive control strategies for solar tracking systems. These data-driven insights empower farmers to make proactive decisions, maximizing both agricultural and energy output while minimizing environmental impacts.

4.3.2.1 Electric tractors

A vehicle that uses electric traction motors for propulsion is called an electric tractor (ET). To electrify farm equipment and accessories, the "International Harvester Company" originally offered the Electrall feature for Farmall 400 tractors in 1954. This method was designed to electrocute insects at night. The first fuel cell tractor was introduced in 1959 by Allis Chalmers. This vehicle had 1008 alkaline fuel cells that allowed it to power a 20 hp DC motor [48,49]. Using solar and wind energy to power farm tractors will significantly reduce CO_2 emissions while being economically feasible [45]. A Dutch agricultural machinery dealer named Schmetz launched an electric tractor in the 1990s that used an electric motor-generator mounted between the ICE and the transmission. In this way, an electric drive could achieve step-less speed transmission without a hydraulic mechanism [48]. To solve the problems with the electrification of farm tractors at the beginning of the twenty-first century, German universities and institutions, in collaboration with Fendt (AGCO), also known as MELA (Mobile Electrical Power

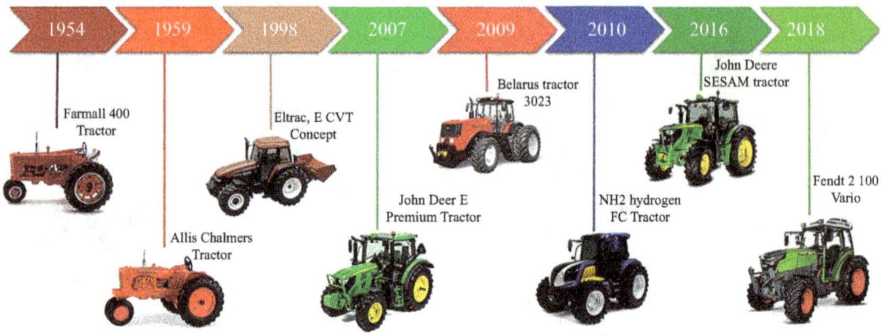

Figure 4.12 Historical development of hybrid tractors [25]

and Drive Technology), introduced a series of auxiliary electrical devices. Universal compatibility of tractors with implements from all manufacturers was ensured two years later with the introduction of the ISO11783 (ISObus) standard, created by the "Agricultural Industry Electronics Foundation" (AEF) [50]. In 2010, New Holland introduced the NH2, a hydrogen-powered tractor with an electric motor that produced about 106 horsepower. German manufacturer Fendt entered the electric tractor market with the introduction of the e100 Vario at Agritechnica in Hannover in 2018 [51]. A timeline of hybrid electric tractors for agriculture from 1954 to 2018 is shown in Figure 4.12. Leading innovators in the development and electrification of automated agricultural tractors include John Deere, AVL, Fendt, Solectrac, Ztractor, Rigitrac, Claas, Case IH, Escorts, Kubota and Schaffer.

4.3.2.2 Agricultural robots

In the modern agricultural industry, automation of agricultural machinery can lead to significant change by enabling farming techniques, increasing the precision of various processes, limiting human intervention, reducing downtime, and ultimately increasing overall production. The health risks associated with machine–human interaction also decrease as human presence in the field decreases and the cost of operating and maintaining machinery decreases [46].

Farmers are using technology to address labor shortages by managing a range of processes remotely. With the introduction of technologies such as wireless connectivity, better bandwidths that enable human-machine interaction (HMI), machine-to-machine (M2M) communications, and unmanned vehicles, agriculture is becoming smarter, faster, and more automated. Apart from the rising trend in labor costs, new developments in agricultural automation are making this technology more accessible in the near future, even though most agricultural robots are still at an early stage of commercialization due to high investment and maintenance costs [52]. Automation and robotics agricultural systems are changing the way work is done today. Some of the leading companies are following this trend and focusing on modern, sophisticated, and automated agricultural issues. One of the biggest obstacles to the development of agricultural robotics is the fact that agricultural work requires specialized methods and more resources than in other industries. As multipurpose and durable robotics technology advances, more progress can be made. This could address the challenges of a wide range of field operations and reduce the problems of changing conditions in agriculture [25]. AgBotII is an agricultural robot unveiled by the Queensland University of Technology (QUT). This autonomous robot has a 90% hit rate in identifying, classifying, and removing weeds. This robot is also capable of fertilizing and seeding. A computer, tablet, or smartphone app controls a variety of sensors, software, and electronic components used to perform weed control. Using a wireless remote control and virtual globe software, the farmer can control and set up the robot both manually and automatically. The ability of this robot to weed using force or herbicides is an important feature (Figure 4.13). The main power source of this robot is a 48 VDC electric motor with a power of 5 kW and an efficiency of 75–85%, coupled with a two-stage planetary gearbox to move the device at a speed of 5–10 km/h. In addition, the robot has integrated trailing batteries with a parallel connection that can store about 10.5 kWh of electricity and provide 7–11 hours of independent operation in the field [53]. An external solar array is used to charge these integrated batteries while the robot is docked at the station, as shown in Figure 4.13(b).

Two Intel i7 quad-core CPUs provide vision and navigation for this robot. An RGB camera and image processing controller are used to detect the weeds while weeding. Based on the image of the weed, the type of weed is determined, and the best weeding tool is selected and used at the right time. The software of the robot is based on the instructions of the robot operating system (ROS), which is divided between the two computers mentioned above [53]. The main steps of the weeding process by the AgBotII are depicted in Figure 4.14.

The use of agricultural robotics in agriculture can have a positive impact on both the design and operation of agrivoltaics. Robotic systems are expected to be smaller than conventional machines in both height and width, which can lead to significant cost reductions in agrivoltaics. In addition, smaller grass strips are required due to the higher driving precision, which means less land loss for the farmer. Overall, the higher precision of robotic systems in both driving and task execution can positively contribute to addressing the operational challenges identified in Section 4.3.1. For example, high driving precision and unmanned driving

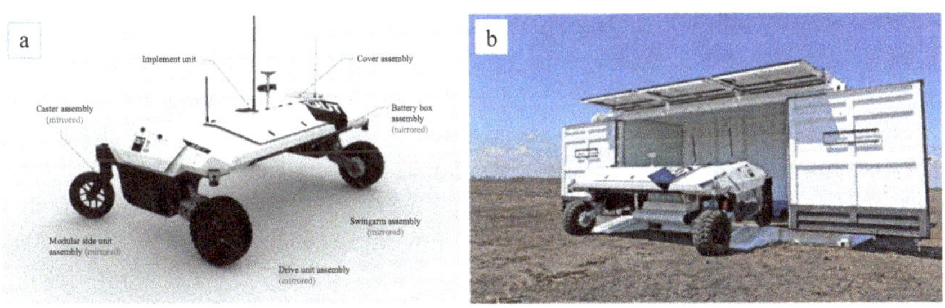

Figure 4.13 Solar-powered AgBotII weeding robot [54]: (a) robot's major outlines and (b) charging process at the solar charge station

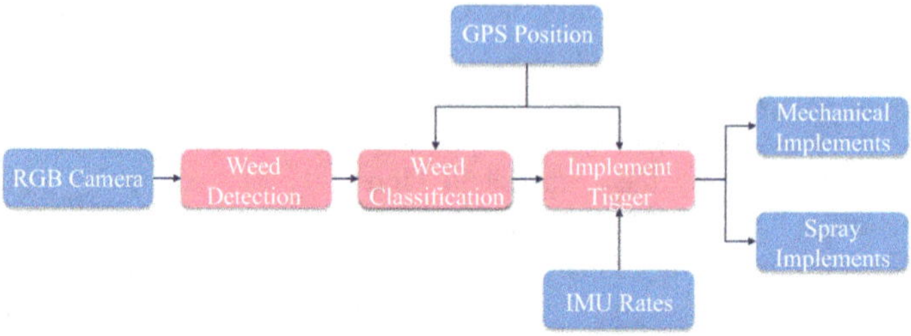

Figure 4.14 Block diagram of the AgBotII's weeding system software, adapted from [53]

significantly increase operational safety. In addition, precise fertilization and seeding can potentially reduce edge effects and chemical buildup in modular components. Finally, agrivoltaic systems can also prove beneficial to the robotic systems themselves by providing a standardized, repetitive workspace. In addition, agrivoltaics can increase the stability of the robot by integrating rails into the structure and providing power for operation. The potential benefits that robotic systems can bring make them one of the most promising areas for future research.

4.4 Smart monitoring of agrivoltaic systems

Measuring climate conditions is a critical challenge for a variety of scientific and practical fields. In tropical regions where there are no seasons, weather variables are highly dependent on local and temporal variations. In addition, due to the increasing prevalence of precision agriculture or smart farming, it is critical to improve measurement systems while placing them widely at crop locations. Therefore, monitoring agrivoltaic systems is critical to optimizing their performance and ensuring sustainable agricultural practices. These integrated systems, which combine agriculture with solar energy production, require careful monitoring to find the right balance between crop growth and energy production. Smart monitoring technology offers immense opportunities in this area. By using sensors to collect real-time data on factors such as soil moisture, temperature, solar radiation, and crop health, farmers can make informed decisions. This data can be analyzed using artificial intelligence (AI) algorithms to predict optimal planting times, irrigation schedules, and solar panel adjustments. The benefits are many: higher crop yields through optimized conditions, improved solar energy production by optimizing the tilt angle of the panels, and resource efficiency through reduced water and fertilizer use. In addition, such intelligent systems enable early detection of problems, minimizing crop damage and downtime [55].

4.4.1 Agronomical and PV monitoring

The agricultural Internet of Things (Ag-IoT) is at the forefront of digitalization in the agricultural sector. The concept of Ag-IoT refers to the collection of multiple agricultural parameters by sensors or drones, which are transmitted wirelessly and stored on the Internet. This data is later processed by specially developed algorithms that support the farmer's decision-making. Ag-IoT is already being used commercially in many cases. Xu *et al.* [56] categorize the application of Ag-IoT into the following five main groups:

(I) **Water-saving irrigation**: Microclimatic and soil data combined with irrigation control enable a precise, water-saving irrigation schedule.
(II) **Crop growth environment monitoring**: This refers to the monitoring of the whole range of microclimatic parameters such as temperature and humidity of the air, leaves, and soil, PAR, atmospheric pressure, wind speed, wind direction, and others via weather stations (Figure 4.15(a)).
(III) **Intelligent agricultural machinery**: In this application, sensors and satellite data enable higher driving accuracy (see Section 4.2).
(IV) **Animal and plant life information monitoring**
   - **Animal life information monitoring**: Sensors can be used to monitor the location and health of animals in real-time.
   - **Plant life monitoring**: It refers to the monitoring of plant physiological parameters, mainly through visual inspections (with cameras or drones) and measurements such as photosynthetic rate, and leaf area index (Figure 4.15(b)).

(V) **Agricultural product quality safety and traceability**: This is about managing the transportation, storage, and distribution of crops after harvest.

One of the essential technologies needed for solar power plants is a measurement system for weather and irradiation parameters. This can be achieved through appropriately calibrated sensors that meet WMO criteria and can consistently and accurately record all key data, including direct and reflected radiation, air temperature, wind direction and speed, and precipitation. AgriSense IoT is a product of Netsens, a series of wireless IoT units with microclimate sensors (air temperature

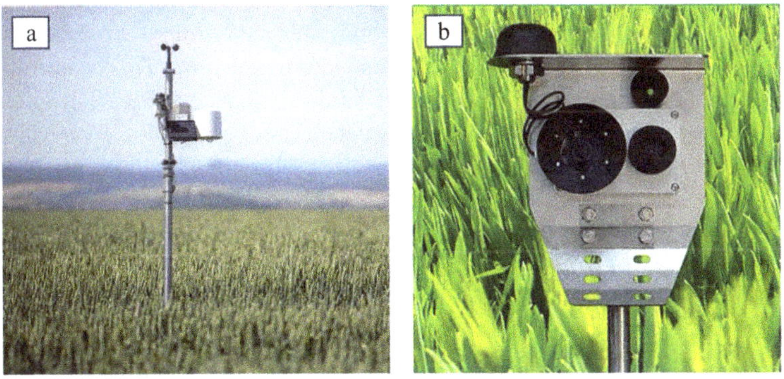

Figure 4.15 Examples of sensors deployed in agriculture: (a) a weather station and (b) a dual view camera system for crop monitoring and analysis © NEXT Farming

Figure 4.16 Experimental setup of the AgriSense IoT [57]

and humidity, leaf wetness, soil moisture, etc.) which are arranged around a main station containing the main weather sensors (rain, wind, solar radiation, air pressure) (see Figure 4.16) [57]. The wireless IoT modules are placed on plots up to 8 kilometers from the station and operate without solar power or an electrical connection. All data collection is accessible at any time through a LiveData cloud interface, which can be accessed through the website or app. The platform provides models for managing and analyzing data for the three main agricultural needs—crop protection, water demand, and frost alarm and prediction—as well as a complete record of all data collected.

According to Botero-Valencia *et al.* [79], an adaptable, autonomous, solar-powered weather station with 3G or WiFi wireless connectivity was designed and manufactured. The station measures relative humidity, temperature, barometric pressure, precipitation, wind speed, and light radiation. The system serves as a reference cell for the potential of solar energy generation and photovoltaics at the installation site and checks the state of charge of the main battery and the energy generated by the photovoltaic module. The station can be configured and controlled remotely. The collected sensor data can be uploaded to the cloud in real-time. This project aims to promote the development of free open-source hardware that can be used in agriculture and help local experts increase production efficiency and crop yield. PV power plant monitoring is a standard in the industry and allows for continuous measurement of energy generated, evaluation of system performance, and early and targeted detection of faults. Therefore, at the time of construction, the plant is equipped with several sensors that monitor in real-time the following main parameters [58]:

(I) **PV parameters**: In-Plane irradiation in the front and back side of the modules (Figure 4.17(a)), module temperature, and soiling sensor (Figure 4.17(b)).
(II) **Weather station**: Horizontal global irradiation, ambient temperature, wind speed, wind direction, relative humidity, air pressure.
(III) **Electricity production**: Current, voltage, and power at the PV array output and at the connection point to the utility grid, durations of system outage.

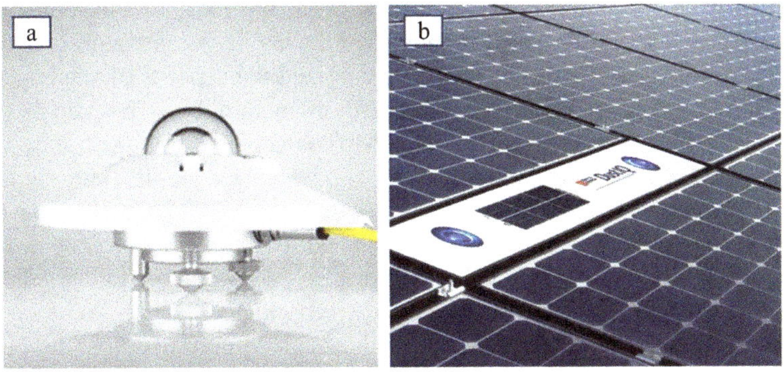

Figure 4.17 Examples of sensors deployed in PV: (a) a pyranometer for measuring horizontal and in-plane global irradiation, and (b) a soil monitoring system © KIPP&ZONEN [59]

As described above, smart monitoring is widely used in both agriculture and PV sectors. Therefore, in the case of agrivoltaics, the hardware for monitoring agriculture and photovoltaics is commercially available. What is interesting for agrivoltaics is to evaluate the extent to which smart monitoring can improve the performance of the system, develop algorithms based on the collected data, support the farmer's decision-making based on the new microclimatic conditions, and combine agronomic and PV monitoring in a single platform.

4.4.2 Advantages and challenges of agrivoltaic monitoring

The use of intelligent monitoring systems can bring many benefits for optimizing the performance of agrivoltaic systems. At the research and development level, intensive monitoring of the agrivoltaic systems around the world has led to a better understanding of the effects of photovoltaics on microclimatic conditions, plant physiology, crop yield and quality, animal welfare, and photovoltaic system performance [60]. Data collected in the field also contributes to the validation and development of various agrivoltaic algorithms that can be used to model crop yield, electricity yield, light availability, water availability, and microclimatic conditions. Monitoring of agrivoltaic systems has shown that many of the microclimatic conditions are significantly affected, introducing new diurnal and spatial distribution patterns [42]. The new conditions add another layer to the already complex equation that determines final crop yield and quality. With the application of Ag-IoT, the farmer can be assisted in quantifying and analyzing the new growing conditions created within the agrivoltaic system. For example, through the use of weather stations, the impact of agrivoltaics on the microclimate can be tracked in real-time, while with plant life monitoring, the response of plants to new conditions can be observed. With this data, the farmer can make agronomic decisions that improve crop growth and quality. In addition, it has been observed that reduced evapotranspiration under agrivoltaics leads to reduced irrigation demand [60], therefore, a smart irrigation monitoring platform can make the most of this potential. In addition, Ag-IoT applications for monitoring animals in systems with grazing can support real-time analysis of animal well-being and the benefits of shading, while geolocation sensors can indicate where animals are within the agrivoltaic park [61]. Overall, the applications and benefits of Ag-IoT in conventional agriculture can be transferred and scaled up to agrivoltaic farming, where decision-making for a more complex farming environment can be supported. A very interesting case where both agronomic and PV monitoring are combined for optimal agrivoltaic operation is given in dynamic agrivoltaic systems. Normally, a PV tracker follows the position of the sun with the goal of maximum electricity production. When used in agrivoltaic systems, custom tracking algorithms can be used to adjust the position of the tracker to achieve specific microclimatic conditions for the plants. This solution is offered at the commercial level by Sun'Agri [62] and is based on internally developed algorithms.

At the research level, the first attempt to deviate the tracker from the position of maximum solar flux was performed by Valle *et al.* [63]. In their study, the trackers were shifted from the electrically suboptimal position during the morning and afternoon hours as part of an agrivoltaic pilot project in southern France. In addition, tracking models that aimed to find an optimal tracker position based on light availability were proposed by Riaz *et al.* [64] for interspace agrivoltaics and by Bruno [65]. Overall, custom tracking systems have great potential because of the flexibility they offer. However, for the optimal tracking algorithms to be applied, advanced crop and PV yield algorithms must be successfully coupled with real-time weather, crop, and PV monitoring data, and state-of-the-art weather forecasts. Overall, the high-tech capabilities offered by Ag-IoT and PV monitoring systems can contribute to a better understanding of the interaction between plants and PV modules within the agrivoltaic system. In addition, robust and validated modeling tools can be developed through data collection. The algorithms can help the system designer, PV operator, and farmer optimize the design and operation of the agrivoltaic system. In addition, Ag-IoT can assist the farmer in adapting and optimizing farming practices under the newly created microclimatic conditions. By providing freely accessible data, the overall learning curve of agrivoltaics can be steeper. In addition, the importance of monitoring is highlighted in the agrivoltaic quality standards published in Germany [24] and Italy [66], which show that monitoring also plays a key role at the policy level.

At the technical level, standardization of the required equipment, communication protocols, and measurement parameters must be developed within agrivoltaics. This standardization will facilitate the comparison of different monitoring results and can help stakeholders in system selection. At the same time, standardization can help policy makers to make clearer how monitoring is envisaged within the quality standards. In addition, the free availability of the collected data is key to accelerating agrivoltaic technology. In addition, it is important to develop a unified monitoring platform for both agronomic and PV parameters. Another challenge is the complexity of crop yield algorithms, which is key to leveraging monitoring data. On an economic level, one challenge is the potentially high initial cost of an advanced Ag-IoT by the farmer.

4.4.3 Research monitoring systems

Plant operation can be tracked, possible malfunctions and failures can be indicated, and this information can be communicated to the plant manufacturer and raw material supplier. This is made possible by using automated systems and sensors based on digital technology. Cho *et al.* [67] evaluated the impact of agrivoltaic crop management using information and communication technologies (ICT). The subject of the study was grapes, and six parts of cropland were used. Solar panels were installed on three test plots, not the other three. A photovoltaic module measuring 1300 × 520 mm was installed on an area with 30% shading. In addition, sensors for cultivation settings and wireless communication devices were used to collect information about cultivation, and habitats and to study electricity generation. Results showed that regular modules generated 25.2 MWh over five months, bifacial modules generated 21.6 MWh, and transparent modules generated 25.7 MWh.

A vertical agrivoltaic hydroponics was studied by Karim *et al.* [68] to maximize land use in urban environments, as well as PV-powered LED grow light irradiation, which could maximize energy production and reduce operating costs. The steps of socialization, installation, collaboration, training, monitoring, and evaluation were used to implement the activity. The results of this activity included: (i) 1.35 kWh of agrivoltaic technology, (ii) 476 planting sites in an area of 8 m^2, (iii) the use of circulation tanks as fish breeding tanks in an area of 6 m^2 with a capacity of 60 fish, and (iv) a 24-hour LED grow light irradiation system. They claimed that this nonprofit initiative can increase agricultural production, provide value-added activities, and reduce agricultural operating costs, which will increase the long-term economic value of the organization. Mengi *et al.* [69] developed a method for agrivoltaic system design based on digital replication and a genomic optimization framework. In this approach, light rays in a procedurally generated agrivoltaic system were simulated at an hourly time step for a given crop, site, and growing season to model light absorption by the photovoltaic modules and underlying plants. The hourly radiation values were then summed to daily radiation values and fed into a crop model to simulate the performance of an agrivoltaic system and a reference crop at a daily time step. The results of the photovoltaic and plant performance metrics for a given design were then used in a genomic optimization algorithm to perform multi-objective optimization across different designs to find an optimal plant-driven solution for a given plant, season, and location. A numerical example was demonstrated using a SunnySD tomato crop in Davis, California. The result indicated a 28.9% optimization of combined crop and energy production using a genomic optimization scheme over 50 generations (Figure 4.18).

A project by Waghmare *et al.* [70] was conducted to intelligently monitor and control greenhouses in India. A powerful combination of agrivoltaics and image processing was used to develop a sustainable and effective method for managing greenhouses. The project was powered by a Raspberry Pi. Plant health monitoring and early disease detection were enabled by real-time data collection from sensors and image processing methods using the PiCam. To improve solar energy production, the system also included solar tracking technology. Based on sensor data,

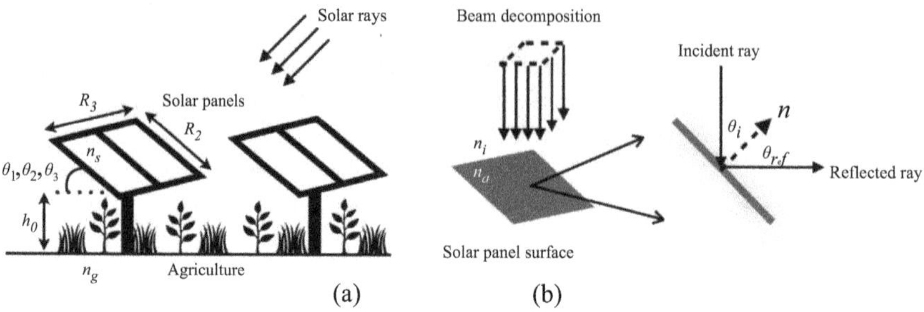

Figure 4.18 (a) Modeled APV system and (b) beam decomposition for the raytracing model [69]

actuators such as fans, heaters, LEDs, and a drip irrigation pump were controlled to ensure perfect environmental control for optimal crop development. Growers could easily access and manage greenhouse conditions from anywhere thanks to remote monitoring and control features provided by a user-friendly Blynk interface. The researchers claimed that this all-encompassing method of greenhouse cultivation promotes environmentally friendly practices, uses less energy, and produces more crop yields.

In Sweden, an IoT sensor station was developed to monitor albedo, temperature, and humidity in an agrivoltaic system. Researching and testing the electrical components to be used in the monitoring station and writing the code that the microcontroller uses to connect to the various sensors were part of the task. The prototype station was then tested three times in Kärrbo Prästgrd, and the information collected was compared with the sensors already installed there. The performance of the IoT stations was evaluated after each test run to identify potential improvements that could be made before further testing. Due to the inherent limitations of the sensitivity of the electrical components used to measure solar irradiance, the final design of the monitoring station showed great accuracy of albedo data throughout the daylight hours with some variation in the early morning and late afternoon [71]. The development of a novel testing method known as the parametric open-source cold-frame agrivoltaic system (POSCAS) was documented in an article by Pearce [72]. POSCAS is a cold frame horticultural system modified for use in cold locations because it functions like a tiny greenhouse for growing food. POSCAS was developed to test partially transparent PV modules intended for agricultural use. It has been claimed that POSCAS can serve as both a conventional cold frame and, when automated, as a full-scale greenhouse. The controls in this design could be powered by the integrated PV module on the roof, or it could be connected to a micro-inverter to generate electricity. The POSCAS could be used in an experimental facility to evaluate food and energy production. Any type of PV module, some of which are transparent, can be easily mounted on it. By adjusting the density of silicon-based cells or the thickness of a thin-film PV material, as well as various types of optical amplification, anti-reflective coatings, and materials to shift the sunlight spectrum in the back sheet, a variety of POSCAS systems can be used to test the impact of agro-photovoltaics. To find the best PV designs for a particular crop, all agrivoltaic variables can be changed.

4.4.4 Commercial monitoring equipment

The Netsens WiSense 2.0 control unit manages the valves of the irrigation system by radio or cable, detecting the flow and pressure of the different sectors, regulating the operation of the pumps and automatically alerting the user in the event of an anomaly (Figure 4.19(a)). In addition, it can automatically schedule irrigation sessions and, thanks to the connection with Netsens IoT technology, read data from all ground sensors (weather, climate and soil), helping to reduce water consumption. Key operations can be controlled remotely through the website and LiveData app. These include activating and controlling irrigation on a schedule (daily or weekly), setting thresholds (soil moisture, precipitation, temperature, leaf wetness, etc.) to dynamically control successive sessions, and variable irrigation rates based on water demand. In addition, you can set automatic alarms for anomalies or missing programming [73]. A web-based user interface called iSolarCloud has been developed by Sungrow Power Supply Co., Ltd (Figure 4.19(b)). It is also used for real-time energy data monitoring in the Tatura Smart Farm experiment for agriculture. Three smart inverters with a total power of 60 kW and two batteries with a storage capacity of 20 kWh are installed in the irrigation pump shed next to the agrivoltaics experiment. The on-site inverters, equipped with Wi-Fi, control energy production, storage, consumption, and grid feed-in, and send data to the iSolarCloud. You can access the data via a smartphone app. The iSolarCloud provides highly reliable data management and is fast. It has proven very useful for tracking and downloading data needed to understand the potential of different solar panel configurations to offset the energy costs of the running irrigation pump and helps identify system faults if problems occur [74].

In agrivoltaic sites with photovoltaic systems, farmers operate heavy equipment and irrigation systems near PV products, making fencing impractical and posing a risk of electrocution. In some installations, animals graze on the PV

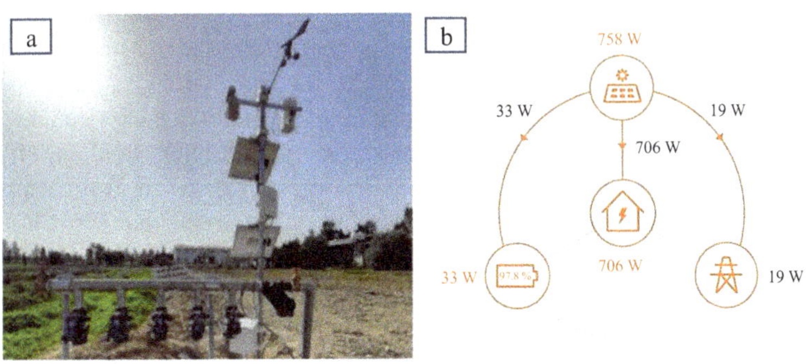

Figure 4.19 (a) WiSens as an automated irrigation and water saving [73], and (b) iSolarCloud operational app [75]

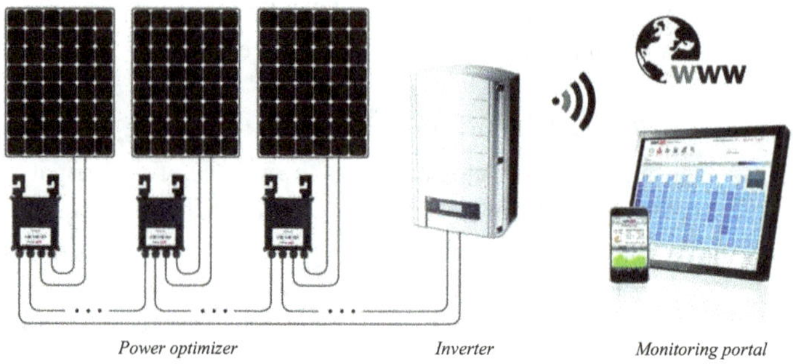

Figure 4.20 SafeDC^{TM}; Shutdown, arc detection, and termination at the module-level [76]

panels, which could damage the cables. This is important because damaged cables increase the risk of fire. With Agri-PV, it is even more important to ensure safety because it is a young business. Equipment is more susceptible to damage at an Agri-PV site, which puts it at increased safety risk. With two built-in features, SafeDCTM and arc fault detection and interruption, SolarEdge provides greater safety and reduces the risk of electrocution and fire (Figure 4.20). As PV systems age and the plugs and wires fail, when animals chew on the wires, or when they are broken or improperly connected, arcing can occur. An electrical arc can occur when wires or connectors are broken. The heat generated by arcing can cause flames. Arcing can also energize the equipment, causing the mounting system to become charged and potentially electrocuting people who touch it. SolarEdge inverters are built to automatically go into safety mode when AC is shut down, lowering the voltage from DC to a safe level. When AC is turned off, our built-in SafeDCTM feature ensures that the output voltage of each module is lowered to a touch-safe level of 1 V.

4.5 Replacement of plastic-based crop protection nets with agrivoltaics

An interesting way to improve is to replace traditional agricultural nets with PV panels in agrivoltaics systems. Traditionally, agricultural plastic nets have been used to protect plants from excessive sunlight, reduce the risk of sunburn, and optimize plant growth. However, integrating PV panels in place of nets is leading to an exciting shift toward more energy production and more efficient land use. By harnessing solar energy through PV panels while shading crops, this innovative approach provides a dual benefit and opens new opportunities for sustainable agriculture and renewable energy production. In this section, the concept of replacing conventional plastic-based crop protection systems with PV panels in agrivoltaics systems will be discussed and the potential benefits and challenges of this transformation will be explored. This section will provide insights into optimizing agrivoltaics designs and pave the way for more efficient and environmentally friendly systems that bridge the gap between agriculture and clean energy production.

4.5.1 Plastic nets in agriculture

By easily regulating the microclimate under the cover, agricultural nets not only help increase productivity but also improve the quality of produce. Specifically, nets are used to improve the microclimate for crops and increase the productivity, quality, and uniformity of plants and fruits by creating a mild greenhouse and windbreak effect. In addition, nets can be used to postpone the normal production period, increasing the market value of produce and spreading the availability of fresh, high-quality produce more evenly over a longer period [77]. The most common uses of plastic nets in agriculture include protection from hail, wind, snow, or heavy rain in fruit growing and ornamental crops, shade nets for greenhouses, and nets that moderately alter the microenvironment around a crop. In addition, nets are often used during harvest and post-harvest operations and for protection against virus-carrying insects and birds [78]. Currently, there are several agricultural nets on the market, and they are well-advertised. These nets are typically utilized in applications that fall under a few distinct categories. The earliest uses of nets in protected cultivation were for ornamentals, such as cut flowers, and for fruit farming, such as grapes, peaches, apricots, apples, and cherries, to shield plants and fruits from hail and wind. These nets are often made of transparent threads, have a white appearance, and have an average net mesh size of roughly 1.5–4 mm [77]. Different structural characteristics, such as the type of material, the type, and dimensions of threads, texture, mesh size, porosity/solidity, and weight, as well as radiometric and physical characteristics, such as air permeability and tensile stress, strength, elongation at break, and durability, are used to describe different types of nets. The most widely used raw material for agricultural nets is high-density polyethylene [79].

4.5.2 Agricultural applications of nets

4.5.2.1 Windbreak nets

Windbreak netting is used to: reduce wind load on agricultural structures; prevent mechanical damage to plants (e.g., breaking of branches and flowers); prevent biological effects on plants (e.g., high evapotranspiration and difficulty with pollination); increase the quality of produce by protecting it from dust, salt, and sand; and minimize heat loss from animals through ventilation in open livestock [79]. Generally, windbreak nets are attached to a support structure of steel, concrete, or wood columns or trusses mounted on a foundation (Figure 4.21). Occasionally, farmers also attach them directly to trees located along the edges of fields. The height of the structure, the permeability of the mesh, and the distance of the fence from the crop affect how much wind is reduced.

4.5.2.2 Anti-hail nets

Anti-hail nets are typically used to protect against hail damage to fruit trees, including grapes, peaches, apricots, and cherries (Figure 4.22). They are either attached directly to the plants or installed with a special support frame. Sometimes these nets are also used as vital protection for greenhouses with glass panes to

Figure 4.21 Different types from Duranet windbreak nets [80]

Figure 4.22 (a) Orchard anti-hail netting on concrete structures and (b) vineyard anti-hail netting on wooden supports [79]

prevent costly economic impact on the materials [78]. In addition, these nets can reduce the amount of sunlight entering the greenhouse in the summer, preventing temperatures from rising. Anti-hail nets on greenhouses require a special steel support system attached to the greenhouse frame [79].

4.5.2.3 Nets for reducing solar radiation

Shading nets are designed to block solar radiation to lower the air temperature in the greenhouse or to reduce the light intensity for some plants that prefer shade, such as some ornamental plants (Figure 4.23). The shade factor of the net affects how effective shading systems are. In sunny locations, they are often used to extend or delay the harvest season. For virus-free production, screen printing companies use shading nets [79]. However, their use is limited by the increase in humidity and the decrease in airflow. During cold winter nights, heat screens are used in greenhouses to reduce convection and radiation heat losses. To improve the reflection of heat radiation from inside the greenhouse, aluminum-colored nets are usually used in this type of application. Steel structures specially designed for greenhouses are usually used to support the shading nets [78].

Figure 4.23 (a) Thermal and shading screens inside a greenhouse, and (b) steel made flat roof shading structure [79]

Figure 4.24 Pollination net system in Wageningen (Nl) [79]

4.5.2.4 Protection against insects

Pesticide alternatives such as insect-proof nets are often used in organic farming because they are considered safer for the environment and human health. For virus-free products, insect-proof nets with one or two layers are used as covers for shelters (Figure 4.24). Insect screening can improve integrated control measures and reduce reliance on chemical pesticides. In addition, insect-proof nets are used to prevent pollinating insects, such as bumblebees, from escaping from the greenhouse. The most important elements for this type of netting are mesh size and color, which can either attract or repel insects. Consequently, very low porosity and permeability are characteristics of insect-proof nets [79].

4.5.2.5 The effect of nets on different factors

The use of nets to protect against the threats posed by abiotic factors such as heat, light, and wind will be critical in the face of climate change. Nets can increase water use efficiency while protecting from the intense light and high temperatures that harm fruit in warm climates. In cold climates, nets protect fruit from hail, wind, and heavy rains, which can result in significant yield losses. Crop response can depend on factors such as type of netting, timing of application, netting material, and establishment system. To improve crop response and ensure quality fruit yield, it is important to better understand the effects of such elements [81]. Most research addressed the timing of netting, with nets placed after full bloom until harvest to allow for better insect pollination. In situations where nets were applied before pollination, colonies were sometimes housed in enclosures during flowering. Both the reproductive and vegetative activities of plants can be affected by nets in different ways. More research is needed to understand the mechanisms underlying the increase in productivity [82]. This knowledge would help scientists and farmers select the appropriate type of netting and shading required for the desired plant response. Given the challenges of climate change and invasive exotic pests, protected cultivation of tree fruit crops is a creative alternative strategy that needs to be further explored and utilized to ensure orchard sustainability. Network systems will also promote the growth of sustainable production techniques, such as organic and low-input methods for growing fruit trees.

4.5.3 Agrivoltaics as a crop protection system

Agrivoltaic systems can be used on farms with open fields as well as in combination with protected growing areas in greenhouses. While overhead PV is best suited for horticulture, interspace PV is used on pasture and cropland farms. Crops grown in the shade of PV modules benefit from improved water/rain distribution, wind mitigation, protection from temperature fluctuations, reduced evapotranspiration, optimal soil moisture, security from climatic uncertainties, and protection from damaging events such as hail [15]. At Germany's largest agrivoltaic research center, located near Lake Constance, Trommsdorff *et al.* [41] studied the electrical yield and the productivity and behavior of four crops. While the dry and hot summer of 2018 showed that the agrivoltaic system can increase land production by almost 90%, the land use efficiency evaluated by the LER showed an increase between 56% and 70% in 2017. Gadhiya and Patel [83] conducted experiments on the performance evolution of agrivoltaic net houses in the experimental field of the department REE at the College of Agriculture Engineering and Technology, JAU, Junagadh (21.5 N, 70.1 E) (Figure 4.25(a)). The experimental agrivoltaic structure was designed and constructed with 24 150-W solar modules with a total power of 3.6 kW and an area of 76.94 m^2. The final yield and the reference yield were highest in February and March, respectively, while they were lowest in December. During the trial period, a power ratio of 0.78 was observed. Prakash *et al.* [84] studied the distribution of radiation and shade on the plant surface for three densities of PV panels: (i) partial density (PD), (ii) half density (HD), and (iii) full density (FD) (Figure 4.25(b)). Wheat cultivar GW 496 was selected for cultivation using an agrivoltaic system with drip irrigation and row seeding. It was found that the percentage of shaded areas in the cultivated area was highest in the full-density

Figure 4.25 *(a) Photo of an agrivoltaic net house [83], (b) shade zones in agrivoltaic array system [84]*

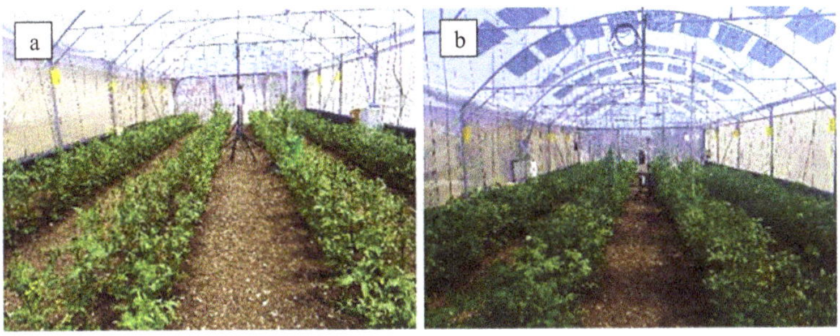

Figure 4.26 *Inside view of greenhouses [85]: (a) the control tunnel and (b) the OPV tunnel*

plots and lowest in the partial-density plots. The percentage of the total area of agrivoltaic systems ranged from 24.1% to 75.4%. The amount of photosynthetically active radiation (PAR) available over the surface of the plants was negatively affected by the shaded areas under the plots with different densities. Directly under the panel, the lowest value of PAR (56 mol m^{-2} s^{-1}) was found.

Friman-Peretz *et al.* [85] studied the microclimate in two greenhouse tunnels in which tomato plants were growing. One served as a control (without shade screens and with 25% shade screens), and the other was covered with flexible semi-transparent organic photovoltaic (STOPV) modules (with 37% roof cover and 23% shade modules) (see Figure 4.26). Less solar radiation entered the OPV tunnel without shading screens, but no discernible differences in microclimate were observed between the OPV tunnel with 25% shading screens and the control tunnel with shading screens. Conclusion: OPV modules can be used in place of shade screens to protect plants and provide clean electricity.

A single-span gable roof greenhouse with STOPV modules installed was the subject of a study by Ravishankar *et al.* [86] using an energy balance model. The greenhouse had typical dimensions (29.4 m long, 7.3 m wide, and 3 m eave height). According to the simulations, organic photovoltaic (OPV) could meet the annual energy demand of the greenhouse in hot and mixed-humid climates with an annual

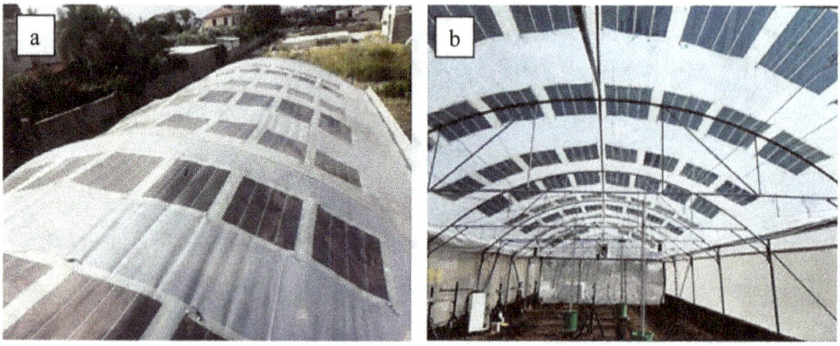

Figure 4.27 Installation of OPV modules on the roof of a tunnel greenhouse [87]: (a) outside view and (b) inside view

PAR reduction of only 10% and 25%, respectively. It was also found that the integration of OPVs reduces the energy demand load. Semi-transparent OPV modules were found to offer tremendous potential for creating zero-energy consumption greenhouses. Magadley *et al.* [87] investigated the feasibility of installing OPV modules on a greenhouse roof in the Mediterranean region (Figure 4.27). To this end, they thoroughly investigated how OPV modules behave outdoors, specifically how environmental factors and panel orientation affect how quickly they degrade and how much electricity they produce. According to the data, the OPVs installed on the roof produce the most yields, efficiencies, and fill factors (FFs), while the OPVs installed on the east and west walls reduce midday peaks, resulting in a more consistent energy supply throughout the day.

An evaluation of the performance of large-area, roll-to-roll printed OPV arrays (3.4 m^2 active area) installed on a Gothic-arch greenhouse roof in Tucson, Arizona, was performed by Waller *et al.* [88] to estimate incident solar radiation on curved surfaces. Curved surface modeling techniques were used to evaluate the performance of six outdoor OPV installations, focusing on the effect of irradiance on electrical performance. The overall power conversion efficiency (PCE) of the OPV arrays was 1.82%, with lower PCE in the afternoon than in the morning and midday. Over the test period, the normalized PCE of the OPV arrays decreased by 32.6% on average. Based on these results, we conclude that to support the use of agrivoltaic OPV greenhouses, more powerful OPV devices that are more durable in outdoor environments are needed, along with accurate performance monitoring systems. Using glass-glass c-Si STPV modules with variable transparency and cell arrangement, the German company BayWa r.e. has installed four experimental agrivoltaic systems on blueberry, blackberry, redberry, and strawberry fields in the Netherlands [89]. A 2.67 MWp agrivoltaic system was established in Babberich, the Netherlands, using STPV modules with a rated power of 258 Wp on a 3.3-ha raspberry field. This was done after the establishment of pilot projects. In Fig. there are pictures of these projects (see Figure 4.28).

Figure 4.28 Orchard agrivoltaics installed in the Netherlands [89]: (a) 60 kWp pilot project over strawberries, (b) 95 kWp pilot project over blueberries, (c) 130 kWp pilot project over blackberries, and (d) 2.67 MWp project over raspberries

Figure 4.29 The ASCA® OSTPV modules installed on (a) vertical greenhouse façade, (b) indoor horizontal shade, and (c) indoor horizontal shade sails [90]

The company "ARMOR" produced 43 m2 of ASCA® OPV films (consisting of 79 modules) that were used in greenhouses in France to test different installation strategies [90]. In this case, as shown in Figure 4.29, the modules were attached to the outer vertical greenhouse facade, the inner horizontal shade sails, and the inner vertical shade curtain. The manufacturers claim that these modules are lightweight, flexible, and up to 30% transparent, making them easily removable. In a study by Willockx *et al.* [91], a novel 3D agrivoltaic tool developed by KU Leuven that can simulate light distribution, crop yields, and electrical efficiency was used for a techno-economic analysis of an agrivoltaic system for orchards. Four different arrangements—opaque PV modules in landscape formation, checkerboard arrangement, PV modules with 40% transparency, and a reference without PV modules—were studied in this context. The results showed that the STPV module arrangement resulted in a much more uniform distribution of light. Therefore, a 13.3 kWp agricultural photovoltaic system was installed on a tiny part (120 m^2) of the pear orchard. In this case, as shown in Figure 4.30, the hail nets were removed and PV modules in the form of a dome structure were installed in their place. 36 PV cells with an efficiency of 21% were used in each of the c-Si STPV modules, which have a nominal power of 185 Wp to achieve 40% transparency. The height of the PV system was set at 4.6 m and a tilt angle of 12.5° was selected to reduce self-shading and reduce wind loads based on the hail net and spray system already in use. According to the results, PV modules can reduce the occurrence of radiative frost because they shield the ambient air from long-wave radiation losses.

Graham *et al.* [92] discovered that the number of flowers increased, and the timing of flowering was delayed in semi-shaded plots, which could be beneficial to late hunters in water-scarce habitats. Both the fully shaded and partially shaded plots had higher abundance, diversity, and greater richness of pollinators than the fully shaded plots, which were comparable. At this level, treatment differences did not affect the frequency of pollinator flower visits. This indicates that pollinators choose habitats under solar arrays despite differences in community structure along shade gradients. When seeking land for pollinator habitat restoration in targeted areas, local farmers, solar developers managing solar shelters, and agricultural and pollinator health advocates should be armed with the findings from these studies. According to Ramos-Fuentes *et al.* [93], the description of plant growth dynamics, distinguishing between fixed and dynamic panels, dealt with the interaction between radiative

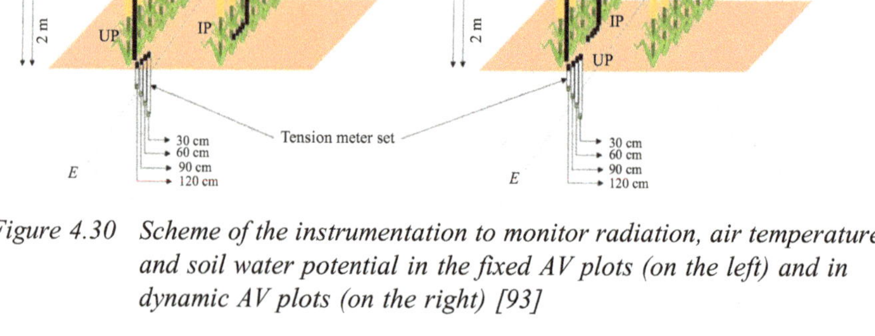

Figure 4.30 Scheme of the instrumentation to monitor radiation, air temperature, and soil water potential in the fixed AV plots (on the left) and in dynamic AV plots (on the right) [93]

transfer, plant development, and irrigation requirements in maize cultivation under field conditions (see Figure 4.30). They discovered that corn leaves respond to shade by reducing stomatal conductance, net CO^2 uptake, and leaf temperature in ways that are related to radiation on a finer temporal scale. This finding opened the possibility of using this behavior to optimize water use and shading techniques. In contrast to the effects of radiation reduction, spatial heterogeneity of radiation in fixed AV systems at the plot level was found to have a second-order effect on leaf area index and phyllochron. In addition, dynamic AV showed its ability to minimize spatial heterogeneities in soil water extraction, highlighting the importance of controlled shading methods in AV systems regarding water use.

Jung and Salmon [94] analyzed the cost of a photovoltaic system on a blueberry orchard and compared it to the cost of plastic covers, introducing a metric for calculating the price of covering cropland with a photovoltaic system. Using a selected case, they showed that the price of covering cropland with an agrivoltaic system is still higher than the price of covering cropland with shade nets, and therefore additional costs or design changes for photovoltaic systems cannot be justified by the economic benefits of shading alone. They asserted that depending on the local cost structure for crop protection with plastic covers, agrivoltaic can achieve significant synergistic benefits. They also concluded that agrivoltaics can further reduce the cost of crop cover if the capacity density of the system is reduced. They asserted that further research is needed to investigate the physical aspects of replacing plastic covers with photovoltaic panels. Apple orchards could be covered with movable solar panels to mitigate climate change and help transition to a clean energy source. Li *et al.* [95] investigated the effectiveness of a blind shade regulator that can automatically rotate semi-transparent PV louvers mounted on the greenhouse roof in response to changes in sunlight. Under cloudy conditions, the PV louvers were rotated so that they were perpendicular to the roof, allowing sunlight to pass through but still partially blocking intense sunlight from entering the greenhouse (Figure 4.31). Annual operation proved that the blinds' electrical

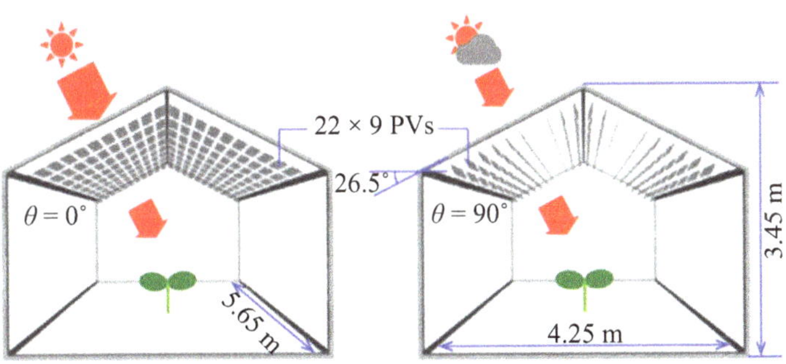

Figure 4.31 Model greenhouse equipped with PV blinds at closed (θ = 0°) and open (θ = 90°) orientations according to the degree of irradiance. Underneath the greenhouse glass roof, 198 PV modules can be installed neatly [95].

Figure 4.32 Photos of (a) control and (b) photovoltaic greenhouses [97]

production could support their function and generate additional electrical energy. The results showed that an overcast sky reduced electrical energy production by 50%. Nevertheless, the PV blinds can meet some or all the greenhouse's electrical energy needs, depending on how electrified the greenhouse is. Under current sky conditions, 8–10 MJ m^2 of solar radiation is expected to irradiate the plants under the PV blinds on day 1. With this solar radiation, important horticultural crops can be cultivated. To enable sustainable energy-food dual production in a greenhouse, the irradiance threshold for turning the PV blinds can be adjusted. In this way, the ratio of solar irradiance for agriculture and power generation is controlled.

A dynamic agrivoltaic system was established over a 10-year-old "Golden Delicious" apple orchard in southern France and explored by Juillion [96] throughout three experimental seasons (2019–2021). According to the results, air temperature decreased by an average of 3.8 °C under shade conditions, while relative humidity increased by 14%. Reduced radiation and harsh environments decreased irrigation by 6–31%, depending on the season. They concluded that agrivoltaic systems could reduce alternate bearing behavior in apple trees, underscoring the value of a protracted study. Ezzaeri *et al.* [97] studied microclimate and tomato yields during the summer and winter seasons as a result of the shading effect of flexible PV panels arranged in a checkerboard pattern on the greenhouse roof surface. The two Canary Island tomato greenhouses used for this study are typical of the southern Mediterranean region (Figure 4.32). The results of our study showed that the solar panels covering 40% of the roof area of the Canarian greenhouse had no discernible effect on the climatic variables. During the hot season, the solar panels lowered the temperature in the greenhouse, occasionally to the range most suitable for the growth of the tomato. In addition, the overall yield of tomatoes is not significantly affected by the utilization of solar panels.

4.6 Conclusions and prospects

The adoption of agricultural processes enhanced by agrivoltaic applications represents a promising way to address the complex challenges of our time, including food and energy security in the face of population growth and climate change. The synergy between solar energy generation and traditional agricultural practices not only optimizes land use but also helps reduce greenhouse gas emissions, providing a sustainable solution for the future. However, this use also presents some technical, operational, and economic challenges that require careful consideration and innovation. As we have seen in this chapter:

- Conventional farm machinery plays a central role in the design and operation of agrivoltaic systems, influencing the final dimensions of the system and bringing additional operational challenges to the agronomic and PV sides. However, if these elements are carefully considered in the early stages of project development in collaboration with the farmer, many solutions can be found for an agrivoltaic system where the agronomic and PV operations are optimally coupled.
- Although the adaptation of agricultural machinery and robots to agro-industrial systems is very promising, several challenges must be overcome. These include high initial investment costs, technological compatibility, and the need for adequate training and support for farmers to effectively use advanced technologies. Overcoming these challenges will require collaboration among researchers, manufacturers, policymakers, and agricultural stakeholders.
- The integration of agricultural machinery and robotics with agrivoltaic systems presents an opportunity to revolutionize modern agriculture. As these technologies evolve, the combination of agrivoltaics and agricultural automation holds the potential to create a sustainable, technologically advanced agriculture capable of meeting the world's growing demand for food and energy. Collaboration between the agriculture and renewable energy sectors will be critical to drive innovation, spread these integrated systems, and ensure a more sustainable and prosperous future.
- Replacing conventional plastic-based crop protection nets with Agrivoltaics represents a breakthrough change in sustainable agriculture and clean energy production. This innovative approach offers a dual benefit: It shades and protects crops while harnessing solar energy, optimizing land use and resource consumption. However, this transformation is not without challenges, including technical considerations and potential adjustments to farming practices.

In conclusion, collaboration between sectors, stakeholders, and innovators will be essential in propelling these integrated systems forward and realizing their full potential on the path toward a brighter tomorrow.

References

[1] Zhang X, Yao G, Vishwakarma S, *et al.* Quantitative assessment of agricultural sustainability reveals divergent priorities among nations. *One Earth* 2021;4:1262–77. https://doi.org/10.1016/j.oneear.2021.08.015.

[2] Gorjian S, Singh R, Shukla A and Mazhar AR. On-farm applications of solar PV systems. In: Gorjian S and Shukla A (eds.) , eds. *Photovoltaic Solar*

Energy Conversion. Elsevier; 2020, p. 147–90. https://doi.org/10.1016/B978-0-12-819610-6.00006-5.

[3] The Sustainable Development Goals Report 2023. The Sustainable Development Goals Report. 2023.

[4] Mulligan M, Neave M and Mulligan M. *In Brief to The State of Food and Agriculture. The State of Food and Agriculture 2022*. FAO; 2022. https://doi.org/10.4060/cc2459en.

[5] Gorjian S, Calise F, Kant K, *et al.* A review on opportunities for implementation of solar energy technologies in agricultural greenhouses. *J Clean Prod* 2021;285:124807. https://doi.org/10.1016/j.jclepro.2020.124807.

[6] Gorjian S, Ebadi H, Jathar LD and Savoldi L. Solar energy for sustainable food and agriculture: Developments, barriers, and policies. *Solar Energy Advancements in Agriculture and Food Production Systems*. Amsterdam, The Netherlands: Elsevier; 2022, p. 1–28. https://doi.org/10.1016/B978-0-323-89866-9.00004-3.

[7] Jaiswal B and Agrawal M. Carbon footprints of agriculture sector. *Carbon Footprints*. 2020, p. 81–99. https://doi.org/10.1007/978-981-13-7916-1_4.

[8] Yousefi M, Khoramivafa M and Damghani AM. Water footprint and carbon footprint of the energy consumption in sunflower agroecosystems. *Environ Sci Pollut Res* 2017;24:19827–34. https://doi.org/10.1007/s11356-017-9582-4.

[9] Zhang D, Shen J, Zhang F, Li Y and Zhang W. Carbon footprint of grain production in China. *Sci Rep* 2017;7:4126. https://doi.org/10.1038/s41598-017-04182-x.

[10] Gorjian S, Ebadi H, Trommsdorff M, Sharon H, Demant M and Schindele S. The advent of modern solar-powered electric agricultural machinery: A solution for sustainable farm operations. *J Clean Prod* 2021;292:126030. https://doi.org/10.1016/j.jclepro.2021.126030.

[11] Bolyssov T, Yessengeldin B, Akybayeva G, Sultanova Z and Zhanseitov A. Features of the use of renewable energy sources in agriculture. *Int J Energy Econ Policy* 2019;9:363–8. https://doi.org/10.32479/ijeep.7443.

[12] European Agricultural Machinery Industry Association (CEMA). The role of agricultural machinery in decarbonising agriculture. 2022.

[13] FAO. The future of food and agriculture – Alternative pathways to 2050. 2018.

[14] CAP Strategic Plans 2023. https://agriculture.ec.europa.eu/cap-my-country/cap-strategic-plans_en (accessed September 9, 2023).

[15] Gorjian S, Bousi E, Özdemir ÖE, *et al.* Progress and challenges of crop production and electricity generation in agrivoltaic systems using semi-transparent photovoltaic technology. *Renew Sustain Energy Rev* 2022; 158:112126. https://doi.org/10.1016/j.rser.2022.112126.

[16] Owley J and Wilson Morris A. The new agriculture : From food farms to solar farms. *Columbia J Environ Law* 2019;44:2:409–77.

[17] Macknick J, Hartmann H, Barron-gafford G, *et al.* The 5 Cs of Agrivoltaic Success Factors in the United States : Lessons from the InSPIRE Research Study. 2022.

[18] Trommsdorff M, Dhal IS, Özdemir ÖE, Ketzer D, Weinberger N and Rösch C. Agrivoltaics: Solar power generation and food production. *Solar Energy Advancements in Agriculture and Food Production Systems.* Amsterdam, The Netherlands: Elsevier; 2022, p. 159–210. https://doi.org/10.1016/B978-0-323-89866-9.00012-2.

[19] Heins BJ, Sharpe KT, Buchanan ES and Reese MH. Agrivoltaics to shade cows in a pasture-based dairy system. AIP conference proceedings, 2635, 2022, p. 060001. https://doi.org/10.1063/5.0103148.

[20] Hermann C, Dahlke F, Focken U and Trommsdorff M. Aquavoltaics: Dual use of natural and artificial water bodies for aquaculture and solar power generation. *Solar Energy Advancements in Agriculture and Food Production Systems.* Amsterdam, The Netherlands: Elsevier; 2022, p. 211–36. https://doi.org/10.1016/B978-0-323-89866-9.00009-2.

[21] Artigianfer | Photovoltaic greenhouses: The advantages. 2023. https://artigianfer.com/en/news/photovoltaic-greenhouses-advantages/ (accessed September 9, 2023).

[22] Schallenberg-Rodriguez J, Rodrigo-Bello J-J and Río-Gamero B D. Agrivoltaic: How much electricity could photovoltaic greenhouses supply? *Energy Reports* 2023;9:5420–31. https://doi.org/10.1016/j.egyr.2023.04.374.

[23] Lewandowski I, Lippe M, Montoya JC, *et al.* Primary production. *Bioeconomy.* Cham: Springer International Publishing; 2018, p. 97–178. https://doi.org/10.1007/978-3-319-68152-8_6.

[24] DIN. *Agri-photovoltaic Systems - Requirements to Primarily Agricultural Use*: English translation of DIN SPEC, 2021, 91434:2021–05.

[25] Gorjian S, Minaei S, MalehMirchegini L, Trommsdorff M and Shamshiri RR. Applications of solar PV systems in agricultural automation and robotics. In: Gorjian S and Shukla A, eds. *Photovoltaic Solar Energy Conversion.* 1st edn. London: Elsevier; 2020, p. 191–235. https://doi.org/10.1016/B978-0-12-819610-6.00007-7.

[26] Bousi, F, Schindele, S and Tegtmeyer MD, M. Land equivalent ratio as a key assessment indicator for utility scale interspace Agri-PV, *Agrivoltaics Conference 2023*, Daegu, South Korea; 2023.

[27] Campana PE, Stridh B, Amaducci S and Colauzzi M. Optimisation of vertically mounted agrivoltaic systems. *J Clean Prod* 2021;325:129091. https://doi.org/10.1016/j.jclepro.2021.129091.

[28] Katsikogiannis OA, Ziar H and Isabella O. Integration of bifacial photovoltaics in agrivoltaic systems: A synergistic design approach. *Appl Energy* 2022;309:118475. https://doi.org/10.1016/j.apenergy.2021.118475.

[29] Riaz MH, Imran H, Younas R, Alam MA and Butt NZ. Module technology for agrivoltaics: Vertical bifacial versus tilted monofacial farms. *IEEE J Photovoltaics* 2021;11:469–77. https://doi.org/10.1109/JPHOTOV.2020.3048225.

[30] Fendt. Fendt Former. 2023.

[31] Fendt. Crop protection sprayers. 2023.

[32] Wild K. Challenges in the planning, construction and farming practices in agrivoltaic systems with vertically mounted panels, *Agrivoltaics Conference 2023*, Daegu, South Korea; 2023.

[33] Li M, Imou K, Wakabayashi K and Yokoyama S. Review of research on agricultural vehicle autonomous guidance. *Int J Agric Biol Eng* 2009;2:1–16. https://doi.org/10.3965/j.issn.1934-6344.2009.03.001-016.

[34] Perez-Ruiz M, Slaughter DC, Gliever C and Upadhyaya SK. Tractor-based real-time kinematic-global positioning system (RTK-GPS) guidance system for geospatial mapping of row crop transplant. *Biosyst Eng* 2012;111:64–71. https://doi.org/10.1016/j.biosystemseng.2011.10.009.

[35] Fendt. Guidance systems. 2023.

[36] Deere, J. RTK precision Ag technology. 2023.

[37] Ortiz B V., Balkcom KB, Duzy L, van Santen E and Hartzog DL. Evaluation of agronomic and economic benefits of using RTK-GPS-based auto-steer guidance systems for peanut digging operations. *Precis Agric* 2013;14:357–75. https://doi.org/10.1007/s11119-012-9297-y.

[38] Eurostat. *Agri-Environmental Indicator - Tillage Practices - Statistics Explained.* European Union; 2017.

[39] Kaan Kurtural S and Fidelibus MW. Mechanization of pruning, canopy management, and harvest in winegrape vineyards. *Catal Discov Pract* 2021; 5:29–44. https://doi.org/10.5344/catalyst.2021.20011.

[40] Schindele S, Trommsdorff M, Schlaak A, *et al.* Implementation of agrophotovoltaics: Techno-economic analysis of the price-performance ratio and its policy implications. *Appl Energy* 2020;265:114737. https://doi.org/10.1016/j.apenergy.2020.114737.

[41] Trommsdorff M, Kang J, Reise C, *et al.* Combining food and energy production: Design of an agrivoltaic system applied in arable and vegetable farming in Germany. *Renew Sustain Energy Rev* 2021; 140:110694. https://doi.org/10.1016/j.rser.2020.110694.

[42] ISE FI for SES. Agrivoltaics: Opportunities for Agri_Culture and the Energy Transition. 2022.

[43] Agostini A, Colauzzi M and Amaducci S. Innovative agrivoltaic systems to produce sustainable energy: An economic and environmental assessment. *Appl Energy* 2021;281:116102. https://doi.org/10.1016/j.apenergy.2020.116102.

[44] Jung D, Gareis GH, Staiger A and Salmon A. Effects of soiling on agrivoltaic systems: Results of a case study in Chile. AIP conference proceedings 2022; 2635:020001. https://doi.org/10.1063/5.0107943.

[45] Malik A and Kohli S. Electric tractors: Survey of challenges and opportunities in India. *Mater Today Proc* 2020;28:2318–24. https://doi.org/10.1016/j.matpr.2020.04.585.

[46] Sunusi II, Zhou J, Zhen Wang Z, *et al.* Intelligent tractors: Review of online traction control process. *Comput Electron Agric* 2020;170:105176. https://doi.org/10.1016/j.compag.2019.105176.

[47] Khatawkar DS, Shaji James P and Dhalin D. Modern trends in farm machinery-electric drives: A review. *Int J Curr Microbiol Appl Sci* 2019; 8:83–98. https://doi.org/10.20546/ijcmas.2019.801.011.

[48] Ghobadpour A, Boulon L, Mousazadeh H, Malvajerdi AS and Rafiee S. State of the art of autonomous agricultural off-road vehicles driven by renewable energy systems. *Energy Procedia* 2019;162:4–13. https://doi.org/10.1016/j.egypro.2019.04.002.

[49] Buning EA. *Electric Drives in Agricultural Machinery-Approach from the Tractor Side*. 2010.

[50] Moreda GP, Muñoz-García MA and Barreiro P. High voltage electrification of tractor and agricultural machinery - A review. *Energy Convers Manag* 2016;115:117–31. https://doi.org/10.1016/j.enconman.2016.02.018.

[51] Caban J, Vrabel J, Šarkan B, Zarajczyk J and Marczuk A. Analysis of the market of electric tractors in agricultural production. *MATEC Web Conf* 2018;244:03005. https://doi.org/10.1051/matecconf/201824403005.

[52] Relf-Eckstein JE, Ballantyne AT and Phillips PWB. Farming reimagined: A case study of autonomous farm equipment and creating an innovation opportunity space for broadacre smart farming. *NJAS - Wageningen J Life Sci* 2019;90–91:100307. https://doi.org/10.1016/j.njas.2019.100307.

[53] Bawden O, Kulk J, Russell R, *et al.* Robot for weed species plant-specific management. *J F Robot* 2017;34:1179–99. https://doi.org/10.1002/rob.21727.

[54] Khasha Ghaffarzadeh Leader in autonomous mobility|IDTechEx Research Article 2020. https://www.idtechex.com/ja/research-article/agriculture-leader-in-autonomous-mobility/11643 (accessed January 25, 2020).

[55] Klokov A, Loktionov E, Loktionov Y, Panchenko V and Sharaborova E. A mini-review of current activities and future trends in agrivoltaics. *Energies* 2023;16:3009. https://doi.org/10.3390/en16073009.

[56] Xu J, Gu B and Tian G. Review of agricultural IoT technology. *Artif Intell Agric* 2022;6:10–22. https://doi.org/10.1016/j.aiia.2022.01.001.

[57] Agricolture 4.0 - Agricoltura 4.0, Stazioni meteo professionali, Sensori meteo climatici 2023. https://www.netsens.it/en/agricolture-4-0/ (accessed August 25, 2023).

[58] Woyte, A, Richter, M, Moser, D, Reich, N,Green, M, and Mau, S. (2014). Analytical monitoring of grid-connected photovoltaic systems: Good practices for monitoring and performance analysis: IEA PVPS Task 13, Subtask 2: Report IEA PVPS T13-03: 2014. International Energy Agency, Photovoltaic Power Systems Programme.

[59] KIPP & ZONEN. OTT HydroMet. KIPP & ZONEN. https://www.kippzonen.com/Products.

[60] Weselek A, Bauerle A, Hartung J, Zikeli S, Lewandowski I and Högy P. Agrivoltaic system impacts on microclimate and yield of different crops within an organic crop rotation in a temperate climate. *Agron Sustain Dev* 2021;41:1–15. https://doi.org/10.1007/s13593-021-00714-y.

[61] Sharpe KT, Heins BJ, Buchanan ES and Reese MH. Evaluation of solar photovoltaic systems to shade cows in a pasture-based dairy herd. *J Dairy Sci* 2021;104:2794–806. https://doi.org/10.3168/jds.2020-18821.
[62] Homepage - Sun'Agri 2023. https://sunagri.fr/en/ (accessed August 25, 2023).
[63] Valle B, Simonneau T, Sourd F, *et al.* Increasing the total productivity of a land by combining mobile photovoltaic panels and food crops. *Appl Energy* 2017;206:1495–507. https://doi.org/10.1016/j.apenergy.2017.09.113.
[64] Riaz MH, Imran H, Alam H, Alam MA and Butt NZ. Crop-specific optimization of bifacial PV arrays for agrivoltaic food-energy production: The light-productivity-factor approach. *IEEE J Photovoltaics* 2022;12:572–80. https://doi.org/10.1109/JPHOTOV.2021.3136158.
[65] Bruno M. Tracking optimization in agrivoltaic systems. A comparative study for apple orchards. *KTH* 2023.
[66] Srl C. CEI PAS 82-93 Impianti Agrivoltaici 2023.
[67] Cho J, Park SM, Park AR, Lee OC, Nam G and Ra I-H. Application of photovoltaic systems for agriculture: A study on the relationship between power generation and farming for the improvement of photovoltaic applications in agriculture. *Energies* 2020;13:4815. https://doi.org/10.3390/en13184815.
[68] Karim MH, Diantoro M, Nasikhudin N and Lestari SR. Implementation of agricultural technology urban farming agrivoltaic based system to increase productivity and empowerment of farmer women's community. *J Community Serv Empower* 2023;4:184–95.
[69] Mengi E, Samara OA and Zohdi TI. Crop-driven optimization of agrivoltaics using a digital-replica framework. *Smart Agric Technol* 2023;4:100168. https://doi.org/10.1016/j.atech.2022.100168.
[70] Waghmare RB, Chimankare RV, Jha NM, Kar AD, Patil TS and Mahadik KM. An Intelligent greenhouse monitoring and controlling system with agrivoltaics optimization. *Int Res J Mod Eng Technol Sci* 2023;5:2397–408. https://doi.org/10.56726/IRJMETS36121.
[71] Stridsman E. Monitoring of microclimate underneath agrivoltaic systems using IOT sensor station. 2023. diva-portal.org
[72] Pearce JM. Parametric open source cold-frame agrivoltaic systems. *Inventions* 2021;6:71. https://doi.org/10.3390/inventions6040071.
[73] Automated Irrigation - Agricoltura 4.0, Stazioni meteo professionali, Sensori meteo climatici n.d. https://www.netsens.it/en/products/automated-irrigation/ (accessed August 25, 2023).
[74] Solar Power Inverter System & Energy Storage System for the World - Sungrow 2023. https://en.sungrowpower.com/ (accessed August 25, 2023).
[75] Home Assistant Community Add-on: iSolarCloud / Sungrow - Share your Projects! - Home Assistant Community n.d. https://community.home-assistant.io/t/home-assistant-community-add-on-isolarcloud-sungrow/411602 (accessed August 25, 2023).
[76] SolarEdge Always Puts Safety First. n.d.

[77] Briassoulis D, Mistriotis A and Eleftherakis D. Mechanical behaviour and properties of agricultural nets—Part I: Testing methods for agricultural nets. *Polym Test* 2007;26:822–32. https://doi.org/10.1016/j.polymertesting.2007.05.007.

[78] Maraveas C. The sustainability of plastic nets in agriculture. *Sustainability* 2020;12:3625. https://doi.org/10.3390/su12093625.

[79] Castellano, S, Scarascia Mugnozza, G, Russo, G, *et al.* Plastic nets in agriculture: A general review of types and applications. *Appl Eng Agric* 2008; 24:799–808. https://doi.org/10.13031/2013.25368.

[80] Windbreak nets | DURANET 2023. https://www.duranet.com/en/a/108-windbreak-nets/ (accessed September 28, 2023).

[81] Lee T-C, Zhong P-J and Chang P-T. The effects of preharvest shading and postharvest storage temperatures on the quality of 'Ponkan' (Citrus reticulata Blanco) mandarin fruits. *Sci Hortic (Amsterdam)* 2015;188:57–65. https://doi.org/10.1016/j.scienta.2015.03.016.

[82] Manja K and Aoun M. The use of nets for tree fruit crops and their impact on the production: A review. *Sci Hortic (Amsterdam)* 2019;246:110–22. https://doi.org/10.1016/j.scienta.2018.10.050.

[83] Gadhiya G and Patel U. Studies on the performance evaluation of agrivoltaic insect net house system. *THINK INDIA J* 2019;22:3811–25.

[84] Prakash V, Lunagaria MM, Trivedi AP, *et al.* Shading and PAR under different density agrivoltaic systems, their simulation and effect on wheat productivity. *Eur J Agron* 2023;149:126922. https://doi.org/10.1016/j.eja.2023.126922.

[85] Friman-Peretz M, Ozer S, Geoola F, *et al.* Microclimate and crop performance in a tunnel greenhouse shaded by organic photovoltaic modules – Comparison with conventional shaded and unshaded tunnels. *Biosyst Eng* 2020;197:12–31. https://doi.org/10.1016/j.biosystemseng.2020.06.007.

[86] Ravishankar E, Booth RE, Saravitz C, Sederoff H, Ade HW and O'Connor BT. Achieving net zero energy greenhouses by integrating semitransparent organic solar cells. *Joule* 2020;4:490–506. https://doi.org/10.1016/j.joule.2019.12.018.

[87] Magadley E, Teitel M, Peretz MF, Kacira M and Yehia I. Outdoor behaviour of organic photovoltaics on a greenhouse roof. *Sustain Energy Technol Assessments* 2020;37:100641. https://doi.org/10.1016/j.seta.2020.100641.

[88] Waller R, Kacira M, Magadley E, Teitel M and Yehia I. Evaluating the performance of flexible, semi-transparent large-area organic photovoltaic arrays deployed on a greenhouse. *AgriEngineering* 2022;4:969–92. https://doi.org/10.3390/agriengineering4040062.

[89] Agri-PV Activities. 2021. https://www.baywa-re.de/en/agri-pv/our-agri-pv-activities/#technology (accessed March 30, 2021).

[90] Greenhouses fitted with the ASCA® organic photovoltaic film n.d. https://en.asca.com/projects/asca-structures-en/greenhouses-fitted-with-the-asca-organic-photovoltaic-film/ (accessed February 24, 2021).

[91] Willockx B, Herteleer B and Cappelle J. Techno-economic study of agro-voltaic systems focusing on orchard crops. *37th EU PVSEC, WIP*; 2020.

[92] Graham M, Ates S, Melathopoulos AP, *et al.* Partial shading by solar panels delays bloom, increases floral abundance during the late – season for pollinators in a dryland, agrivoltaic ecosystem. *Sci Rep* 2021;11:1–13. https://doi.org/10.1038/s41598-021-86756-4.

[93] Ramos-Fuentes IA, Elamri Y, Cheviron B, Dejean C, Belaud G and Fumey D. Effects of shade and deficit irrigation on maize growth and development in fixed and dynamic agrivoltaic systems. *Agric Water Manag* 2023; 280:108187. https://doi.org/10.1016/j.agwat.2023.108187.

[94] Jung D and Salmon A. Price for covering cropland with an agrivoltaic system: PV panels replacing shading nets in Chilean blueberry cultivation. *AIP Conf. Proc.*, vol. 2635, 2022, p. 030001. https://doi.org/10.1063/5.0107946.

[95] Li Z, Yano A and Yoshioka H. Feasibility study of a blind-type photovoltaic roof-shade system designed for simultaneous production of crops and electricity in a greenhouse. *Appl Energy* 2020;279:115853. https://doi.org/10.1016/j.apenergy.2020.115853.

[96] Juillion P, Lopez G, Fumey D, Lesniak V, Génard M and Vercambre G. Shading apple trees with an agrivoltaic system: Impact on water relations, leaf morphophysiological characteristics and yield determinants. *Sci Hortic (Amsterdam)* 2022;306:111434. https://doi.org/10.1016/j.scienta.2022.111434.

[97] Ezzaeri K, Fatnassi H, Wifaya A, *et al.* Performance of photovoltaic canarian greenhouse: A comparison study between summer and winter seasons. *Sol Energy* 2020;198:275–82. https://doi.org/10.1016/j.solener.2020.01.057.

Chapter 5
The agronomy of crops in agrivoltaic systems

Christian Dupraz[1]

5.1 Introduction

What crop species are suitable in agrivoltaic (Agri-PV or AV) systems? How do we perform rigorous crop yield assessment in AV systems? Are there really 'shade' species and 'sun' species? Is the shade of AV systems in winter and summer similar? Do annual crops, perennial crops and permanent pastures behave the same way in the shade of AV systems? What services could photovoltaic (PV) panels provide to crops that could help them to perform better, in particular when climate change is concerned? These are some of the questions that will be addressed in this chapter. The question of crop productivity in AV systems is central and deserves a careful approach. The equation is simple: if the crops are not profitable due to low yields or high production costs, the farmer will set aside, the system will no longer be an AV system, and the crop production will decrease. This is why crop productivity in AV systems is a central question. It's a matter of life and death for agrivoltaism, including with grazed systems or aquaculture systems. The PV panels productivity is not depending on the crops, but on the optimisation of the electrical part of the system (use the right orientation, tilting, spacing of the panel and many other specifications of the electrical system). Conversely, the crop productivity is a challenge and depends directly on the PV panels. Although the crop income is usually 10–100 times less than the electricity income per hectare, crops are a key component of AV systems. Growing crops in an AV system is a challenge; crops face shade that reduces photosynthesis and alter morphogenesis. But crops can also benefit from several services provided by the AV system. The balance between competition (mainly for light) and facilitation (through a modified microclimate) is key to understand the functioning of AV systems, in a similar way as it is in mixed cropping or agroforestry (AF) [1]. While competition (for light) is a compulsory feature of AV systems, facilitation (due to a modified microclimate) is an optional and stochastic one. Facilitation may happen or not. Competition (mainly for light, but also for water in some cases) always affect crops. The agronomical evaluation of AV systems can therefore only be probabilistic, and should include several years

[1]INRAE, University of Montpellier, UMR ABSYS, France

with climate variability. Long-term experiments are difficult and costly to manage. Using crop models may allow to explore more options of crop management in AV systems.

5.2 Growing crops in the shade: from agroforestry to agrivoltaism

The idea of mixing PV panels and crops was originally formulated by PV physicists [2], and there were several attempts to grow crops or let animal grazing under PV panels during the early years of the XXI century in Europe and Japan. However, these trials did not include detailed full sun control sites and/or the results were not published in scientific journals. It is not a surprise that the first monitored experimental sites of AV were elaborated by agronomists specialised in AF research [3] or on farm research [4] (Figure 5.1).

Crops and pastures in AF are also facing light competition from trees. However, the shade of trees is much more complex than the shade of PV panels. The leaf area of trees is constantly evolving, the canopies of trees are moving with the wind, and last but not least, trees have roots that compete with crops for water and nutrients. We could also cite many more aspects of trees that affect soils and crops: their litter recycle nutrients, their structural roots hold the ground and facilitate rain infiltration, their deep fine roots capture leached nitrogen. In contrast AV systems are simpler: fixed structure, fixed area, predictable movements with mobile panels, no below-ground competition.

Plants react to shading in two ways: shade avoidance and shade tolerance (Figure 5.2). The shade of PV panels and the shade of trees are quite different [6]. Tree canopies are green and absorb red and blue radiations selectively, while most

Figure 5.1 Pictures of various crops under AV systems worldwide (top, Japan, France and Italy) and of the historical first experimental site at Montpellier, France (bottom)

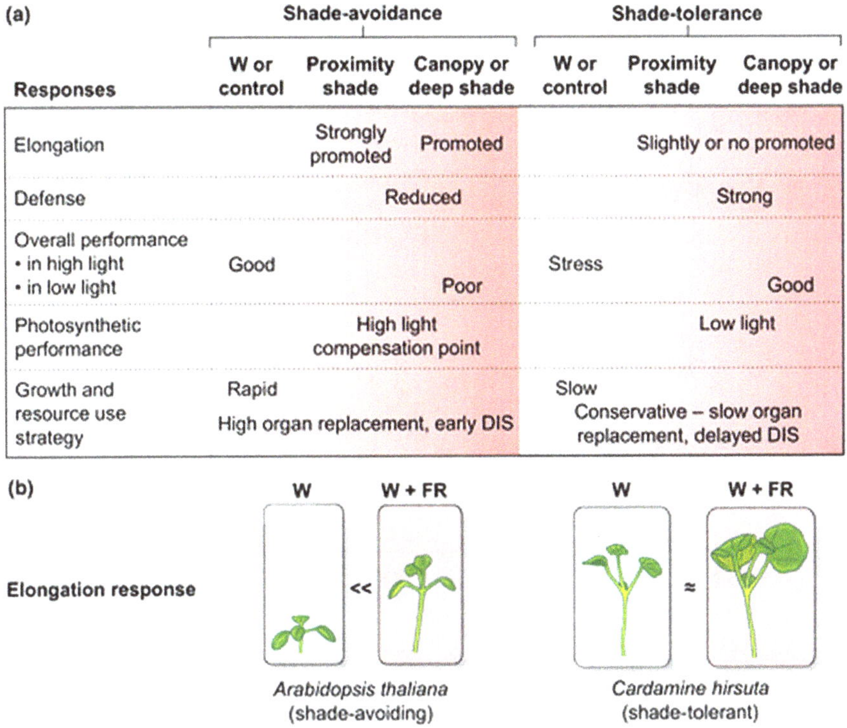

Figure 5.2 Summary of some differential responses of plants associated with either shade avoidance or shade tolerance strategies [5]

PV cells are opaque and thus do not alter the spectral composition of transmitted radiation (Figure 5.3). In sunlit areas (flecks of light in AF, rows or patches of light in AV), the radiation transmitted is nearly identical, qualitatively and quantitatively, to the incident radiation above both systems. In shaded areas, the irradiation is reduced, and includes only diffuse radiation. The Red:Far Red ratio under panels remains close to the incoming radiation value (1.35), while under the trees it decreases to 0.35. The Blue:Red ratio increases in the shade in both systems, but more so in the PV panel system. Therefore, radiation quality in the shaded areas differs substantially between AF and AV systems. The higher Red:Far Red ratio and Blue fraction under PV panels may fail to induce shade adaptation responses in plants, unless low radiation level signals prevail over radiation quality signals in inducing such a response [6].

While some forerunner papers explore how PV panels could convert into electricity only the radiations that are not used by photosynthesis [7,8], currently available PV panels are using the whole sun radiation spectrum. Semi-transparent panels are also developed [9], but they usually do not alter the spectral composition of the transmitted radiation to the crops. Some tinted semi-transparent panels are also under consideration [10]. Under such panels, the shade may be more alike the shade of trees.

(a)

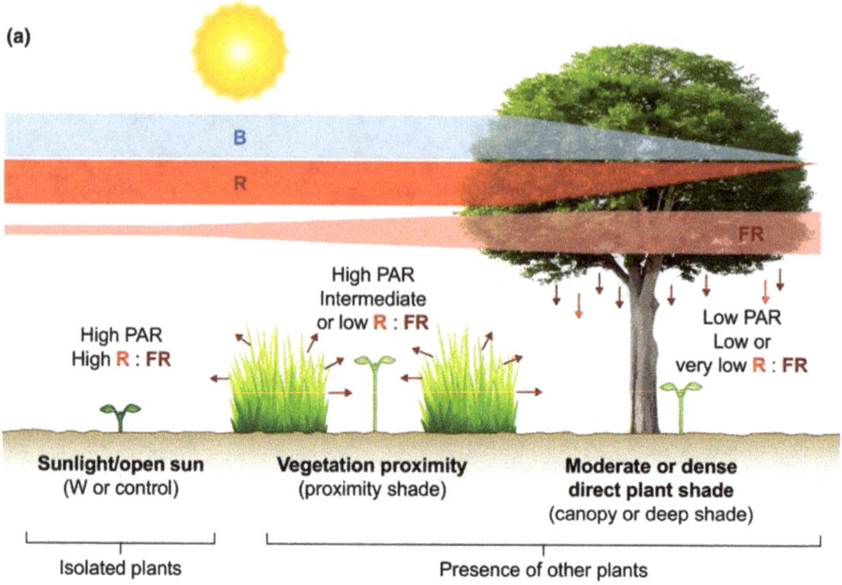

(b)

Natural condition	Relative R : FR value	How it is mimicked	Laboratory term
Sunlight/open sun (unshaded)	High		W or control
Vegetation proximity (close neighbors without direct shading)	Intermediate or low		Proximity shade
Moderate direct plant shade	Low or very low		Canopy shade
Dense direct plant shade	Very low PAR reduced		Deep shade
Vegetation proximity	Very low when starting the dark period	Light Dark	End-of-day FR (EOD-FR)

Figure 5.3 Impact of trees on the light availability for crops in agroforestry systems [5]

However, the primary impact of light competition is the reduction of the photosynthetically active radiation, and hence the reduction in plant growth. This impact will be different if the harvested part of the crop is a vegetative one (leaf vegetables, roots and tubers, pasture) or a reproductive one like flowers or fruits (most cereals and annual crops, edible fruits, fruit vegetables). The sexual part of a plant cycle is usually the most sensitive to a deficit of light, that may induce flower abortion or fruit fall. It is also the most sensitive stage to many stresses like heat or drought that may be mitigated by the shade of panels. This is where AV systems may have a bright future: alleviating stresses on the reproductive organs of the crops, and especially fruit trees [11]. The productivity issue of the underlying crop is also concerning grazed systems [12] or aquavoltaic systems [13,14]. If the plot under a grazed system does not provide forage, the system is a shelter building, not an AV system. If the water below an aquaculture PV system does not provide feed for fishes or shells from photosynthesis, it is also a shelter building, not an AV systems. In aquaculture, shelters are sometimes used to prevent algae growth and all the feed is imported. This is battery farming comparable to feedlots for cattle. In both cases, there is no link to the ground (or the water) and the system is a PV shelter, similar to any PV installation on farm buildings. An AV system requires, by definition, solar sharing between the PV panels and the agricultural production.

5.3 The shade pattern under AV systems

A number of papers have measured or modelled the shade pattern of various AV systems [3,15–18]. This shade pattern is highly variable depending on the height of the panels, their orientation and tilting, and the season. The shading pattern can be averaged for various periods of time such as one year, the crop's growing season or specific phenological phases of those crops that are considered to be the most sensitive to shade. Depending on the AV system design, the shading pattern at the crop level may be very homogeneous or heterogeneous in both space and time. Any heterogeneity of irradiation under the panels may be a constraint to crop farming. Plants are sensitive to the accumulation of irradiation and plant organs temperature, and any heterogeneity in irradiation will induce heterogeneity in phenology and productivity. As an example, if the harvesting is delayed in the relatively shadier zone, it could be a no go for farmers. It is usually not acceptable to harvest a crop at different times in the same plot, for practical and financial reasons. Strongly shaded areas in any AV plot could therefore be set aside, and this would result in a significant decrease of the land productivity. This is why many AV projects aim at finding designs of the system that allows for a reduced but homogeneous irradiation of the crop.

The heterogeneity of the shading pattern at crop level is increased with a low elevation of the panels above the ground and/or a small distance between panels and crop canopy, with clustered panels (panels arranged in lines or blocks) and with opaque panels. The homogeneity of the shading pattern at crop level is conversely increased with a high elevation of the panels above the ground, resulting in a large distance between panels and crop canopy, with a diffuse positioning of the panels (in a quincunx, checkered, etc.) and with semi-transparent panels. Trommsdorff *et al.* [18] designed a system with fixed panels rows not aligned east-west as usual, but north-west to south-east. Using a ray-tracer radiation interception simulator, they showed that this angle will increase the homogeneity of radiation at the crop level under the system. However, this deviation from the east-west usual orientation reduces the electricity production, and very few AV projects adopted this design later.

For the same ground coverage ratio (GCR) (Area of panels/Area of land), mobile panels on trackers with pure solar tracking cast a heavier shade on crops than fixed panels. For the same PV panel area, fixed tilted or vertical panels aligned east-west cast a more heterogeneous shade than fixed tilted or vertical panels aligned north-south. However, it is not easy to design systems with fixed panels rows aligned north-south, as the cost of the support structure would be more expensive. Similar results have been obtained in AF alley-cropping systems at temperate latitudes, where north-south tree lines cast a more homogeneous shade on crops than east-west tree lines [19]. Most mobile panels on 1 axis trackers rows are aligned always north-south, and provide an homogeneous irradiation of crops, while fixed panels are usually aligned east-west. Vertical panels can also be aligned north-south and provide an homogeneous irradiation of the crop irrespective of the pitch [20], but the orientation of the panel rows is not always easy to choose. The slope of the field, the agricultural machinery available on the farm, or the shape of the parcel may impose other orientations.

5.3.1 Reducing the shade heterogeneity in AV systems

The shade pattern may also be modified by various innovations. Some systems include lamps under the panels to provide more light at some key phenological stages of the plants [21], or add specific (e.g., red band) wavelength radiation to stimulate crop production under the panels [22]. Some recent innovations include also semi-transparent PV panels with narrow PV cells that provide an almost homogeneous shading, or narrow cylindrical PV units that also provide an homogeneous shade with multiple sun/shade transitions at the hour time scale (Figure 5.4).

Figure 5.4 The seek for a more homogeneous shading of crops had led to systems with small size PV cells such as those of Tubesolar (A), Insolight (B) or the Japanese systems with narrow cell rows (C, D). The narrower the panels, the more shade/sun transitions per day.

5.3.2 The physiological impact of alternate shade and full sun periods

The shade impact on plants is well understood [23]. Plants sense shading and activate adaptive morphological and physiological responses [5]. Simple morphological responses (e.g., internode elongation) to a single light cue (low red to far-red (R:FR) ratio) are well documented, but multiple responses elicited by a broad spectrum of neighbour proximity signals have also been described [24]. The architectural responses to shade include accelerated elongation of the hypocotyl, internodes and petioles, upward leaf movement (hyponasty), and reduced shoot branching and are collectively referred to as the shade avoidance syndrome. Shade avoidance has become a textbook example of adaptive plasticity. Unfortunately, shade avoidance is of no use for plants grown in AV systems. They have no chance to grow high enough to top the PV panels. The situation is even worse for plants grown right under the panels: the closer they get to the panels, the higher the shade ratio is. The carbon affected to this plasticity reaction is used in vain by the plant. Therefore, the selection of crops or varieties with no shade avoidance strategy would be recommended for AV systems.

Shade tolerance and adaptation is more useful for crops in AV systems, as it is in AF systems [25]. A number of reviews (Figure 5.5) of the shade adaptation of crops are available [26,27]. Sun and shade leaves are a well-known example of an acclimation to light intensity, but what kind of leaves will crops in AV systems produce? Most of the experiments cited in these review papers used various shade nets that provided an homogeneous shade pattern, which is quite different from the

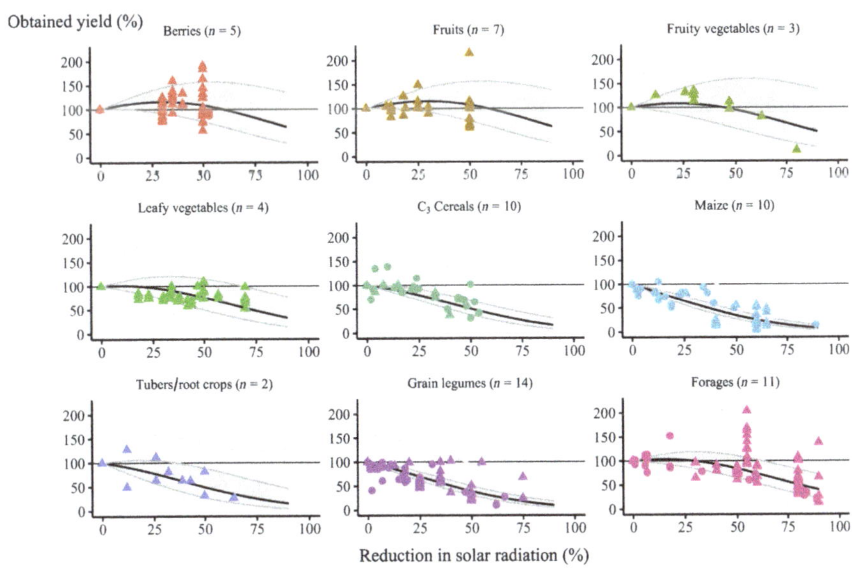

Figure 5.5 Impact of shading on the relative yield of various types of crops (from Laub et al. [26])

shade strips or large shade patches cast by an AV system. This means that one should refrain from directly extrapolating such results to AV systems, unless evidences are produced that on/off shades and homogeneous shades have the same impact on plants. Miyashita and Tateno [28] showed that the leaf lifespan is a key trait for plants to succeed in the shade, and could maintain positive relative growth rates even in deep shade allowing for higher survival rates for shade tolerant species. But in AV, we are not dealing with the survival of plants. We are considering plant productivity and this is another issue. This is why the conclusions of ecological studies of shade adaptation are not relevant for AV. The impact of shade on crops was also evidenced in alley cropping AF and included some etiolation (higher plants), larger leaves and a much longer leaf lifespan, but this was not sufficient to compensate for the low irradiation level due to the tree canopies [29].

The rigid pattern of intermittent shade of AV systems is indeed unusual and almost never experienced by plants in nature. In monocropping, two crucial mechanisms of plant acclimation to changes in light conditions comprise thylakoid protein phosphorylation and dissipation of light energy as heat by non-photochemical quenching [30], allowing the plant to face excess light intensity. In AV systems, plants will still experience these excess light intensity phases, when they are in full sun. But they will also experience several full sun/shade and shade/full sun sharp transitions per day. While the transition from sun to shade is quite neutral for the physiology of the crop, the reverse transition is more an issue. These transitions are quite different from those induced by clouds (progressive transitions from full sun to full diffuse radiation) or tree canopies (high frequencies of transition due to tree movement). Tiffon-Terrade *et al.* [31] monitored the behaviour of vineyard plants during sharp transitions of irradiation. When the plants went from sun to shade, net assimilation was dropping as expected, but the stomas remained open (hysteresis effect) allowing some assimilation at a significant (5 μmol m^{-2} s^{-1}) level during the shade event. When the leaves passed from shade to sun, both net assimilation and stomatal conductance of AV plants were identical or higher as compared to the control full sun plants. The main explanation was the better water comfort of the AV plants after the shaded period, allowing for a larger stomatal aperture and net photosynthesis during sunny times. This process compensates partially the reduction of photosynthesis due to the shade. This is why the usual stomatal closure during the afternoon is less pronounced in AV systems.

The same level of accumulated irradiation may be received by plants with various timings. For example, a 50% relative irradiation may be the result of only one full sun/shade transition per day (a garden that is in the shade of a building after noon), 4–5 full sun/shade transition per day (typically under a fixed panel AV facility), 50–100 full sun/shade transitions per day (in a Tubesolar AV system) or with no full sun/shade transitions (under a shade net). This issue is quite new for plant ecophysiologists (Figure 5.6).

Some recent studies on fruit trees [32] go even further: they suggest that in dynamic AV systems, where the farmer can decide when he provides more light to the plants, it could be useful to provide high irradiation levels during the leaves formation, as sun leaves are more photosynthetically active in the shade later.

Figure 5.6 Shade of AV systems with few sun/shade (fixed panels, left) daily transitions or a lot of sun/shade daily transitions (Tubesolar, right)

This is counter-intuitive and deserves careful investigations. It makes sense as the adaptation of plants to shade (thin and large leaves with low maximum photosynthetic activity) was an evolution consequence of situations in nature that are quite different from what crops experience under AV facilities.

An empirical answer to this issue is now shared by many AV promotors. In Japan, narrow panels are preferred to large panels, although this is more binding on construction costs of the facility (Figure 5.4). In Europe, the Tubesolar design allows for a more homogeneous insolation pattern, but the technology is expensive. In AV greenhouses, semi-transparent panels may also homogenise the irradiation pattern. But the best answer to this heterogeneity of light on the crops is the elevation of the panels. The higher the panels are, the more sun/shade transitions per day. Unfortunately, this increases the cost of construction of the system, and many AV developers prefer not to build such high clearance systems. However, while radiation quality and intermittency may affect plant growth in AV systems, the major limitation in AV systems is definitely the irradiation quantity. Deep shades strongly reduce photosynthesis and growth, and moderate shade seems to be the optimum [31].

5.4 The potential benefits of AV systems for crop productivity and quality

5.4.1 Will AV systems protect crops from climate change?

A number of papers insists on positive aspects of crop protection against several hazards such as hail, frost, excess heat and drought in AV systems [11,33]. These hazards are increasing with climate change in many regions and motivate a number of studies on AV systems as an adaptation tool to climate change. This positive impact could compensate the light competition, and in some cases result in higher yields under AV panels than in an unshaded crop. It could also reduce the costs of production for the same yield. While these facilitation aspects are exciting, it must be kept in mind that light competition is a permanent feature of AV systems while microclimate facilitation is a stochastic one, that happens only in some circumstances.

Positive impacts of the shade may improve not only the yield of the crop, but also its quality. A good example is vineyard AV, sometimes called vitivoltaics. Climate change has a detrimental impact on grape quality. The phenology is fastened, leading to very early harvest dates and the berries are too rich in sugar and poor in secondary metabolites and acidity that are key for the wine quality. Vineyards under AV systems are harvested later, with an improved quality of the berries [34], especially on acidity [35], but the impact on the reduction of the sugar content of the berries under AV systems is not always significant [36]. Current studies on fruit trees also focus on the quality of peach, apricot, apple, cherries under AV systems [37]. Mobile systems with agronomical tracking allow to provide light to the fruit trees when they need it most, and shade when excess light is a problem. This may avoid sunburned fruits that is a worrying event, e.g. in apple production. Less alternate bearing of apple trees was observed under shading [37], which emphasises the need of multi-year studies. Better frost protection resulted in a higher proportion of trees bearing fruits under PV panels during some years. Fruit size was less sensitive than fruit number to shading. However, in this study the yield of apples was too low under the AV system which indicates that sustainable apple orchard shading strategies are still to be identified.

5.4.2 Microclimate management under AV systems

The shade of PV panels affects the microclimate of the area cultivated under and between the panels. In open air AV systems, turbulence of air and wind results in a very high renewal rate of the air under the panels and in the vegetation. Impacts on air temperature and air humidity are usually very small, and often within the limits of detection of usual sensors. By contrast, the soil temperature and the plant temperature are significantly reduced in AV systems (as well as the perceived temperature by humans and animals). Apart from the radiation impact, these are the major microclimate modifications of AV systems on the crop environment [38].

5.4.2.1 Air temperature under AV systems

Measuring the air temperature in contrasted environments like a full sun area and an intermittently shaded area is not as easy as it seems. Several studies documented some small differences in air temperature, but many also concluded to the absence of difference [39]. Al-agele *et al.* [40] documented some degrees of air temperature difference while the wind speed was reduced between the rows of panels. According to Noor and Reeza [39], the microclimate under the panels varies differently depending on the area's climate and the land's topography. Their study in the tropics showed that the air temperature was the same in the open field and under the panels for areas on high topographical locations, while air temperature was

Figure 5.7 No white frost under the Fruitvoltaics Apple Research Project in Haidegg, Graz, Austria, as a consequence of a higher soil and plant temperature under the protection of the panels

lower under panels in the lowlands. A common observation in open air AV systems is that the average daily temperature is the same in full sun and under the AV system. This results from a compensation between lower maximum day-time temperature and higher minimum night-time temperature under AV systems. AV systems display a reduced thermal amplitude day/night. In vineyards, the slow and gradual maturation of the grapes for the beneficial aromatic complexity and freshness of wines requires amplitudes above 20°C, and this may not always happen under AV systems.

Frost is also a major concern for many crops, including fruit trees and vineyards. With climate change, may species open their buds earlier in the year, and may suffer from late frosts. PV panels will reduce radiative frost by limiting the night emission of infra-red radiation of the soil and plants (Figure 5.7). Mobile panels placed horizontally during the night will be the more efficient to prevent heat loss at night. Regarding convective frost, when cold winds below zero degrees flow over the landscape, it is very difficult to combat the frost. Mobile panels placed vertically may create turbulence that will mix the cold air flowing in on the ground with milder air that is above. But this impact is highly unpredictable and would require an adequate wind speed. Some positive impacts of AV systems to protect the crops against radiative frost have been evidenced with vineyards and fruit trees in Southern France, but they were not yet documented in reviewed publications.

5.4.2.2 Air humidity under AV systems

The impact on air relative humidity (RH) is even more difficult to evaluate. Most papers fail to evidence any change in the RH of air under PV panels. Studies that conclude to differences as low as 2% are not very convincing. Noor and Reeza [39] measured 65% RH under panels and 53% RH in full sun, but methodological difficulties should not be overlooked when measuring RH in full sun. The use of mechanically ventilated sensors is recommended for such measurements, but very rarely performed.

5.4.2.3 Wind under AV systems

Wind patterns in AV systems are complicated. Vertical panels have a clear windbreak effect if the wind direction is at right angle with the lines of panels, but vertical panels are usually raised above the ground, and strong winds may occur below the panels as a venturi effect. Fixed tilted and moving panels (that are usually maintained horizontally during windy episodes) will not protect the crop much against raging winds (Figure 5.8).

5.4.2.4 Soil temperature in AV systems

PV panels have a strong impact on soil temperature as they reduce the incoming heat during days and the loss of radiative heat during the night. Soil temperature is decreased during the warm season (long days and short night), and increased during the cold season (short days and long nights), as compared to a full sun site. In spring, the soil may remain colder under the panels, and this may slow the germination of spring crops. Conversely, crops sown in autumn may benefit from a

Offset positioning of vertical panels adapted to hurricane winds (South-Korea)

Raised vertical panels with a strong wind blowing under the panels and into the pasture (Japan)

Mobile 1-axis (left, Sun'Agri technology, France) or 2-axis (Right, RemTec technology, Italy) panels are set horizontal in case of strong winds, annulling the windbreak effect.

Figure 5.8 Managing wind in AV system: facing the wind rage and preventing wind damage to crops is not easy

warmer soil. This is useful for winter cereals [38,41,42]. A number of studies have documented the soil temperature under ground-mounted PV systems. In the UK, the panels reduced the soil temperature (measured at 10 cm soil depth) by an average 5.2°C in summer and increased it by 1.7°C in winter [43]. The impact on soil temperature can vary depending on the installation technology. Fixed panels mitigated more soil temperature than mobile panels in China [41] as they were intercepting more radiation due to their higher ground cover rate.

5.4.2.5 Plant temperature in AV systems

Plant temperature is a key driver of plant phenology. A reduced accumulated plant temperature under PV panels will induce a slower phenology and a longer life cycle of the plants. Very cold or very hot plant temperatures are also a threat to plant physiology. Plants under AV system will experience less extreme (both cold and hot) temperatures, and this may prove beneficial when the extreme temperatures are damaging photosynthesis and growth processes. However, the main effect of shade is a reduced temperature of plant organs in the shade, resulting in a very significant reduction in cumulated plant temperatures. Leaves senescence is also delayed, resulting in a longer life cycle of leaves.

5.4.2.6 Water use efficiency in AV systems

The shade of PV panels reduces the amount of water loss through soil evaporation and therefore increases water use efficiency [44–46]. This advantage will be more expected in drought-prone areas during hot periods [47]. Yue *et al.* [41] described the soil humidity increase under PV panels but failed to report the role of vegetation in the observed differences between the panels and the open control. Barron-Gafford *et al.* [47] showed that applying water to the crops based on a two-day irrigation frequency (instead of a daily frequency) was a better strategy to maintain higher soil moisture, and resulted in an increased water use efficiency under the panels for chili (157%) and tomato (65%). Marrou *et al.* [46] evidenced a slightly increased (12%) water use efficiency for lettuce under 4 m elevation panels, but a reduced water use efficiency for cucumber, showing that this advantage is not systematic. Some AV systems are able to provide shade in the afternoon and full sun during the morning: this is mainly the case for dynamic systems with mobile panels [11] or fixed panels facing the south-west [18]. While this seems an optimal design to avoid water stress in the afternoon, the overall impact on the water use and the water use efficiency is difficult to predict. In the full sun, crops with water stress close their stomata during the afternoon; in the shade, crops close their stomata because of the shade. If the shaded crops have less water stress as a result of reduced soil evaporation or reduced biomass, they will maintain photosynthesis in the shade during the afternoon. However, this may be compensated by different rooting patterns in the shade and in full sun, making the final result difficult to predict. This may explain why a crop species may benefit from the afternoon shade and another crop species will suffer at the same site. Crop models may help to disentangle these different and opposite processes that are very difficult to measure in the field. Crop models were very helpful to interpret tree-crop interactions in AF systems [48,49] and should also be efficient for AV systems.

5.4.3 Dynamic AV systems as a key innovation for crops

Moving panels with an agronomic control to provide crops with light when they need it most is an attractive version of AV. The key question to solve is to identify when the crops need full lighting, and when they can tolerate shading, or even better benefit from shading. All mobile systems could be monitored to provide more light to crops, but in fact very few systems are piloted with this wiping option. The Sun'Agri technology is the most advanced system so far [11,34,37,50,51], but other technologies are also including some agronomical tracking with panels wiping during the crop sensitive stages to light deficit. By providing the right level of irradiation to the crops during the key sensitive stages, the agronomical tracking of mobile panels could boost the crop yields in AV systems.

This is especially true with perennial crops such as fruit trees and vineyards in dry and hot climates. In that case, climate protection could be very positive, but will require high GCR to provide enough shade during stress events. The wiping option of panels is then required to allow a satisfactory growth of the crops during the rest of the growing cycle of the crops. A balance needs to be struck between yield reduction due to shade and crop protection thanks to the panels during extreme weather events. If the GCR is too low, the protection may not be adequate. The optimal GCR will depend on the frequency of weather events when panel protection is positive. This is a clear illustration of the competition/facilitation balance in mixed systems documented by [52]. Some PV developers argue that with some specialty crops (e.g., berries, fruit trees and vineyards), a high GCR is compatible with high relative yields without the need to wipe the panels out during some crop development stages [53]. High GCRs are required to protect the crop efficiently against climate hazards, and this protection will increase the crop relative yields. While this may be true, such systems with a high GCR will not allow the farmer to replace the fruit crops with species that require more light. This would in fact reduce the farmer's options to change crop rotation in the future.

5.5 Measuring the crop productivity in AV systems

Crop and pasture yields under AV systems have been measured by a number of studies worldwide, mostly in the temperate zone so far (Appendix A). Measuring crop yields under AV systems is not as easy as it seems.

5.5.1 Relative plot yields and relative field yields

Usually, the crop production in AV systems is compared to crop production in monoculture. But a key issue is often forgotten: the yield of the reference monocropping is at the field scale, while the yield in the AV system is usually measured as the plot yield of the cropped area. But in AV system, a part of the field cannot be cropped any longer. Therefore when assessing the relative crop yield, it is necessary to provides two values of the yield in AV systems [54]:

- The relative plot yield (RPY) that is the relative yield of the cropped area in the AV system

- The relative field yield (RFY) that relates the production to the full AV system area, including uncropped areas, i.e., areas where no crop production can be obtained (lines of posts, electric systems, areas for support structures, e.g. guy cables, access tracks for maintenance). These uncropped areas may be useful (e.g., for biodiversity protection), but they reduce the agricultural production of the field. The RFY and RPY are linked by the following equation

$$\text{RFY} = \text{RPY} * (1 - \text{Uncropped Area rate}) \tag{5.1}$$

with the uncropped area rate = area unavailable for cropping/total area of the AV system.

As an example, a system where the crop would produce a normal yield per m^2 (i.e., 100% of the control yield) on only 50% of the field area would have a RFY of 50%. The RPY is a measure on how the crop is doing on the cropped area, while the RFY is a measure of land productivity, and should be used when comparing AV and monocropping systems. As an example, policies that refer to the crop yield in AV systems always refer to the RFY. All papers on the productivity of crops in AV systems should provide the two figures, but [54] found less than 10% of the papers that did so.

5.5.2 Methodological issues for measuring crop yields in AV systems

Edge effects on small-sized systems, soil heterogeneity between AV and control zones, questionable extrapolation of measures done under isolated panels, unsuitable control systems, and bias due to weeds and diseases are the most common issues met by AV researchers.

5.5.2.1 Small-sized systems induce bias as the shade impact on crops is underestimated

When evaluating this bias, we must hinge on the distance between the panels and the top of the crop canopy and on the maximum sun elevation in the sky. Low sun elevation as in winter exacerbates the bias. A small system is acceptable when there is a small distance between the crop canopy and the panels. This is usually the case with tall fruit trees. But when the distance between the panels and the crop canopy is large (e.g., low annual crops under panels at a 4–5 m elevation), the edge effect will be significant. This effect is negligible with large systems, where measurements can be performed far from the edge the system, but it can be a concern when the width of the system is narrow. On the winter solstice, the sun is at its lowest daily maximum elevation in the sky. Radiation will penetrate deeply under the AV system. If the system is very narrow, the entire area under the system will get full sun at noon on the winter solstice, and possibly during long periods of the year around the winter solstice. Dupraz [54] suggested a rule for determining the minimum width Wmin of an experimental AV system. This rule takes into consideration the latitude, the maximum height of the crop, and the clearance of the system. It can also be computed for different periods of the year (for winter, spring, summer or autumn crops). At the 40–60° latitude, a Wmin/H ratio of 5 was

	A maize crop in the oldest experimental AV prototype in the world (Lavalette, Montpellier, France, [3]) with both fixed and mobile panels. With low crops such as salad, the area suitable for crop yield measures is small. The control area is on the right (south)
	One of the highest AV prototype in the world. With such a clearance (5m) and narrow width, the area suitable for winter crop yields measurement is limited. The control is located at the North-West. Heggelbach, Germany [42]
	Very recent prototype by RemTec in a paddy field in Camargue, France. High clearance, narrow width, inducing potential large edge effects. The control area is located South of the prototype.

Figure 5.9 Small-sized AV prototypes that need special attention to avoid overestimating the relative crop yields

assumed to be a sensible minimum if winter crops are considered. A more site-specific calculation can be done at each experimental site. Sites at low latitudes may be narrower. If there are no winter crops, the calculation can be done with the sun elevation during the actual crop growing season. It is therefore recommended to make sure that the width of the system is larger than five times the clearance height of the system. This was particularly concerning for some very narrow experiments (Figure 5.9) such as those of [55] on maize or at the Tanzanian site of [56].

5.5.2.2 Soil heterogeneity across AV and control plots

Soil heterogeneity was observed lately at sites including the world's first dynamic AV farm in Tresserre, France, and in a vertical panel experiment with cereals described by [34]. A good assessment of soil homogeneity before the setting up of the AV facility is always a good idea, but in practice, AV systems are located with other criteria (e.g., close to a grid connecting station). Actually, a very critical moment regards soil heterogeneity is during the AV facility construction, as the installation process – if not considered by the project developer – will lead to soil compaction, soil damage and thus have an impact especially on the agricultural yields in the first years after the installation. The usual way to control soil heterogeneity in agronomic studies is to replicate the experiments on different fields or plots (factorial experiments), but this approach is complicated in AV research due to the size and cost of the systems, and due to the difficulty for the farmer to manage small distant and replicated units. The same difficulties are met in AF research [57]. Not a single factorial design experiment could be found so far in the literature for AV systems.

5.5.2.3 Extrapolation of measures under isolated panels

When the experimental device consists of isolated, small-sized panels, with only a limited number of yield measurements made under the panels or close to the panels, it is not easy to calculate a plot yield, and even less a field yield [10,35,47].

5.5.2.4 Unsuitable or unrealistic control systems

Concerning AV greenhouses, some studies did not compare a standard greenhouse with a PV greenhouse [21]. Cossu *et al.* even added light under the PV panels to increase the very low irradiation in winter in a greenhouse with half the roof covered with panels. Dramatic drops in crop yields were recorded, but could not be compared to a true control as it was missing.

Some studies explore whether it might be possible to grow a crop in full summer under an AV system, in sites where the climate and soils make it impossible to do so in full sun. A good example for this is the work of [58] on growing lettuce during the summer in Italy. The results are convincing, with a higher yield of fresh lettuce under the AV system. But an increased RPY for a short duration in summer may be compensated by low RPYs during the other growing seasons. The impact of an AV system should be assessed at least on a timescale of one year, including the full rotation of crops over the year.

Extremely low yields in the control indicate that solar radiation is not a limiting factor, and usually result in very high RPYs. Such systems, which cannot provide farm revenue, are often designed by researchers to explore the limits of the system. A good example occurs when researchers include unirrigated maize in their experiments [50]. The same very low yield in AV and in full sun are usually recorded, with some advantage to AV maize that may result, as a consequence of the almost null yield in full sun, to huge RPYs tangenting infinity. But these yields are so low (less than 1 T of grain $DM.ha^{-1}$) that they make no economic sense. Another example concerned pastures with an annual Dry Matter productivity lower than 3 T $DM.ha^{-1}.year^{-1}$ [59]. In the area studied by Madej *et al.*, pastures usually have an annual productivity of 5–8 T of $DM.ha^{-1}.year^{-1}$. In such conditions, very high RPYs are evidenced, with very low absolute yields.

5.5.2.5 Biotic stresses that do not result from the panels

A contrasted impact of weeds/pests/diseases between the AV system and the control may also occur. In an experiment in Montpellier, barley had much higher yields under the panels than in the full sun. But the key explanation was that the control plot was overrun with weeds, whereas the barley located under the panels was almost weed-free. There was no way to prove that the absence of weeds under the panels resulted from a positive impact of the panels, as there was no replication on different sites.

5.5.2.6 The issue of perennial crops

Measurements on perennial plants during a single year are not enough. For perennial plants, year-to-year depletion of carbon reserves due to the shade is crucial, and several years of measurements are required to reach a solid conclusion. Moreover, yield measurements made on perennial plants after only one year of shade may not reflect the true impact of an AV system, especially if the system was constructed on a previously existing perennial crop (e.g., orchard or vineyard) that had been growing in full sun conditions for many years. Conversely, measurements made during a one-year period but on perennial crops that have already been under the AV system for several years could be considered, given that the impact of the system on the plant reserves is probably stabilised.

5.5.2.7 Beware of land equivalent ratios in agrivoltaics

Land equivalent ratios (LER) are often used to illustrate the productivity of AV systems, in a similar way as their use for AF systems [57,60]. They have been first introduced to AV systems by Dupraz *et al.* [3]. The LER of an AV system is defined as

$$\mathrm{LER} = \mathrm{RFYc} + \mathrm{RFYe} \tag{5.2}$$

where c and e stand for crop and electricity respectively. It is a measure of the synergistic effect of collocating PV panels and crops together. If an AV system has a LER of 1.4, it means that a 100 ha AV farm would produce as much crop products and electricity as a 140 ha farm where PV panels and crops are located on different plots.

To account for the uncropped areas in the system, this equation should always be used with the RFY of crops. Trommsdorff *et al.* [18] and Edouard *et al.* [61] modified the equation into

$$\mathrm{LER} = \mathrm{RFYc} + \mathrm{RPYe} - \mathrm{UA} \tag{5.3}$$

where (UA stands for the Uncropped Area rate), which is not correct algebraically. It should be

$$\mathrm{LER} = \mathrm{RFYc} + \mathrm{RPYc} - \mathrm{UA} * \mathrm{RPYc}. \tag{5.4}$$

Some very high LERs of AV systems have been published, resulting from RFYc values higher than 1. Such high LER values are always explained by high values of the crop RFY, as the electricity RFY is always lower than 1. RFYe may be slightly higher than 1 if the AV system has the same GCR as the standard Ground Mounted PV system, and if the elevation of the panels allows a cooler temperature of the panels that increases the electricity productivity of the panels.

But usually, AV systems have a lower GCR than the standard ground mounted PV systems, and RFYe is usually in the range [0.5; 0.9].

This is why high values (>2) of AV LERs are always the consequence of very low yields in the control monocropping system. These values may be deceptive, and must be considered with care. As all ratios values, LERs are very sensitive to the denominator value. Ratio distributions are heavy-tailed, and it may be difficult to work with such distributions and develop an associated statistical test. When the monocropping control crop yield comes close to zero, infinite Relative Yields can be measured, resulting in infinite LERs. It could happen that both values of crop productivity in the monocropping control and under the AV system are very low and economically not sensible, but this may result in a very high LER if the low yield under the panels is still higher than the very low yield in full sun conditions. This is why it is not recommended to use LERs when crop yields are not both in the range of sensible values to make the crop profitable. A very high value of the LER obtained with very low absolute yield values will hide the major point: AV systems are not efficient in this case.

In some extreme weather conditions, crops could be successful under the panels, and not in the full sun. This is what happened in the study by [47] in Arizona, USA. Chiltepin pepper and tomato crops resulted in extremely high values of relative yields (close to 3 and 2, respectively). Indeed, this was done under desert conditions, where growing crops in full sun is a challenge, even with irrigation. We may question the significance of the control crops, that should probably be managed under shading nets, if this would have represented the standard way of cultivating in that area without PV panels. When the shade of panels allows crops to grow in very harsh environments, such as deserts where growing crops in full sun is impossible, infinite relative yields might be expected. These huge LERs do not indicate if the system is profitable or not. Such scenarios relate to places where agriculture is a challenge, and they differ from mainstream agriculture.

Another limitation to the use of LERs is that a fair estimate of the LER of a system would require many years of crop monitoring. Single year values should never be extrapolated. This is even more important for perennial crops. Another example of the deceptive use of LERs in AV with perennial plants is illustrated by the work of [61] on alfalfa. These authors measured a RFYc of 0.7 and 1.4 during the two first years of the experiment. The very high RFYc during the second year was explained as the consequence of the decay of the control crop during the first year after a severe summer drought. The same pattern was again observed during the third year (Edouard, pers. com.). In that case, any farmer would have replaced the alfalfa crop in the control area, and the comparison during the following years would have been different. In that example, the AV system had the good impact of protecting the perennial crop from decaying, saving the cost of seeding it again. A fair comparison should have included reseeded alfalfa, in parts of both the control and the AV system, to elucidate what the best strategy would have been for the farmer. Alfalfa is usually cultivated for 3–5 years, and the system should be assessed for this duration.

5.6 A synthesis of crop potential yields in AV systems

The yield of crops in both AV and agroforestry (AF) systems is difficult to predict. Most countries allow AV systems on croplands only if the crop productivity is maintained (e.g., in France) or slightly reduced, as in Japan and Germany, with 80% and 66% minimum relative yield required respectively. The number of published papers with yield measurement under AV systems is increasing rapidly (see the Appendix A). The variety of systems (open air and greenhouses; with fixed or mobile panels; with opaque or semi-transparent panels; with tilting or vertical panels) make the synthesis a perilous target. When climate conditions, soil conditions and crops species and varieties are considered, each assessment is unique and making a synthesis is tricky. Dupraz [54] suggested to use the GCR of the system as an explanatory variable to predict crop yield in the shade of AV systems. The GCR is simple and defined as

$$\text{GCR} = \text{Area of solar panels/Area of the land used for the AV system} \tag{5.5}$$

This calculation assumes that the panels are optimally positioned (azimuth and tilt angle). It can be corrected to account for the transparency of panels, for the deviation of the tilting and orientation from the optimum, or for the deviation of the solar tracking when agronomical tracking is performed with mobile panels [62]. The GCR can easily be computed and controlled for all kinds of AV systems with panels that are either fixed (horizontal, tilted or vertical) or mobile (on 1 or 2 axis trackers).

Tonita *et al.* [63] have shown that at latitudes ranging from 17°N to 75°N, the efficiency of fixed-tilt arrays peaks for GCRs comprised between 50% and 70%. Detailed measurements of the irradiation available under the panels of several AV power stations have been published. For example, under an experimental site in Lavalette, Montpellier, France, annual shade was 28% under fixed panels with a 25% GCR, and 56% under fixed panels with a 49% GCR [3,64]. The shade cast by the mounting structure adds to the shade of the panels and may account for discrepancies in the results obtained on different sites using different technologies. The order of the difference between the shade ratio and the GCR is quite stable across the various experimental sites. The additional shade due to the mounting structure decreases when the GCR increases, as more parts of the structure tend to be in the shade of the panels.

A number of papers investigated both the system's averaged shade ratio and its GCR [3,15,18,51,64]. In some facilities, the design of the system is not optimal for electricity production. This is the case at the Heggelbach facility [18] where fixed panels are not facing the south, but the south-west. In that case, the GCR should be corrected to account for the sub-optimal positioning of the panels. Many authors assume that the shade level under an AV system is close to the GCR of the system [65], but most field measurements show that usually the average annual shade level is slightly higher than the GCR. The GCR appears to be a good predictor of the average relative crop productivity under AV systems (Figure 5.10).

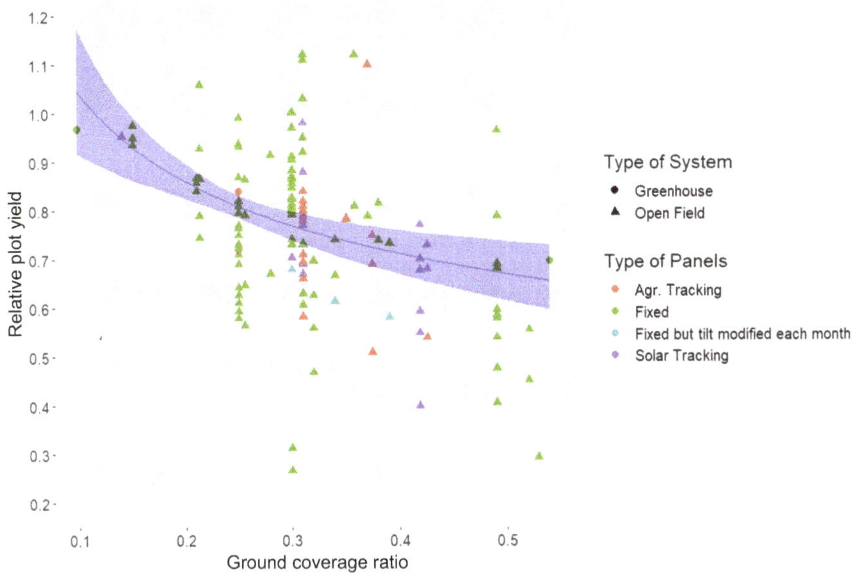

Figure 5.10 *Decrease in the relative plot yield in agrivoltaics as a function of the system's ground coverage ratio adjustment: RPY =* $aGCR^b$ *with* $a = 0.5348$ *and* $b = -0.2724$ *after [54] but modified to add more recent experimental points. As compared to [54], the GCR data from Heggelbach [42] were corrected to account for the sub-optimal orientation of the panels.*

RPYs decrease rapidly when GCRs increase and a GCR <25% is required to ensure that most crop RPYs stay >80%. Assuming an average 10% lost area rate in an AV project, a GCR <18% is necessary to ensure a RPY >89% that results in a RFY >80%. These results are consistent with a recent meta-analysis examining the impact of shade on crops [26]. Most AV projects so far have GCRs >30% and will not comply with this rule. The use of the GCR criterion to validate AV projects is a simple and cost-effective alternative to the tricky control of crop yields in the fields [54].

5.7 What crop species are recommended in AV systems?

The flexibility in the choice of crops and varieties depends on the GCR of the system.

5.7.1 Low GCR systems

The safer approach to crop choice in AV is probably the choice of a low GCR system. This allows all crops to be grown, and will permit the farmer to adapt to market issues, pest issues, labour availability issues. AV systems with low shade ratio (GCR <=10%) are suitable for any kind of crops. When the shade ratio is higher, limitations become significant, and the choice of the right crop is an issue. Several authors tried to classify crop species regarding their compatibility with AV systems.

5.7.2 High GCR systems

When the GCR >25%, significant yield reductions can be expected. For intermediate values (10%<=GCR<=25%), the prediction is highly difficult.

5.7.2.1 Classifying crop species with regard to their shade tolerance

Obergfell *et al.* [66] divided Germany's most important crops into three categories (Figure 5.11) regarding their harvestable yield under moderate shading conditions (up to 40% light reduction):

- Category *PLUS* (+) positive effects are dominant, crop yield will increase (e.g., potatoes, salads and spinach)
- Category *ZERO* (0) no significant change in crop yield (e.g., Rape, rye and oats).
- Category *MINUS* (−) negative effects dominate, crop yield will decrease (e.g., maize, wheat and horticulture).

Mamun *et al.* [67] provided another list including crops grown in Southeast Asia (Table 5.1).

Such classifications are highly controversial. For example, maize is usually considered as a light demanding species not adapted to shaded environments. But

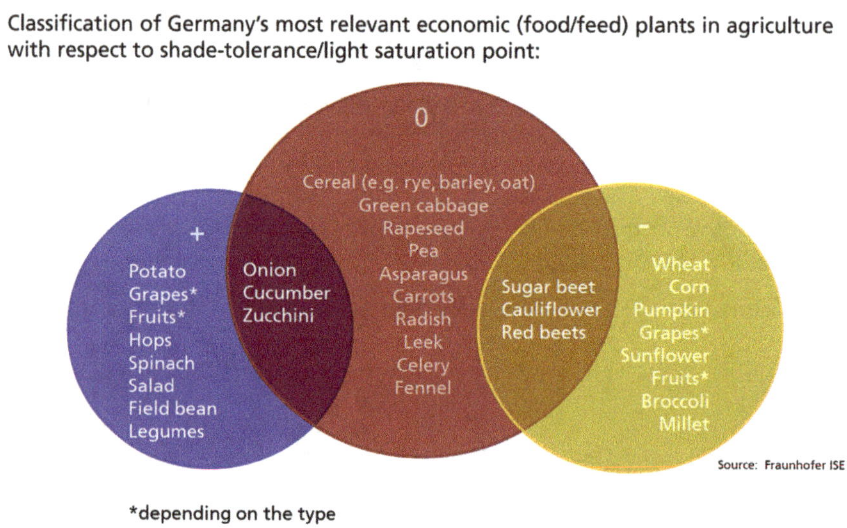

Figure 5.11 Classification of the main crops in Germany according to their assumed tolerance to shade [66]

Table 5.1 Shade-tolerant crops for potential use in agrivoltaic systems [67]

Major commercial shade-intolerant crops	Shade-tolerant-crops		
	Full shade	**Moderate light**	**Low light**
Cabbage, corn, cucumber, pumpkin, rice, tomato, turnip and watermelon	Alfalfa, arugula, Asian greens, broccoli, cassava, chard, collard greens, hog peanuts, kale, kohlrabi, lettuce, mustard greens, parsley, scallions, sorrel, spinach, sweet potatoes, taro and yam	Beans, carrots, cauliflower, coriander, green peppers and onions	Mushroom

Sekiyama and Nagashima [55] measured a very high productivity of maize in the shade of PV panels. Kiwifruits are often cited for their tolerance to shade (as the kiwi plant is originating from a forest ecosystem), but Jiang *et al.* [68] measured significant decreases in kiwifruit production under the shade of AV system that are quite similar to those reported for many crop species. Similarly, many AV systems in Japan are grown with rice, a very light demanding species.

Shade tolerance is always a site-specific behaviour that will depend on other limiting factors. If nitrogen or water are the limiting factor of growth, a ‘false’ shade tolerance will be evidenced. The GCR is an indicator of the potential relative yield, when the crop is properly managed. The use of crop models may be instrumental in determining what are the limiting factors for crop growth at a particular site. Besides, many crops have not been evaluated for AV applications. Many tropical shade-tolerant crops are often grown in AF systems such as coffee, cocoa, tea and vanilla. These have not been included in any published AV experiment so far, but some data are probably available in Japan with tea production. They would offer a new avenue for AV, given that they may tolerate shade better than temperate annual crops do. Future work is needed to test alleged shade tolerant greens including: arugula, Asian greens, chard, collard greens, kale, mustard greens, parsley, sorrel, spinach and scallions or brassicas such as broccoli, kohlrabi and cabbage. Other crops such as hog peanut, alfalfa [69] yam, taro, cassava and sweet potato [70] should also be investigated for AV applications. The shade tolerance depends on the radiation interception efficiency (RIE) of the leaves and is independent of the level of shading. Hence, when lettuce is grown under shading, it compensates for the constant RIE by increasing its leaf area to maximise its ability to tap the most of the incoming solar radiation [30]. There is currently a large dearth of information on the shade tolerance of crops and those with data are not overly promising. Ravi *et al.* [71] suggested that accumulating evidences of the behaviour of many crop species and varieties in the shade could be a useful application of citizen science. The issue is even more complicated, as varieties of the same crop species may respond differently under the same conditions.

5.7.2.2 Special systems that may accept high GCRs

When the crop is heterotrophic and does not rely on photosynthesis, shade is not a problem, and high GCRs can be accepted. This is the case mainly for mushrooms, but also for fish farming if the phytoplankton productivity is not used to feed the fishes. In oyster farming, phytoplankton productivity is essential to feed the shellfishes, but if the oyster park is open to the sea or the lagune, sea currents will bring the required food for shellfish from the surroundings, provided that the area ratio between covered waters with PV panels and uncovered water is low. In all these cases, if the sun radiation is not used for the plant or animal component production, the system is closer to a shelter building, and no longer to an AV system.

Regarding fodder productivity in grazed systems, the reduction with increasing GCR applies fully. In some grazed system, the fodder production of the plot is not required as the animals are fed with external resources. This is often the case in poultry production. In that case, if the fodder production is no longer required, the system is no longer an AV system. The panels are a shelter for the animals, and should be classified as a PV barn.

Some crops are reputed to tolerate heavy shade ratios. Red berries (raspberries, strawberries, etc.) or some tree fruits (including kiwifruit) are often cited for their success in the shade. This may be explained by the low amount of biomass harvested and exported in such crops. However, more experimental results are needed to validate this point that is controversial.

5.8 Practical constraints for growing crops and pastures in AV systems

An AV facility is not only PV panels and crops or pastures. PV panels need a superstructure to be maintained and serviced. This structure may be a hindrance to the agricultural activity on the plot, while its cost may also be a disadvantage as compared to ground-mounted PV facilities.

5.8.1 Maintaining the soil quality for cropping

Building an AV system may destroy the soil structure through compaction or pollution. Compaction should be avoided as it will reduce the crop productivity for years. It requires a careful planning of the construction site. The use of cranes to provide the structure and panels is recommended but not always possible, depending on the technology. The use of concrete to anchor the support structure is not recommended and several systems are now available to support the AV structure without any concrete. However, in some particular cases, concrete pillars will still remain indispensable, such as in flooded paddy fields, where the soil strength is very loose in saturated conditions. For rice AV, two options are available: concrete foundations if the posts are widely spaced, or a large number of posts, in order to reduce the weight on each individual post. Isolated large size trackers will always need a heavy anchoring, and concrete is the only option in that case. The reversibility of the system is often a condition for obtaining permission to build an AV facility. Removing the foundations of the system is a concern that should be addressed during the design stage of the system.

5.8.2 Obtaining a clearance that allows easy mechanisation of crops and pastures

While the clearance is essential to provide an homogenous irradiation of the underlying crops, it is also essential to provide room for tractors and machineries. Some crops do not require heavy machinery, but a change in the crop choice or crop rotation should always be considered as a possibility, even with perennial crops or pastures. Market changes may force to envisage such modifications that are more than probable during the long time life (25 to 40 years) of an AV facility. The system should be as versatile as possible, and versatility depends mostly on the support structure. A high clearance is a key aspect of versatility. Some machineries are very tall, like the grape harvesting machines for vineyards or combine harvesters for most cereal and oil crops. A 4 m clearance is a safe option for crops. With grazing animals, it is recommended to elevate the lower part of the PV panels at the height at the withers plus 1 m. This would be at least 1.5 m for sheep and goats, and 2.5 m for cattle and horses.

5.8.3 Providing the right fertilisation

In most cases, the potential yield of crops under AV systems is reduced by the shade. If standard fertilisation schemes are applied, there is a risk that part of it may be wasted. This is especially concerning with nitrogen fertilisation, as leaching and watertable pollution may result from excess fertilisation.

5.8.4 Harvesting nonhomogeneous crops?

If the AV system provides an nonhomogeneous irradiation of the crop, this may result in different maturity dates in different parts of the field. For some crops, the harvest should be done right on time and not delayed, as quality concerns may occur (e.g., for fruits), or seed dropping may reduce the yield (e.g., for rapeseed). Doing several delayed harvests on the same field is not always an option and will encumber the profit. This is why a high clearance, narrow panels, or a checkered design of the panels location are highly recommended for a successful AV system.

5.8.5 Rainfall distribution in AV systems

Detailed investigations have evidenced that the rainfall distribution at the ground level may be highly disturbed under fixed panels facilities [45,46]. The wider and the closer from the ground the panels are, the more areas will stay dry or underwatered. In very humid climates, this may be an advantage, but AV systems are mostly set up in dry areas. Irrigation may be a solution to this issue. Vertical panels or moving panels that are positioned vertically during the rainfalls provide uniform watering of the field, as well as systems with very narrow PV cells (Figure 5.4). Plants under the panels will not be exposed directly to rain which could increase dust deposition on leaves and decrease photosynthetic effectiveness. This problem is particularly detrimental in tea plantations in that it affects leaf quality [67].

5.8.6 Runoff and erosion from panels

This is usually not a concern with vertical panels or mobile panels, as the mobile panels can be maintained in the vertical position during the rain. A refinement of this management includes to position the moving panels parallel to the rain drops trajectory, taking into consideration the wind direction and speed during the rain. Very narrow fixed panels such as those used today in Japan (Figure 5.4) do not collect much water either. But wide fixed panels with no collection of the collected water may induce soil gully erosion or plant damage, especially for young plantlets (Figure 5.12).

5.8.7 Negative impacts of cropping on panels

Dust is reducing significantly the productivity of PV panels, especially in dry and desert areas [41,72]. But soil tillage is also a substantial source of dust. Cleaning the panels during dry seasons would significantly increase the electricity

Figure 5.12 Soil erosion under large mobile panels that moved during a moderate rainfall event. This may destroy plantlets and induce gully erosion, especially on sloping grounds.

production cost, and may not be easy in dry areas where water is scarce. No tillage farming is therefore recommended to avoid dust deposition on panels.

5.9 Conclusion: AV is not elevated ground-mounted PV

The ecology of crops and pastures indicate that there are no miracle crops that can grow normally with low irradiation under PV panels. Regarding light, crops have two essential needs: enough light, and an homogeneous cumulated radiation pattern at the day time step in the cropped zone. This leads to the two agronomical dilemmas of AV.

5.9.1 The PV panel density dilemma

Should we design systems with low panel coverage (e.g., GCR <20%) that always allocate enough radiation to the crops and ensure sustained average yields, or with high panel coverage (GCR >20%) that may protect the crops from extreme events, but reduce their average yield? This is a very common compromise between average and variance of yields. To solve this issue, we need longer time series of crop yield in AV systems, or validated models of PV panels/crop interaction that can predict the crop yield for long durations, including the impact of climate change. Mobile systems that allow to provide more radiation to the plants during their sensitive phases offer an elegant solution to this dilemma. With climate change, the occurrence of negative events will increase, which might lead to systems with a higher panel coverage becoming acceptable, providing that the protection against extreme events is effective.

5.9.2 The panel elevation dilemma

A second dilemma relates to the elevation of the PV panels. High clearance is required for a versatile system where the farmer can change crops easily and use the machinery without difficulty, but also for providing an homogeneous irradiation to the crops. Unfortunately, the higher the structure is, the higher the electricity production costs are. This is why AV producers are not happy with high support structures. Vertical panels aligned north-south or large 1-axis trackers that can be placed vertically are elegant solutions to this dilemma as they don't need to be raised high above the ground. Cylindrical solar cells or semi-transparent panels also provide homogeneous irradiation of crops without needing to be elevated.

However, so far mainstream AV uses mostly fixed and tilted panels. For these systems, the two dilemmas must be solved. If the average crop yields in AV systems are not attractive to make a living, crops will no longer be cultivated. This already happened in AV greenhouses with a 50% GCR and had a disastrous impact on the reputation of AV. In contrast, if the panel density is too low, the leveraged cost of electricity will be high as the cost of the elevated structure is not proportionally reduced when the panel density is reduced.

5.9.3 The need for consistent regulations

In the few AV regulations available, high and often unrealistic yields of crops in AV systems are expected. This places the farmer in a very uncomfortable position.

In Japan, the law requires an 80% minimum RFY of the crops, which is almost impossible to achieve. Consequently, new AV projects in Japan are no longer bankable. The banks refuse to finance new projects, in fear that the control of the RFY will induce a cut on the feed in tariffs for electricity that are crucial for the economic profitability of the projects. The German regulations are more pragmatic, and require a 66% relative yield in AV systems. In France, the law suggests that the yields in AV should be 'normal', which is even more difficult to achieve. To achieve 'normal' yields, very low GCRs (<20%) should be recommended. Our recommendation is that the regulations should recognise that the yields will decrease, and suggest thresholds for GCRs that are consistent with this accepted reduction. AV systems with low GCRs could be exempted from measuring crop yields on a control area, and this would simplify the control of AV systems and reduce the costs of management.

Appendix A: Publications included in the final analysis of the relationship between the GCRs and RPYs (sorted by date of publication)

Authors (sorted by year of publication)	Reference	Country	Crop	Year of experiment	GCR	Panel type and movement
Marrou (2012) Ph. D. thesis	[73]	France	Durum wheat; beans; cucumber	2010	0.25; 0.49	Fixed
Marrou *et al.* (2013b)	[64]	France	Lettuce	2010; 2011	0.49; 0.25	Fixed
Dupraz (2014, unpub.data)		France	Durum wheat	2014	0.25; 0.49	Fixed
Valle *et al.* (2017)	[51]	France	Lettuce	2015	0.25; 0.31	Fixed; ST; AT
Aroca-Delgado *et al.* (2019)	[74]	Spain	Tomato	2010–2012	0.09	Fixed
Thompson *et al.* (2020)	[10]	Italy	Basil; spinach	2016; 2019	0.43	Fixed, tinted, semi-transparent
Andrew *et al.* (2021)	[12]	USA	Grass	2019–2020	0.28	Fixed
Trommsdorff *et al.* (2021) Weselek *et al.* (2021)	[18,42]	Germany	Potato; wheat; celeriac; Clover grass	2017; 2018	0.36	Fixed
Al-agele *et al.* (2021)	[40]	USA	Tomato	2019	0.52	Fixed

(Continues)

(*Continued*)

Authors (sorted by year of publication)	Reference	Country	Crop	Year of experiment	GCR	Panel type and movement
Gonocruz *et al.* (2021)	[75]	Japan	Rice	2014–2017	0.21; 0.3; 0.39; 0.34	Fixed
Hudelson and Lieth (2020)	[76]	USA	Kale; chard; broccoli; peppers; tomato; spinach	2018	0.42	ST
Kim *et al.* (2021)	[65]	South Korea	Sesame; mung bean; red bean; maize; soybean	2020	0.21; 0.26; 0.32	Fixed
Potenza *et al.* (2022)	[77]	Italy	Soybean	2021	0.14	ST
Lee *et al.* (2022)	[78]	South Korea	Potato; sesame; soybean; rice	2021	0.25–0.3	Fixed; ST
Jiang *et al.* (2022)	[68]	China	Kiwifruit	2018–2020	0.15; 0.25; 0.31	Fixed, semi-transparent
Jo *et al.* (2022)	[79]	South Korea	Rice; rye; soybean; adzuki bean; silage maize; garlic; onion	2018–2020	0.30	Fixed
Juillion *et al.* (2022)	[37]	France	Apple	2022	0.43	ST, AT
Kumpanalaisatit *et al.* (2022)	[80]	Thailand	Bok Choi	2018	0.53	Fixed
Edouard *et al.* (2023)	[61]	France	Alfalfa	2020; 2021	0.37	Fixed; AT
Ramos-Fuentes *et al.* (2023)	[50]	France	Maize	2019–2021	0.25; 0.31; 049	Fixed; ST; AT
Prakash *et al.* (2023)	[81]	India	Wheat	2018	0.30; 0.34; 0.39	Fixed but tilt modified each month

References

[1] Vandermeer, J.H., *The Ecology of Intercropping*. Cambridge, UK: Cambridge University Press; 1989. p. 237.

[2] Goetzberger, A. and A. Zastrow, On the coexistence of solar-energy conversion and plant cultivation. *International Journal of Solar Energy*, 1982; **1**:55–69.

[3] Dupraz, C., Marrou, H., Talbot, G., *et al.*, Combining solar photovoltaic panels and food crops for optimising land use: Towards new agrivoltaic schemes. *Renewable Energy*, 2011;**36**(10):2725–2732. https://doi.org/10.1016/j.renene.2011.03.005.

[4] Nagashima, A., *Solar Sharing: Changing the World and Life*. Tokyo: Access International Ltd.; 2020, 208 pages.

[5] Martinez-Garcia, J.F. and M. Rodriguez-Concepcion, Molecular mechanisms of shade tolerance in plants. *New Phytologist*, 2023;**239**(4):1190–1202. https://doi.org/10.1111/nph.19047.

[6] Adolfo, R., Kyle, P., Azad, D., *et al.*, Agroforestry versus agrivoltaic: Spectral composition of transmitted radiation and implications for understory crops. *Agroforestry Systems*, 2023. https://doi.org/10.1007/s10457-023-00914-3.

[7] Yano, A., M. Onoe, and J. Nakata, Prototype semi-transparent photovoltaic modules for greenhouse roof applications. *Biosystems Engineering*, 2014; **122**:62–73. https://doi.org/10.1016/j.biosystemseng.2014.04.003.

[8] Liu, W., Liu, L., Guan, C., *et al.*, A novel agricultural photovoltaic system based on solar spectrum separation. *Solar Energy*, 2018;**162**:84–94. https://doi.org/10.1016/j.solener.2017.12.053.

[9] Uchanski, M., Hickey, T., Bousselot, J., and Barth, K.L., Characterization of agrivoltaic crop environment conditions using opaque and thin-film semi-transparent modules. *Energies*, 2023;**16**(7):11. https://doi.org/10.3390/en16073012.

[10] Thompson, E.P., Bombelli, E.L., Shubham, S., *et al.*, Tinted semi-transparent solar panels allow concurrent production of crops and electricity on the same cropland. *Advanced Energy Materials*, 2020;**10**(35). https://doi.org/10.1002/aenm.202001189.

[11] Lopez, G., Chopard, J., Persello, S., *et al.*, *Agrivoltaic Systems: An Innovative Technique to Protect Fruit Trees From Climate Change*. Leuven, Belgium: International Society for Horticultural Science (ISHS); 2023. https://doi.org/10.17660/ActaHortic.2023.1366.20.

[12] Andrew, A.C., Higgins, C.W., Smallman, M.A., Graham, M., and Ates, S., Herbage yield, lamb growth and foraging behavior in agrivoltaic production system. *Frontiers in Sustainable Food Systems*, 2021;**5**. https://doi.org/10.3389/fsufs.2021.659175.

[13] Pringle, A.M., R.M. Handler, and J.M. Pearce, Aquavoltaics: Synergies for dual use of water area for solar photovoltaic electricity generation and aquaculture. *Renewable and Sustainable Energy Reviews*, 2017;**80**:572–584. https://doi.org/10.1016/j.rser.2017.05.191.

[14] Hsiao, Y.-J., J.-L. Chen, and C.-T. Huang, What are the challenges and opportunities in implementing Taiwan's aquavoltaics policy? A roadmap for achieving symbiosis between small-scale aquaculture and photovoltaics. *Energy Policy*, 2021;**153**:112264. https://doi.org/10.1016/j.enpol.2021.112264.

[15] Amaducci, S., X. Yin, and M. Colauzzi, Agrivoltaic systems to optimise land use for electric energy production. *Applied Energy*, 2018;**220**:545–561. https://doi.org/10.1016/j.apenergy.2018.03.081.

[16] Tahir, Z. and N.Z. Butt, Implications of spatial-temporal shading in agrivoltaics under fixed tilt and tracking bifacial photovoltaic panels. *Renewable Energy*, 2022;**190**:167–176. https://doi.org/10.1016/j.renene.2022.03.078.

[17] Weselek, A., Ehmann, A., Zikeli, S., *et al.*, Agrophotovoltaic systems: Applications, challenges, and opportunities. A review. *Agronomy for Sustainable Development*, 2019;**39**(4):35. https://doi.org/10.1007/s13593-019-0581-3.

[18] Trommsdorff, M., Kang, J., Reise, C., *et al.*, Combining food and energy production: Design of an agrivoltaic system applied in arable and vegetable farming in Germany. *Renewable and Sustainable Energy Reviews*, 2021;**140**:110694. https://doi.org/10.1016/j.rser.2020.110694.

[19] Dupraz, C., Blitz-Frayret, C., Lecomte, I., *et al.*, Influence of latitude on the light availability for intercrops in an agroforestry alley-cropping system. *Agroforestry Systems*, 2018;**92**(4):1019–1033. https://doi.org/10.1007/s10457-018-0214-x.

[20] Campana, P.E., Stridh, B., Amaducci, S., and Colauzzi, M., Optimisation of vertically mounted agrivoltaic systems. *Journal of Cleaner Production*, 2021;**325**:129091. https://doi.org/10.1016/j.jclepro.2021.129091.

[21] Cossu, M., Murgia, L., Ledda, L., *et al.*, Solar radiation distribution inside a greenhouse with south-oriented photovoltaic roofs and effects on crop productivity. *Applied Energy*, 2014;**133**:89–100. https://doi.org/10.1016/j.apenergy.2014.07.070.

[22] Camporese, M. and M. Abou Najm, Not all light spectra were created equal: Can we harvest light for optimum food-energy co-generation? *Earth's Future*, 2022;**10**(12):e2022EF002900. https://doi.org/10.1029/2022EF002900.

[23] Pierik, R. and M. de Wit, Shade avoidance: Phytochrome signalling and other aboveground neighbour detection cues. *Journal of Experimental Botany*, 2013;**65**(11):2815–2824. https://doi.org/10.1093/jxb/ert389.

[24] Ballaré, C.L. and R. Pierik, The shade-avoidance syndrome: Multiple signals and ecological consequences. *Plant, Cell & Environment*, 2017;**40**(11):2530–2543. https://doi.org/10.1111/pce.12914.

[25] Bayala, J., J. Sanou, Z. Teklehaimanot and F. Sinclair, Adaptation of crops to partial shade in mixed cropping systems. *Tree-Crop Interactions: Agroforestry in a Changing Climate*, 2nd Edition, ed. C.K. Ong, C.R. Black, and J. Wilson. 2015, London: CABI Publishing. 309–325.

[26] Laub, M., Pataczek, L., Feuerbacher, A., Zikeli, S., and Hogy, P., Contrasting yield responses at varying levels of shade suggest different suitability of crops for dual land-use systems: A meta-analysis. *Agronomy for Sustainable Development*, 2022;**42**(3):13. https://doi.org/10.1007/s13593-022-00783-7.

[27] Aroca-Delgado, R., Pérez-Alonso, J., Callejón-Ferre, Á.J., and Velázquez-Martí, B., Compatibility between crops and solar panels: An overview from shading systems. *Sustainability*, 2018;**10**(3):743. https://doi.org/10.3390/su10030743.

[28] Miyashita, A. and M. Tateno, A novel index of leaf RGR predicts tree shade tolerance. *Functional Ecology*, 2014;**28**(6):1321–1329. https://doi.org/10.1111/1365-2435.12290.

[29] Dufour, L., Metay, A., Talbot, G., and Dupraz, C., Assessing light competition for cereal production in temperate agroforestry systems using experimentation and crop modelling. *Journal of Agronomy and Crop Science*, 2013;**199**(3):217–227. https://doi.org/10.1111/jac.12008.

[30] Grieco, M., Roustan, V., Dermendjiev, G., *et al.*, Adjustment of photosynthetic activity to drought and fluctuating light in wheat. *Plant Cell and Environment*, 2020;**43**(6):1484–1500. https://doi.org/10.1111/pce.13756.

[31] Tiffon-Terrade, B., *Analyse et modélisation de l'effet d'un ombrage intermittent sur le développement de la vigne, la maturation et la qualité du raisin*. Montpellier; 2022. http://www.theses.fr/2022NSAM0018/document.

[32] Juillion, P., Fumey, D., Génard, M., Lesniak, V., Lopez, G., and Vercambre, G., *Specific Leaf Area and Photosynthesis of Apple Trees Under a Dynamic Agrivoltaic System in Agrivoltaics2023*. Daegu, South Korea; 2023.

[33] Walston, L.J., Barley, T., Bhandari, I., *et al.*, Opportunities for agrivoltaic systems to achieve synergistic food-energy-environmental needs and address sustainability goals. *Frontiers in Sustainable Food Systems*, 2022. **6:932018**. https://doi.org/10.3389/fsufs.2022.932018.

[34] Tiffon-Terrade, B., Simonneau, T., Caffarra, A., *et al.*, Delayed grape ripening by intermittent shading to counter global warming depends on carry-over effects and water deficit conditions. *OENO One*, 2023;**57**(1):71–90. https://doi.org/10.20870/oeno-one.2023.57.1.5521.

[35] Ferrara G., Maurizio, B., Palasciano, M. and Mazzeo, A., Effect of shading determined by photovoltaic panels installed above the vines on the performance of cv. Corvina (Vitis vinifera L.). *Scientia Horticulturae* 2023. **308**. https://doi.org/10.1016/j.scienta.2022.111595.

[36] Cho, J., Park, S.M., Park, A.R., Lee, O.C., Nam, G., and Ra, I.-H., Application of photovoltaic systems for agriculture: A study on the relationship between power generation and farming for the improvement of photovoltaic applications in agriculture. *Energies*, 2020;**13**(18):4815. https://doi.org/10.3390/en13184815.

[37] Juillion, P., Lopez, G., Fumey, D., Lesniak, V., Génard, M., and Vercambre, G., Shading apple trees with an agrivoltaic system: Impact on water relations, leaf morphophysiological characteristics and yield determinants. *Scientia Horticulturae*, 2022;**306**:111434. https://doi.org/10.1016/j.scienta.2022.111434.

[38] Marrou, H., Guilioni, L., Dufour, L., Dupraz, C., and Wery, J., Microclimate under agrivoltaic systems: Is crop growth rate affected in the partial shade of solar panels? *Agricultural and Forest Meteorology*, 2013;**177**(0):117–132. http://dx.doi.org/10.1016/j.agrformet.2013.04.012.

[39] Noor, N.F.M. and A.A. Reeza, Effects of solar photovoltaic installation on microclimate and soil properties in UiTM 50MWac Solar Park, Malaysia. *IOP Conference Series: Earth and Environmental Science*, 2022;**1059**(1):012031. https://doi.org/10.1088/1755-1315/1059/1/012031.

[40] Al-agele, H.A., Proctor, K., Murthy, G., and Higgins, C., A case study of tomato (*Solanum lycopersicon* var. Legend) production and water productivity in agrivoltaic systems. *Sustainability*, 2021;**13**(5):2850. https://doi.org/10.3390/su13052850.

[41] Yue, S., Guo, M., Zou, P., Wu, W., and Zhou, X., Effects of photovoltaic panels on soil temperature and moisture in desert areas. *Environmental Science and Pollution Research*, 2021;**28**(14):17506–17518. https://doi.org/10.1007/s11356-020-11742-8.

[42] Weselek, A., Bauerle, A., Hartung, J., and Zikeli, S., Agrivoltaic system impacts on microclimate and yield of different crops within an organic crop rotation in a temperate climate. *Agronomy for Sustainable Development*, 2021;**41**(5):15. https://doi.org/10.1007/s13593-021-00714-y.

[43] Armstrong, A., N.J. Ostle, and J. Whitaker, Solar park microclimate and vegetation management effects on grassland carbon cycling. *Environmental Research Letters*, 2016;**11**(7):074016. https://doi.org/10.1088/1748-9326/11/7/074016.

[44] Hassanpour Adeh, E., J.S. Selker, and C.W. Higgins, Remarkable agrivoltaic influence on soil moisture, micrometeorology and water-use efficiency. *PLOS ONE*, 2018;**13**(11):e0203256. https://doi.org/10.1371/journal.pone.0203256.

[45] Elamri, Y., Cheviron, B., Lopez, J.M., Dejean, C., and Belaud, G., Water budget and crop modelling for agrivoltaic systems: Application to irrigated lettuces. *Agricultural Water Management*, 2018;**208**:440–453. https://doi.org/10.1016/j.agwat.2018.07.001.

[46] Marrou, H., L. Dufour, and J. Wery, How does a shelter of solar panels influence water flows in a soil–crop system? *European Journal of Agronomy*, 2013;**50**:38–51. https://doi.org/10.1016/j.eja.2013.05.004.

[47] Barron-Gafford, G.A., Pavao-Zuckerman, M.A., Minor, R.L., *et al.*, Agrivoltaics provide mutual benefits across the food–energy–water nexus in drylands. *Nature Sustainability*, 2019;**2**(9):848–855. https://doi.org/10.1038/s41893-019-0364-5.

[48] Reyes, F., Gosme, M., Wolz, K.J., *et al.*, Alley cropping mitigates the impacts of climate change on a wheat crop in a mediterranean environment: A biophysical model-based assessment. *Agriculture, Special Issue Impact of Climate Change on Agriculture*, 2021;**11**(356):19. https://doi.org/10.3390/agriculture11040356.

[49] Dupraz, C., Wolz, K.J., Lecomte, I., *et al.*, Hi-sAFe: A 3D agroforestry model for integrating dynamic tree–crop interactions. *Sustainability*, 2019;**11**(8):2293. https://doi.org/10.3390/su11082293.

[50] Ramos-Fuentes, I.A., Elamri, Y., Cheviron, B., Dejean, C., Belaud, G., and Fumey, D., Effects of shade and deficit irrigation on maize growth and

development in fixed and dynamic agrivoltaic systems. *Agricultural Water Management*, 2023;**280**:108187. https://doi.org/10.1016/j.agwat.2023.108187.

[51] Valle, B., Simonneau, T., Sourd, F., *et al.*, Increasing the total productivity of a land by combining mobile photovoltaic panels and food crops. *Applied Energy*, 2017;**206**(Supplement C):1495–1507. https://doi.org/10.1016/j.apenergy.2017.09.113.

[52] Vandermeer, J., van Noordwijk, M., Anderson, J., Ong, C., and Perfecto, I., Global change and multi-species agroecosystems: Concepts and issues. *Agriculture, Ecosystems & Environment*, 1998;**67**(1):1–22. https://doi.org/10.1016/S0167-8809(97)00150-3

[53] Macdonald, J., T. Reher, and B. Van de Poel. Comparing different agrivoltaics systems and protective film covers: which holds the edge? In *AgriVoltaics2023*. Daegu, South Korea: TIB Open Publishing; 2023.

[54] Dupraz, C., Assessment of the ground coverage ratio of AgriVoltaic systems as a proxy for potential crop productivity. *Agroforestry Systems*, 2023. https://doi.org/10.1007/s10457-023-00906-3.

[55] Sekiyama, T. and A. Nagashima, Solar sharing for both food and clean energy production: Performance of agrivoltaic systems for corn, A typical shade-intolerant crop. *Environments*, 2019;**6**(6):12. https://doi.org/10.3390/environments6060065.

[56] Randle-Boggis, R.J., Mbele, A., Maro, J., Kimaro, A., and Hartley, S.E., Food-energy-water results from raised-system agrivoltaics at sustainable agriculture Tanzania. In *Agrivoltaics2023*. Daegu, South Korea; 2023.

[57] Dupraz, C., Adequate design of control treatments in long term agroforestry experiments with multiple objectives. *Agroforestry Systems*, 1998;**43**(1):35–48. https://doi.org/10.1023/A:1026495002991.

[58] Dal Prà, A., Genesio, L., Miglietta, F., *et al.*, Salad yields under agrivoltaics: A field test. In *Agrivoltaics2023*. Daegu, South Korea; 2023.

[59] Madej, L., Michaud, L., Bouhier de L'Ecluse, C., *et al.*, *Synthèse de la dynamique végétale sous l'influence de panneaux photovoltaïques et du pâturage sur deux sites prairiaux pâturés. Etude des effets sur une période annuelle*. INRAE; 2022, p. 35.

[60] Dupraz, C., L. Dufour, and G. Talbot. A field assessment of the actual land equivalent ratio of a temperate agroforestry system. In *Second World Congress of Agroforestry*. Nairobi, Kenya; 2009.

[61] Edouard, S., Combes, D., Van Iseghem, M., Ng Wing Tin, M., and Escobar-Gutiérrez, A.J., Increasing land productivity with agriphotovoltaics: Application to an alfalfa field. *Applied Energy*, 2023;**329**:120207. https://doi.org/10.1016/j.apenergy.2022.120207.

[62] Dupraz, C., Utilisation du taux de couverture pour qualifier et encadrer les projets agrivoltaiques. 2023:6. https://www.dropbox.com/scl/fi/174od9i-eamhxcwxt3gbsh/Utilisation-du-taux-de-couverture-pour-qualifier-et-encadrer-les-projets-agrivoltaiques-VF.docx?rlkey=liw9jquktyf9vmz3isdfl0csn&dl=0.

[63] Tonita, E.M., Russell, A.C.J., Valdivia, C.E., and Hinzer, K., Optimal ground coverage ratios for tracked, fixed-tilt, and vertical photovoltaic

systems for latitudes up to 75°N. *Solar Energy*, 2023;**258**:8–15. https://doi.org/10.1016/j.solener.2023.04.038.

[64] Marrou, H., Wery, J., Dufour, L., and Dupraz, C., Productivity and radiation use efficiency of lettuces grown in the partial shade of photovoltaic panels. *European Journal of Agronomy*, 2013;**44**:54–66. https://doi.org/10.1016/j.eja.2012.08.003.

[65] Kim, S., S. Kim, and C.-Y. Yoon, An efficient structure of an agrophotovoltaic system in a temperate climate region. *Agronomy*, 2021;**11**(8):1584. https://doi.org/10.3390/agronomy11081584.

[66] Obergfell, T., Bopp, G., Goetzberger, A., Reise, C., and Schindele, S., Agrophotovoltaik – landwirtschaft unter Photovoltaikanlagen: ressourceneffiziente Flächennutzung als Alternative zu Monokulturen. In *Ecotrinova Samstags-Forum*. Freiburg; 2013, p. 17.

[67] Mamun, M.A.A., Dargusch, P., Wadley, D., Zulkarnain, N.A., and Aziz, A.A., A review of research on agrivoltaic systems. *Renewable and Sustainable Energy Reviews*, 2022;**161**:112351. https://doi.org/10.1016/j.rser.2022.112351.

[68] Jiang, S., Tang, D., Zhao, L., *et al.*, Effects of different photovoltaic shading levels on kiwifruit growth, yield and water productivity under 'agrivoltaic' system in Southwest China. *Agricultural Water Management*, 2022; **269**:107675. https://doi.org/10.1016/j.agwat.2022.107675.

[69] Lin, C.H., McGraw, R.L., George, M.F., and Garrett, H.E., Shade effects on forage crops with potential in temperate agroforestry practices. *Agroforestry Systems*, 1998;**44**:109–119. https://doi.org/10.1023/a:1006205116354.

[70] Johnston, M. and I.C. Onwueme, Effect of shade on photosynthetic pigments in the tropical root crops: Yam, taro, tannia, cassava and sweet potato. *Experimental Agriculture*, 1998;**34**(3):301–312. https://doi.org/10.1017/S0014479798343033.

[71] Dinesh, H. and J.M. Pearce, The potential of agrivoltaic systems. *Renewable & Sustainable Energy Reviews*, 2016;**54**:299–308. https://doi.org/10.1016/j.rser.2015.10.024.

[72] Ravi, S., D.B. Lobell, and C.B. Field, Tradeoffs and synergies between biofuel production and large solar infrastructure in deserts. *Environmental Science & Technology*, 2014;**48**(5):3021–3030. https://doi.org/10.1021/es404950n.

[73] Marrou, H., *Conception and Optimisation of Agrivoltaic Systems* (Ph.D. Thesis). Montpellier: Montpellier SupAgro; 2012.

[74] Aroca-Delgado, R., Pérez-Alonso, J., Callejón-Ferre, Á.-J., and Díaz-Pérez, M., Morphology, yield and quality of greenhouse tomato cultivation with flexible photovoltaic rooftop panels (Almería-Spain). *Scientia Horticulturae*, 2019;**257**:108768. https://doi.org/10.1016/j.scienta.2019.108768.

[75] Gonocruz, R.A., Nakamura, R., Yoshino, K., *et al.*, Analysis of the rice yield under an agrivoltaic system: A case study in Japan. *Environments*, 2021;**8**(7):18. https://doi.org/10.3390/environments8070065.

[76] Hudelson, T. and J.H. Lieth. Crop production in partial shade of solar photovoltaic panels on trackers. In *AGRIVOLTAICS2020: Launching Agrivoltaics*

World-wide. Perpignan, France, Online: AIP Conference Proceedings; 2020. https://doi.org/10.1063/5.0055174.

[77] Potenza, E., Croci, M., Colauzzi, M., and Amaducci, S., Agrivoltaic system and modelling simulation: A case study of soybean (*Glycine max* L.) in Italy. *Horticulturae*, 2022;**8**(12):1160. https://doi.org/10.3390/horticulturae8121160.

[78] Lee, H.J., Park, H.H., Kim, Y.O., and Kuk, Y.I., Crop cultivation underneath agro-photovoltaic systems and its effects on crop growth, yield, and photosynthetic efficiency. *Agronomy*, 2022;**12**(8):1842. https://doi.org/10.3390/agronomy12081842.

[79] Jo, H., Asekova, S., Bayat, M.A., *et al.*, Comparison of yield and yield components of several crops grown under agro-photovoltaic system in Korea. *Agriculture*, 2022;**12**(5):619. https://doi.org/10.3390/agriculture12050619.

[80] Kumpanalaisatit, M., Setthapun, W., Sintuya, H., and Jansri, S.N., Efficiency improvement of ground-mounted solar power generation in agrivoltaic system by cultivation of Bok Choy (*Brassica rapa* subsp. chinensis L.) under the panels. *International Journal of Renewable Energy Development*, 2022;**11**(1):103–110. https://doi.org/10.14710/ijred.2022.41116.

[81] Prakash, V., Lunagaria, M.M., Trivedi, A.P., et al., Shading and PAR under different density agrivoltaic systems, their simulation and effect on wheat productivity. *European Journal of Agronomy*, 2023. **149.126922**. https://doi.org/10.1016/j.eja.2023.126922.

Chapter 6

Crop and power yield modelling

Brecht Willockx[1,2] *and Jan Cappelle*[1]

The goal of yield modelling is to determine and quantify the most optimal design and operating conditions for the system, taking into account factors such as land use, shading, irrigation and weather patterns. The models can be based on simulations, field and satellite measurements or a combination of all. The results of the yield modelling help to evaluate the feasibility, profitability and sustainability of the agrivoltaic system, and inform decision-making regarding its optimal design, implementation and management [1].

6.1 The different parts of agrivoltaics yield modelling

Yield modelling in agrivoltaic systems refers in the first place to the calculation and prediction of the amount of energy (electricity) and crops (biomass and quality) produced by a combined agriculture and PV energy production system [2] and consist of following main model blocks shown in Figure 6.1.

- Microclimate analysis: involves studying the local climate conditions that affect plant growth and development within the system. This includes factors such as solar radiation, air temperature, humidity, wind speed and soil moisture levels, which can all be affected by the presence of the solar panels.
- Crop production: This involves the estimation of the amount of crops and its quality parameters that can be produced on the same land, taking into account factors such as plant species, shading, irrigation and management.
- Generated electricity: taking into account factors such as PV module orientation, tilt, shading, albedo and weather conditions.

Besides these main effects, which are considered as the key parameters [3], other important interactions based on the initial crop and energy impact can be

[1]Strategic Department, Storm, Borsbeeksebrug 22, 2600 Antwerpen, Belgium
[2]ELECTA Gent, Faculty of Technology Engineering, KU Leuven, Gebroeders De Smetstraat 1, 9000 Gent, Belgium

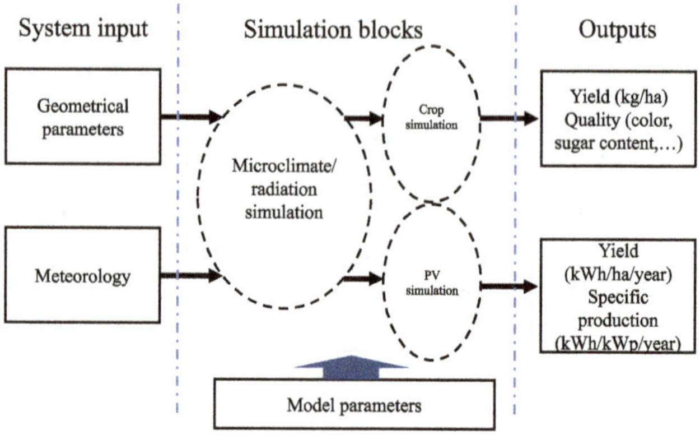

Figure 6.1 The flow of agrivoltaics yield modelling. The inputs are the geometrical parameters and meteorology, and by use of combined radiation, crop and electricity production, the output yields are calculated.

considered in the decision-making and are therefore handled in detail in other chapters of this book:

- Water analysis [4]: This involves the estimation of the water requirements of the crops and the impact of the photovoltaic modules on the precipitation distribution, irrigation need and water stress.
- Financial analysis [5]: This involves the calculation of the costs and benefits of the agrivoltaic system, including the costs of construction, maintenance, operation and the revenues from the sale of electricity and crops.
- Environmental analysis [6]: This involves the evaluation of the environmental impacts of the agrivoltaic system, including the effects on land use (land equivalent ratio), biodiversity, water usage and greenhouse gas emissions (carbon sequestration, . . .).
- Landscape analysis: The calculation of the visual impact of the agrivoltaic system and the evaluation of landscape compatibility and spatial planning.

6.2 Input parameters

6.2.1 System design: geometrical parameters

The performance of agrivoltaic systems is affected by the interaction between the crops and the PV system, and can be influenced by the configuration of the height, orientation, tilt angle, row spacing (ground coverage ratio), panel spacing, panel transparency level and structural elements of the components within the system [7]. To accurately represent the shading and energy production level in agrivoltaic systems, it is important to consider these geometrical parameters in the simulation and design of the system as shown in Figure 6.2. In most cases, the shading

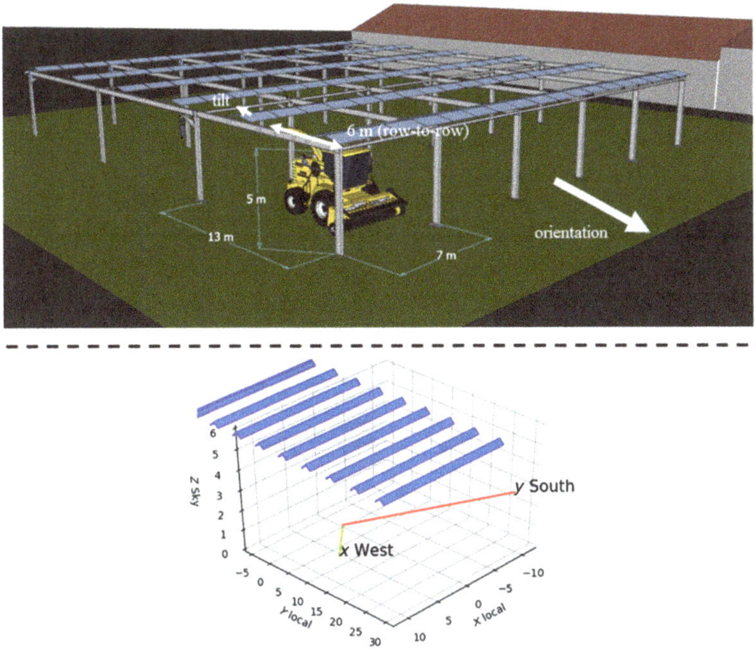

Figure 6.2 The geometrical parameters of agrivoltaic systems: the height, row-to-row, tilt, orientation, define the impact on the crop and energy yield. In most cases, the structural elements are not considered as shading objects.

calculation is based solely on the size and location of the PV modules themselves, and not the structural elements of the construction, which can underestimate the total amount of shading that occurs. Of all geometrical parameters, the Ground Coverage Ratio (GCR) is one of the most impacting variables in an agrivoltaic design. The GCR is defined as the ratio between the surface area of the PV modules and the cultivated ground surface, thereby impacting directly the light sharing between the solar modules and the crops.

6.2.2 Meteorology inputs

In addition to geometrical parameters, another influential input parameter is the meteorology (solar radiation, wind speed, dry bulb air temperature, air pressure, …). Annual PV energy and biomass production can be misinterpreted in case of wrong data [8]. Nevertheless, it is not clear which weather source is the most representative in reality, and no one can predict the weather for the next 20 years. In PV simulation, Typical Meteorological Year (TMY) weather data is often used [9]. This is a set of meteorological data with data values for every hour in a year for a given geographical location. The data are selected from hourly data in a longer time period (normally 10 years or more). For each month in the year the data has been selected from the year that was considered most 'typical' for that month to provide a good representation of the average weather conditions for a particular location. However, TMY data may not capture the inter-annual variability of weather patterns, which can have a significant impact on crop yields, as weather events (droughts, heatwaves, rain, ...) throughout different growing stages (from seeding to harvest) can impact the final yield [8]. To account for inter-annual variability, some models use multi-year weather data, such as 10-year or 30-year weather records from the NASA Prediction Of Worldwide Energy Resources (POWER) Project [10] or the National Solar Radiation Database (NSRDB) [11], to generate weather scenarios for multiple years (based on satellite and weather station data). This can help to capture the range of weather conditions that a crop may experience over a multi-year period, and improve the accuracy of crop yield predictions such as the study of Amaducci [12].

6.3 Microclimatic/radiation analysis

The presence of PV modules in an agrivoltaic system can have several effects on the microclimate of the area [13]. These effects include changes in precipitation [14], wind [15], temperature [16], humidity [17] and (spectral composition of) the available solar radiation [16].

In both overhead and interspace agrivoltaic systems, PV modules can act as a shelter and impact the amount of rain and (direct/diffuse) sunlight that reaches the plants and soil. The reduction in solar radiation on the crops can then lead to lower temperatures in the area below (overhead) or next to the panels (interspace) as well as reduced evapotranspiration rates, which can result in impacted humidity levels [4]. In addition, the presence of PV modules can alter wind patterns in the area. Especially in interspace systems, the modules can act as a barrier to wind, reducing wind speeds and changing the direction of the prevailing wind, preventing crop wind erosion losses [15]. Finally, an aspect that is less studied is the spectral composition of the solar radiation. The photosynthetically active radiation (PAR) and R:FR ratio can be altered due to the presence of modules that absorb and reflect different wavelengths of light [18]. When PV modules are installed, they create a shaded environment that reduces the amount of direct and diffuse red light (wavelengths around 600–700 nm) reaching the plants, while increasing the amount of far-red light (wavelengths around 700–800 nm) that is reflected by (unshaded) neighbouring plants.

This shift in the R:FR ratio can have several effects on plant growth and development as shown in Figure 6.3. One of the most significant effects is the stimulation of stem elongation (also known as shade avoidance) or the stimulation of leaf growth, known as shade adaptation [19]. This can be beneficial for leafy crops such as lettuce [20] or spinach but it can also not be desirable for other crop types resulting in taller and thinner plants.

6.3.1 *The solar radiation as primary element*

The primary changing parameter of the microclimate is considered solar radiation [16], affecting the amount of available photosynthetic active radiation, leaf

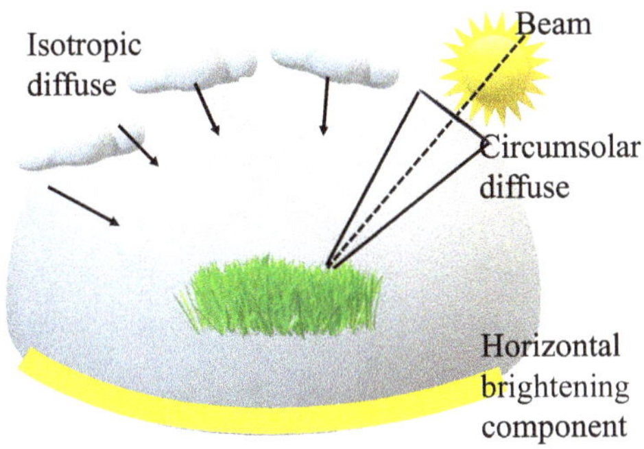

Figure 6.3 Components of the diffuse irradiance following the Perez model, including the isotropically diffuse, circumsolar and horizontal brightening components

temperature and evapotranspiration, and R:FR ratio and is therefore the key element that is discussed in this chapter.

In order to estimate the amount of solar radiation below or next to the PV modules, the agricultural area is divided in different small (cm scale) elements, called the evaluation area. Next, the amount of solar radiation for each element is estimated, composed out of a direct, diffuse and reflected component. The direct component is the light that comes directly from the sun and is not scattered by the atmosphere. It is characterised by high intensity and a high proportion of ultraviolet and infrared radiation. Diffuse sunlight, on the other hand, is the light that is scattered by the atmosphere and is therefore less intense but harder to estimate. In most cases, the anisotropic sky model from Perez [21] is used, assuming the sky as a hemisphere and separating the diffuse light into an isotropic, circumsolar and horizon components explicitly, visualised in Figure 6.3. Finally, the reflected component consists out of the direct and diffuse sun component reflected between the crops and agrivoltaic system.

6.3.2 The calculation of the shading factors

As shown in Figure 6.4, the evaluation surface is meshed in small elements. For each timestep, during the simulation and each element of the evaluation surface, the direct shading factor (a value between zero and one) is calculated [7]. This factor refers to the fact that the direct sunlight is blocked by the PV modules or not and depends on the geometrical lay-out/system design of the agrivoltaic system, the transparency level as well as the position of the sun in the sky.

The diffuse shading factor [7], also unique for each element of the evaluation surface in Figure 6.4 refers to the amount of diffuse sunlight that is available to the crops and is not blocked by the PV modules. Diffuse shading is more complex to calculate than direct shading. In the most simple model, it assumes that all objects are a Lambertian surface, which means that it reflects light equally in all directions.

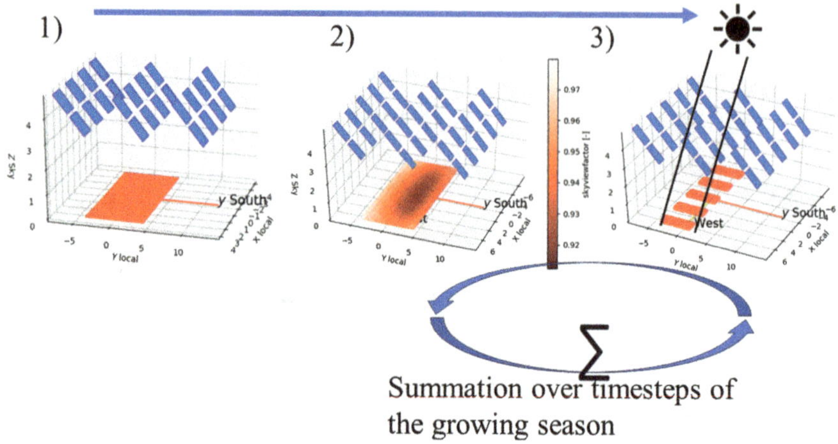

Figure 6.4 Steps in the simulation of the radiation maps. In 1) the evaluation surface below the PV module is meshed in small areas; In 2) the sky view factors are calculated and; In 3) the direct shade factors are calculated depending on the position of the PV modules and sun. Step 2 and step 3 are repeated and accumulated for each timestep, with varying sun position (and direct shade factors) and changing direct and diffuse solar irradiance.

It also assumes that the medium through which the sunlight passes has no effect (scattering, emission and absorption) on the transfer of radiation between surfaces and that secondary (and higher) reflections are negligible. In that case, the 'view factor' or 'configuration factor' between two surfaces (i.e. the sky that is not blocked by the PV modules and the element of the evaluation surface) can be calculated by a mathematical expression [22]. For the reflected diffuse and direct component between the crops and the agrivoltaic system, the same assumptions and view factor model can be applied.

Once the diffuse, direct shading and reflected radiation are estimated for each element of the evaluation surface and this accumulated over the growing period of the crops, the radiation maps in Figure 6.5 can be created. Based on these radiation maps, new indicators can be calculated such as [23]:

- The minimum, maximum and average solar radiation level
- The coefficient of variation, a measure to show the dispersion of the solar radiation

Both an interspace design Figure 6.5a and overhead design Figure 6.5b (low density) and Figure 6.5c (high density) are simulated, showing that a larger cultivation area can be impacted by the shade of an overhead design compared to an interspace design.

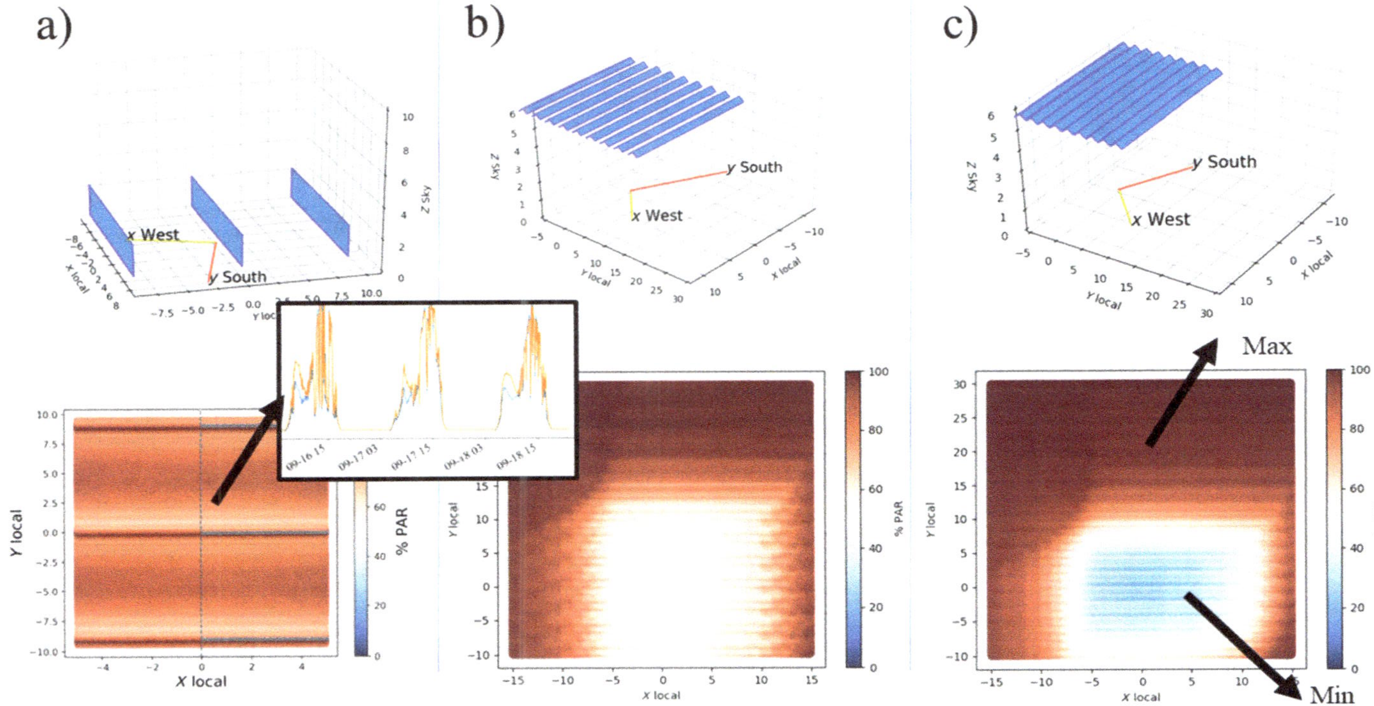

Figure 6.5 *Radiation maps created by a view factor model. Three different Agri-PV designs are simulated in a) (vertical interspace system), b) (overhead low density) and c) (overhead high density). In each situation, the minimum, maximum and average solar radiation can be calculated. Note that these maps show the average solar radiation over the growing season, and that also the temporal impact of the shading moments (shown in a), must be evaluated.*

6.3.3 Sophistication levels in radiation/microclimate simulation

In the simple view factor model, all surfaces are considered as Lambertian, which is a huge simplification. Another model, better representing the reality, is ray tracing [22]. This is a numerical method used to simulate the propagation of light through a material. It involves tracing thousands of individual rays of light through the material and calculating the amount of energy absorbed, transmitted and reflected at each point. Ray tracing can take into account the complex geometries of objects and the interactions between light and the object (forward ray-tracing) or interactions between object and the light (backward ray-tracing) between different wavelengths and angles. The resulting simulation is more accurate but also requires more computational resources. Ray tracing is particular interesting in detailed engineering questions such as the design new PV modules, for example to know the impact of the cell lay-out of silicon semi-transparent modules [24].

6.4 Crop yield modelling

Crop yield modelling and forecasting is a common and well-established practice in agriculture. Farmers routinely use these models to predict and estimate their crop yields, enabling them to make informed decisions about selling their produce, organising logistics and securing favourable prices even before the actual harvest begins.

They are tried-and-true practices that have long been integral to agriculture. In the context of agrivoltaics, the existing tools need not be reinvented but rather adapted to suit the unique microclimate conditions, environmental factors and practical management. First of all, the specific microclimate of agrivoltaic systems can have both positive and negative effects on crop growth [25]. On the one hand, shading can reduce the intensity of sunlight and lower the temperature, which can be beneficial for crops that are sensitive to high temperatures or sunlight, such as lettuce or spinach. Shading and reduced air flows can also reduce water loss from the crops, helping to maintain soil moisture levels during periods of drought. On the other hand, shading can also reduce the overall amount of sunlight available for photosynthesis, which can reduce the growth and yield of some crops, such as tomatoes or peppers. Finally, the reduced air flow and increasing humidity can lead to increased disease pressure and pest problems. Crop modelling involves predicting the growth, development and yield of crops under different environmental and shading conditions. The process of crop modelling typically involves three high-level steps: light interception, light conversion and partitioning [26], which dynamically evolves through the different grow stages.

- Light interception: The first step in crop modelling is to estimate how much sunlight (i.e. solar radiation) is intercepted by the crop canopy. This is typically done using mathematical models that take into account the angle of the sun, the location of the crop and the characteristics of the crop canopy (e.g. leaf area, orientation and density). The amount of intercepted light is an important factor that determines the amount of energy available for crop growth and development.

- Light conversion: The second step in crop modelling is to estimate how much of the intercepted light is converted into chemical energy through photosynthesis. This involves modelling the complex biochemical and physiological processes that occur within the plant, including the absorption of light by photosynthetic pigments (e.g. chlorophyll), the conversion of carbon dioxide and water into carbohydrate and respiration. The efficiency of photosynthesis is influenced by a variety of factors, including temperature, humidity, water availability, nutrient availability and plant genetics.
- Partitioning: The third step in crop modelling is to estimate how the energy produced through photosynthesis is partitioned among different plant organs and used for growth and development. This involves modelling the processes of biomass accumulation, partitioning of biomass among different plant organs, and the development of reproductive structures (e.g. flowers and seeds). The partitioning of resources is influenced by factors such as plant genetics, nutrient availability, water availability and environmental stresses (e.g. drought or heat stress).

There is a misunderstanding that the Light Conversion (mostly represented by the light response curve) is the only indicator to define the suitable of a specific crop type or variety [18,27]. The light response curve shows how the rate of photosynthesis (the process by which plants convert light into energy) changes as the intensity of absorbed light increases. However, designing an agrivoltaic system is not only optimising the light conditions for plant growth alone (the light response curve), also the crop variety choice and possible crop's improved Light Interception or Partitioning to the harvestable organs (e.g. not partitioning biomass to larger leaves but to the potatoes) must be considered during the system's design.

6.4.1 Empirical crop models

The first method to quantify the impact of shading on the crop yield is the use of empirical 'Shadow Response Curves' [28,29]. Shadow response curves show the relationship between the average percentage of shading and the corresponding decrease in crop yield as shown in Figure 6.6. This means that the shadow response curve average the multiplication of the light interception efficiency, light conversion and partition over the growing season. This implies shadow response curves are not able to consider temporal effects by the dynamics of the shade and consequently location and system specific. In addition, other non-linear factors such as irrigation, temperature and soil fertility are not included and may also have a significant impact on crop growth and yield. As a consequence, results from a shadow response curve can act as an initial pre-design or investment decision but should carefully be interpreted and always used in combination with other methods and data, such as field measurements, system configuration and weather information, to provide a more complete and accurate picture.

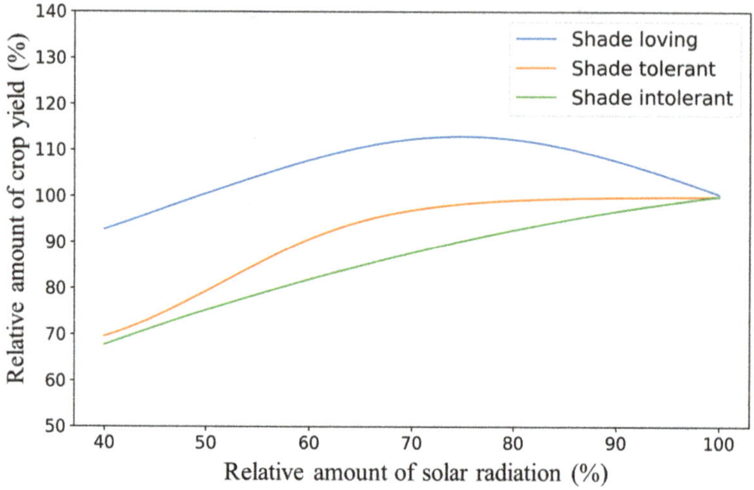

Figure 6.6 A graphical representation of shade response curves (SRC) for shade benefit, tolerance and susceptibility crop types. The shade response curve shows the relation between the average reduction in solar radiation over the growing period and the obtained crop yield [28].

6.4.2 Mechanistic crop models

In contrast to the empirical shadow response curves, a mechanistic crop model is a mathematical complex model that describes light interception, light conversion and partitioning based on physiological and physical processes that occur in a crop, in response to environmental conditions such as climate, soil properties, management practices, and pests and diseases on a much higher spatial (cm) and temporal (daily) scale [30]. These models are typically used for crop yield forecasting and management decision-making, such as optimising irrigation and fertilisation schedules or predicting the effects of climate change on crop yields [30] and run on daily or hourly level to also include the impact of inter-annual events during the growing season such as heat waves and droughts.

The disadvantage of this model is the use of many input variables that must be calibrated based on site-specific conditions [31]. Most models are able to consider the partial shading of the solar modules by including the dynamics in photosynthetically active radiation (for photosynthesis) and water evapotranspiration on a minute or hourly level [12] if they are linked to the microclimatic/radiation model.

Nevertheless, as most crop models consider full light conditions, the impact of the changed R:FR ratio (which is constant in full light conditions) and dynamic shading are not included. This results in the need for incorporating phenological and morphological adaptations into the simulation to accurately predict the crop yield as the dynamic shading can impact the harvestable yield (more biomass partitioned to stems and leaves), water use efficiency (stomata opening and closing)

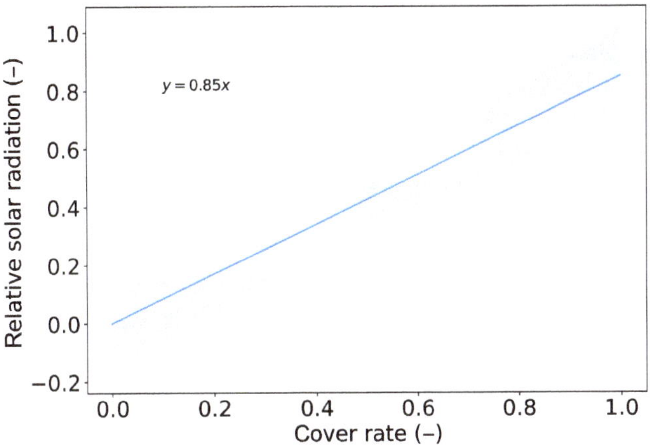

Figure 6.7 Illustration of (linear) increase in cover rate of lettuce (proxy for projected leaf area), with respect to the relative solar amount [20]. These important relations that impact the light interception and the allocation to harvestable organs are not yet considered in the standard mechanistic crop models.

and light interception (specific leaf area). The data on morphological adaptations can be included into the simulation by developing equations or algorithms that describe the changes in morphological traits as a function of the growing conditions. For example, equations can be developed that describe how changes in shading or light intensity affect the height and width of the crops (light interception) and partitioning between the different organs. An example is given by the experimental setup of Marrou [21] where the leave cover rate of lettuce is clearly related to the relative PAR amount as shown in Figure 6.7.

6.4.3 Practical management and aspects

Practical management in agriculture holds a profound influence on crop yield, and this impact becomes even more pronounced when we consider the specific challenges faced by agrivoltaic systems. The most common critical aspects of practical management that significantly affect agricultural yield are:

- Soil compaction: The use of heavy machinery during installation and maintenance can lead to soil compaction. This compaction hinders the healthy development of crop roots by reducing the space and aeration necessary for root growth. As a result, crops struggle to access essential nutrients and moisture, leading to diminished yields. To mitigate this, careful planning and the use of appropriate machinery can help minimise soil compaction and promote optimal root development.
- Land loss: Machinery requirements and the need for mandatory spacing to prevent collisions with the structure can result in significant land loss in

interspace agricultural systems. This land is essentially rendered unproductive for crop cultivation. This inefficiency can lead to reduced overall yield potential as less land is available for planting. Effective management strategies should aim to maximise land utilisation while ensuring the safety of agricultural equipment.

- Weed growth: In agrivoltaic systems on arable land, there is often a greater likelihood of weed growth around structures. Weeds not only compete with crops for resources but also create shading effects, which can reduce the amount of sunlight reaching the crops. Managing weed growth through manual or mechanical means is essential to mitigate these effects.

It is important to note that these challenges are typically more pronounced in interspace systems on arable land as compared to overhead systems with permanent crops like fruits. In the latter, the permanence of fruit crops reduces the need for frequent machinery access, minimising land loss and its associated challenges. Therefore, practical management strategies must be tailored to the specific agricultural system in use to optimise yield while minimising these adverse effects.

6.4.4 Sophistication levels in crop simulation

Typical crop simulators such as WOFOST [30], STICS [32] and GRECOS [33] include a general understanding of crop growth and development. These models simulate the overall effects of environmental factors on crop growth and development, such as temperature, precipitation, radiation and nutrient availability based on empirical relations and simplified representations. These models are typically used for crop yield forecasting and management decision-making, such as optimising irrigation and fertilisation schedules or predicting the effects of climate change on crop yields [30].

In contrast to the typical crop simulators, Functional-structural plant models (FSPM), combine the representation of three-dimensional (3D) plant structure with selected physiological functions with a detailed understanding of the plant's growth and development. These models simulate the underlying physiological processes that determine plant growth and development, such as light interception, carbon assimilation, nutrient uptake and photosynthesis in much more detail. FSPM models such as OpenAlea [34] are typically used for research purposes and require detailed input data on plant structure such as leaf area, stem diameter and root distribution. One example is the detail in calculation of the light interception. Typical crop growth simulators such as WOFOST, STICS and GECROS apply the Beer-Lambert equation [35], which applies that the absorption of light will be more or less decline exponentially with increasing intercepting area down through the canopy given by the extinction coefficient. In contrast, FSPM includes 3D canopy models that include the spatial solar radiation distribution. This is particularly useful for plants with multiple leaf layers such as (fruit and nut) trees and bushes [23] as shown in Figure 6.8, where the radiation distribution within pear trees of a checkerboard PV configuration is shown.

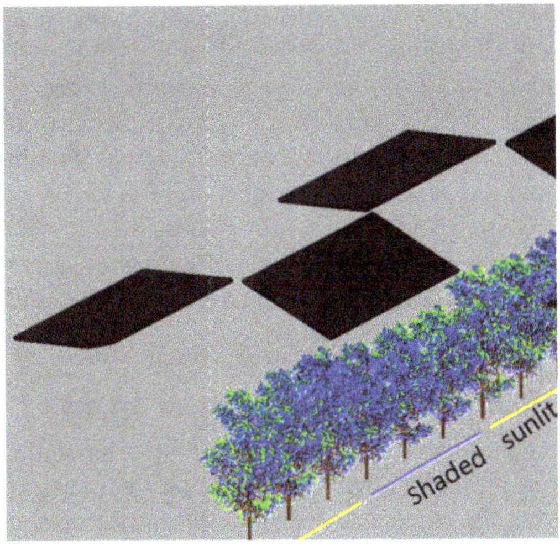

Figure 6.8 Illustration of the light interception for 3D virtual plant architectures created with the OpenAlea framework and VPlants package

6.5 PV power yield modelling

Similar to the crop yield modelling, PV production modelling is a common practice in the field of solar energy. It involves predicting and estimating the electricity production of photovoltaic (PV) solar panels. This practice is widely used by solar industry professionals and researchers to optimise energy generation, assess system performance, plan for efficient energy utilisation and to provide a bankable project with well-estimated yield and revenues.

There are several different types of PV technologies that are used to convert sunlight into electricity. In terms of maturity, crystalline silicon (c-Si) is the most common and widely used PV technology, accounting for more than 90% of the global PV market [36]. Thin-film PV is also commercially available and has been used in various applications, but its market share is smaller than c-Si. Organic PV [37] and perovskite PV are still in the early stages of development and are not yet widely used in commercial applications. The main considered PV technology in this chapter is bifacial crystalline silicon cells, generating electricity from both sides of the PV module, capturing sunlight that is reflected off the ground or other surfaces as well as the direct sunlight that falls on the front side of the panel. These modules have a transparent (plastic or glass) backsheet that allow light to pass through and reach the solar cells on the backside of the panel. The key parameters that impact the electricity production of a PV module are:

- Solar irradiance on the front surface: The amount of solar radiation that falls on the module's front affects the amount of electricity that is produced. The solar irradiance and consequently specific yield (kWh/kWp) can be drastically increased if the PV array follows the sun, for example by a solar tracker system, described in Chapter 3.

- Solar irradiance on the rear side: The reflectivity of the soil, crops or other surfaces beneath the PV modules can impact the amount of light that is reflected onto the back surface of the panel. Surfaces with higher albedo, such as snow or white gravel, can reflect more light onto the back surface of the panel, increasing energy yields.
- Temperature: The operating temperature of the panel can affect its efficiency and energy production. Higher temperatures can reduce the efficiency of the solar cells and decrease energy yields.

6.5.1 The irradiance on the front plane of array

Similarly to the radiation/microclimate model, the first step in the simulation of the electricity yield is to project the solar radiation on the plane of the PV modules (plane of array, POA). The direct sun component is projected on the front's surface given the position of the Sun and the position of the PV modules. Next, the received diffuse radiation can be estimated given the 'view factor' between the sky and PV module or the use of ray-tracing software. Finally, also part of soil/crops can be reflected on the front's surface (especially for a vertical system lay-out).

6.5.2 The irradiance on the rear plane of array

To facilitate the solar radiation on the rear side of the PV modules, the presence of structural elements that block the reflected light and the reflection coefficient (albedo) of the crops and soil must be know. Currently, in most (simplified) simulation programs, no impact of structural shading on the rear side is considered and the albedo of crops is assumed isotropic and fixed throughout the growing season with a reflection value around 20%. However, the value of the albedo as well as its directional components can change as the crops grow and develop, due to changes in the height, structure and surface area of the leaves and stems. For example, young crops with small leaves and stems tend to have a lower albedo than mature crops with larger leaves and stems. Additionally, the albedo and its spectral composition of crops can be affected by the angle of the sun, the time of day and the amount of moisture in the soil and on the leaves (water stress), which can all change throughout the growing season [38]. Also the spatial distribution of the albedo can change, by considering flower strips between the crops and PV modules. These changes in albedo can have an impact on the performance of the solar panels but are currently less addressed in modelling tools.

6.5.3 The cell temperature

The operating temperature of the PV module can affect its efficiency and power production. The nominal efficiency of PV cells is given in standard test conditions at 25°C cell temperature. Higher temperatures can reduce the efficiency of the solar cells and decrease energy yields. The operating temperature of the PV modules depends on the energy balance of the PV module, meaning that the received plane of array irradiance, surrounding air temperature (conduction), wind speed (convection) and leaf and sky temperature (emission) must be considered. This is a complex task considering that the microclimate below the PV modules (wind, air temperature) is also impacted by the soil/crop energy balance [39]. This can leads to interesting effects:

- At some moments, the cell temperature can be lower in agrivoltaic systems given that wind speed at less dense PV structures is higher and able to cool down the modules better.
- At some moments, the cell temperature can be lower due to reduced solar radiation and the evaporative cooling of the air by the crops below the PV system.
- At some moments, the cell temperature can be higher, due to the increased air temperature given the sheltered effect of the PV system that blocks the release of the long wave radiation from soil and plants.

Much depends on the system design (PV density) and the current state of the agrivoltaic system such as current wind speed and how much energy the crops release as latent heat compared to sensible heat. A basic estimation of the PV module and cell temperature can be made given the plane of array irradiance, wind speed at the height of the modules and surrounding air temperature, assuming that the difference in air temperature by evaporative cooling and/or greenhouse shelter effect compared to the surrounding air is limited (1–2°C) [13]. Given that the convective cooling by the wind is better in the case of agrivoltaic systems compared to dense ground-mounted PV systems, an increase in the specific yield (kWh/kWp) is expected.

6.5.4 Sophistication levels in PV simulation

Similarly to microclimatic and agricultural modelling, PV simulation can be at multiple sophistication levels. The most basic model (PV Watts [40]) considers a linear relation between the power production, average plane of array radiation and average cell temperature. Given a calibrated losses factor, which includes module and cell mismatch losses, line losses, soiling losses and inverter losses, the annual electricity can be well estimated. However, to be more precise at a higher temporal level such as day, hour or minute, more complex models must be taken into account [41]. The PV system must be subdivided in PV module models that are subdivided in PV cell models. Each cell includes a non-linear electrical model (such as single diode model), specific irradiance (simulated by ray-tracing or view factor model) and cell temperature (which is calculated based on a 2D/3D thermal model). The behaviour and impact of the heterogeneous distribution of irradiance, air temperature and windspeed on module and cell mismatch losses can now be estimated. Furthermore, DC and AC line losses as well as detailed inverter models can be considered. Moreover, degradation models can be coupled that include temperature and humidity effects (increase the risk of condensation and corrosion) to estimate the possible production over the lifetime of the system [42]. An example of a sophisticated power yield model is PVCASE.

6.6 Towards optimal agrivoltaic systems

The system design of traditional large-scale PV farms is mostly based on PV simulation software where the aim is in most cases to maximise the energy production at the lowest possible cost (levelised cost of energy, LCOE) regarding technical constraints (grid congestion, self-consumption). In contrast, the objective of agrivoltaic systems is to ensure the balance between energy and crop production, which is a complex task due to the many desired objectives (agricultural, electrical, land use, etc.) and practical constraints such as the farming practices and machinery dimensions. This is also highlighted by the study conducted by Pietro Campana [43] evaluating a vertical bifacial system in Sweden based on a genetic algorithm. The results showed that the optimal land use value (land equivalent ratio, LER) underestimates the synergies between crop and electricity production. In other (practical) techno-economic studies from Trommsdorff [29] and Willockx [44] the optimal design are based on a maximum solar radiation reduction (%) in a homogeneous distribution given the practical farming constraints such as machines dimensions.

The design of the optimal agrivoltaic systems depends on the viewpoint of the stakeholder (farmer, PV developer, policy maker) and the inclusion of specific boundary conditions such as fluence of practical (farming) constraints (land loss, soil compaction), electric consumption profile, crop choice, crop quality requirement and (ethical) limits on the crop yield losses sustaining food security. Therefore, a multi-objective optimisation framework is needed including practical constraints evaluating the optimal agrivoltaic design regarding the (simultaneous) optimisation of financial (LCOE), environmental (CO2), spatial (LER) and energetic (MWh/ha) objectives [45].

These considerations are also shown in national agrivoltaic standards, such as DIN SPEC 91434 in Germany, or in Italy and France. The standards stipulate that projects must achieve a minimum yield of 66% in Germany or above 80% of crop yield in order to qualify for financial support mechanisms like auctions. This underscores the significance of maintaining a certain level of crop yield to ensure that permanent and economically feasible agricultural activities can coexist with PV projects. Striking the right balance between energy production and crop cultivation becomes even more critical within the framework of these standards, highlighting the need for a holistic, multi-objective optimisation approach.

6.7 Need for a free public available tool

Given the multidisciplinary character and many variables of agrivoltaic systems, a publicly available simulation tool would facilitate collaboration and knowledge sharing among stakeholders (farmers, policy makers, PV developers) and could help to promote the development and optimal implementation of agrivoltaic systems on a larger scale based on the boundaries of each stakeholder. Moreover, a simulation tool could be used to provide training and education on agrivoltaic systems and their benefits, and to support decision-making regarding the planning and management of agrivoltaic projects. The first free radiation simulation tool, where three regions in the United States can be selected is (https://massgov.github.io/DOER/doer.html). However, besides its geographical limitation, the many input parameters and the visual evaluation of only direct sun shading (and not its diffuse component) are limiting its usefulness across other regions. Another free tool developed by KU Leuven [46], where the user can apply dynamic sliders of the inputs and output values of the shading level, power and energy production are shown. However, no visual output is given and the input data is limited. There is still room for improvement with respect to user experience functionality and the integration of a (complete) financial toolbox where business models and financial data will help the stakeholders in the complex agrivoltaics decision process.

6.8 Conclusion and future work

Crop yield modelling and PV (photovoltaic) modelling are established practices in agriculture and solar energy, respectively. However, when implementing these models in a new combined system, such as agrovoltaics, adaptations become necessary. This is because the unique conditions and interactions require tailored adjustments to ensure accurate predictions and optimal performance. In essence, while the individual practices are common, their integration demands customised approaches to address the specific challenges and opportunities.

Currently, the current focus of agrivoltaic systems modelling is physically and process based driven models (mathematical equations) for shadow calculation, crop simulation and PV electricity generation, where crop simulation is considered as the most complex part. The optimal system design is in most cases determined based on practical considerations such as the farming practices and machinery dimensions, the levelised cost of electricity of the system and the maximal reduction in solar crop yield (given by legislation in e.g. Italy, Germany or France) and can be determined by offline and online simulation tools available for the different stakeholders.

However, as the field of agrivoltaics continues to grow, it is likely that the simulation of more performance indicators will be included in a tool for design, optimisation and performance evaluation [3] and that the sophistication level will increase:

- Crop impact (yield and quality): calibrated mechanistic crop models and meaningful shade response curves based on empirical data from pilot projects [31] that include shade responses and adaptations as well as practical management aspect such as soil compaction and land loss.
- Environmental impact: researchers will need to develop simulation tools to assess factors such as carbon sequestration, soil impact, water use efficiency, biodiversity, up and downstream of materials.
- Microclimatic simulation: there is growing interest to besides solar radiation modelling, evaluate the impact on wind speed (by use of computational fluid

dynamics (CFD) simulations) and air temperature and air humidity (including coupled land surface – crop growth models). Researchers are working to develop more accurate models that can capture the complexity of these systems, and to improve data collection and validation [15].
- Financial modelling: including operational and capital expenditures, both from a farming and investor perspective. New business models such as energy cooperatives must be included.

Future work can also be the development of data-driven models can help simulate agrivoltaic systems by using real-world data to inform the inputs and predictions of the model [47,48]. For example, data on crop growth patterns, weather patterns and energy usage can be used to make more accurate predictions about how a specific agrivoltaic system will perform. These models can also help identify areas for improvement and optimise the design and operation of agrivoltaic systems, leading to increased efficiency and cost savings. By incorporating data from existing agrivoltaic systems, data-driven models can continually learn and improve over time, helping to advance the field of agrivoltaics.

Key aspect of yield prediction and simulation will be the development of standardised measurement protocols [49]. By this, the complexity of these systems can be captured, the data collection will be improved and proposed models can be validated, making decision-making more reliable and robust.

References

[1] P. Burgess, A. Graves, S. G. de Jalón, J. Palma, C. Dupraz, and M. van Noordwijk, "Modelling agroforestry systems," vol. 1, no. 11, 2021, pp. 209–238, doi:10.19103/as.2018.0041.13.

[2] A. Weselek, A. Ehmann, S. Zikeli, I. Lewandowski, S. Schindele, and P. Högy, "Agrophotovoltaic systems: applications, challenges, and opportunities. A review," *Agron Sustain Dev*, vol. 39, no. 4, 2019, pp. 1–20, doi:10.1007/s13593-019-0581-3.

[3] M. A. Al Mamun, P. Dargusch, D. Wadley, N. A. Zulkarnain, and A. A. Aziz, "A review of research on agrivoltaic systems," *Renewable and Sustainable Energy Reviews*, vol. 161. pp. 1–20, Jun. 01, 2022., doi:10.1016/j.rser.2022.112351.

[4] Y. Elamri, B. Cheviron, J. M. Lopez, C. Dejean, and G. Belaud, "Water budget and crop modelling for agrivoltaic systems: Application to irrigated lettuces," *Agric Water Manag*, vol. 208, no. April, pp. 440–453, 2018, doi:10.1016/j.agwat.2018.07.001.

[5] S. Schindele, M. Trommsdorff, A. Schlaak *et al.*, "Implementation of agrophotovoltaics: Techno-economic analysis of the price-performance ratio and its policy implications," *Appl Energy*, vol. 265, pp. 114737, May 2020, doi:10.1016/j.apenergy.2020.114737.

[6] M. Wagner, J. Lask, A. Kiesel *et al.*, "Agrivoltaics: The Environmental Impacts of Combining Food Crop Cultivation and Solar Energy Generation," *Agronomy*, vol. 13, no. 2, pp. 299–313, Feb. 2023, doi:10.3390/agronomy13020299.

[7] S. Zainali, S. Ma Lu, B. Stridh *et al.*, "Direct and diffuse shading factors modelling for the most representative agrivoltaic system layouts," *Appl Energy*, vol. 339, p. 120981, Jun. 2023, doi:10.1016/j.apenergy.2023.120981.

[8] J. Van Wart, P. Grassini, and K. G. Cassman, "Impact of derived global weather data on simulated crop yields," *Glob Chang Biol*, vol. 19, no. 12, pp. 3822–3834, Dec. 2013, doi:10.1111/gcb.12302.

[9] U. Laetitia, H.-G. Kim, J.-Y. Kim, B. Kim, and C. K. Kim, "Multi-Year and Typical Year Data for Solar Energy System in South Korea," *Journal of the Korean Solar Energy Society*, vol. 40, no. 6, pp. 135–149, Dec. 2020, doi:10.7836/kses.2020.40.6.135.

[10] T. Zhang, P. W. Stackhouse, W. S. Chandler, J. M. Hoell, D. Westberg, and C. H. Whitlock, "A GLOBAL PERSPECTIVE: NASA's Prediction of Worldwide Energy Resources (POWER) Project." *Proceedings of ISES World Congress 2007* vol. 1-5, pp 2636–2640, 2007 https://doi.org/10.1007/978-3-540-75997-3_532

[11] M. Sengupta, A. Habte, A. Lopez, and Y. Xie, "The National Solar Radiation Data Base (NSRDB)." *Renewable and Sustainable Energy Reviews*, vol. 89, pp. 51–60, Jun 2018, doi:https://doi.org/10.1016/j.rser.2018.03.003

[12] S. Amaducci, X. Yin, and M. Colauzzi, "Agrivoltaic systems to optimise land use for electric energy production," *Appl Energy*, vol. 220, no. February, pp. 545–561, 2018, doi:10.1016/j.apenergy.2018.03.081.

[13] A. Weselek, A. Bauerle, J. Hartung, S. Zikeli, I. Lewandowski, and P. Högy, "Agrivoltaic system impacts on microclimate and yield of different crops within an organic crop rotation in a temperate climate", *Agronomy for Sustainable Development*, vol. 41, no. 59, pp. 41–59, 2021, doi: 10.1007/s13593-021-00714-y/Published.

[14] Y. Elamri, B. Cheviron, A. Mange, C. Dejean, F. Liron, and G. Belaud, "Rain concentration and sheltering effect of solar panels on cultivated plots," *Hydrol Earth Syst Sci*, vol. 22, no. 2, pp. 1285–1298, Feb. 2018, doi:10.5194/hess-22-1285-2018.

[15] S. Zainali, S. Ma Lu, B. Stridh *et al.,* "Computational fluid dynamics modelling of microclimate for a vertical agrivoltaic system," *Energy Nexus*, vol. 9, pp. 100173, Mar. 2023, doi:10.1016/j.nexus.2023.100173.

[16] H. Marrou, L. Guilioni, L. Dufour, C. Dupraz, and J. Wery, "Microclimate under agrivoltaic systems: Is crop growth rate affected in the partial shade of solar panels?," *Agric For Meteorol*, vol. 177, no. August, pp. 117–132, 2013, doi:10.1016/j.agrformet.2013.04.012.

[17] Barron-Gafford, G.A., Pavao-Zuckerman, M.A., Minor, R.L. *et al.* Agrivoltaics provide mutual benefits across the food–energy–water nexus in drylands. *Nature Sustainability*, vol. 2, pp. 848–855, 2019, doi:https://doi.org/10.1038/s41893-019-0364-5

[18] S. Ma Lu, S. Zainali, B. Stridh *et al.,* "Photosynthetically active radiation decomposition models for agrivoltaic systems applications," *Solar Energy*, vol. 244, pp. 536–549, Sep. 2022, doi:10.1016/j.solener.2022.05.046.
[19] C. M. M. Gommers, E. J. W. Visser, K. R. S. Onge, L. A. C. J. Voesenek, and R. Pierik, "Shade tolerance: When growing tall is not an option," *Trends in Plant Science*, vol. 18, no. 2. pp. 65–71, Feb. 2013. doi:10.1016/j.tplants.2012.09.008.
[20] H. Marrou, J. Wery, L. Dufour, and C. Dupraz, "Productivity and radiation use efficiency of lettuces grown in the partial shade of photovoltaic panels" *European Journal of Agronomy*, vol. 44, no. 2013, pp. 54–66, 2013, doi: https://doi.org/10.1016/j.eja.2012.08.003
[21] R. Perez, R. Seals, R. Stewartt, and D. Menicucci, "A NEW SIMPLIFIED VERSION OF THE PEREZ DIFFUSE IRRADIANCE MODEL FOR TILTED SURFACES," *Solar Energy*, vol. 39, no.3, pp.221–231, 1987, https://doi.org/10.1016/S0038-092X(87)80031-2.
[22] D. Berrian and J. Libal, "A comparison of ray tracing and view factor simulations of locally resolved rear irradiance with the experimental values," *Progress in Photovoltaics: Research and Applications*, vol. 28, no. 6, pp. 609–620, Jun. 2020, doi:10.1002/pip.3261.
[23] S. Edouard, A. J. Escobar-Gutierrez, M. Van Iseghem, R. Barillot, G. Louarn, and D. Combes, "Is the turbid medium-based approach pertinent for estimating light interception when simulating the growth of a crop in an agriphotovoltaic system?," *Biosyst Eng*, vol. 224, pp. 131–142, Dec. 2022, doi:10.1016/j.biosystemseng.2022.10.006.
[24] D. Munoz and H. Colin, "HAL Id: cea-03700287 https://hal-cea.archives-ouvertes.fr/cea-03700287 Agrivoltaics: a review of PV technologies and modelling methods." [Online]. Available: https://hal-cea.archives-ouvertes.fr/cea-03700287
[25] C. Campillo, R. Fortes, and M. del Henar Prieto, "Solar Radiation Effect on Crop Production," *Solar Radiation*, vol. 1, pp. 167–194, 2012, doi:10.5772/34796.
[26] J. L. Monteith, "SOLAR RADIATION AND PRODUCTIVITY IN TROPICAL ECOSYSTEMS," *Journal of Applied Ecology*, vol. 9, p. 747, 1972, doi:https://doi.org/10.2307/2401901
[27] T. Sekiyama and A. Nagashima, "Solar Sharing for Both Food and Clean Energy Production: Performance of Agrivoltaic Systems for Corn, A Typical Shade-Intolerant Crop," *Environments*, vol. 6, no. 6, p. 65, 2019, doi:10.3390/environments6060065.
[28] M. Laub, L. Pataczek, A. Feuerbacher, S. Zikeli, and P. Högy, "Contrasting yield responses at varying levels of shade suggest different suitability of crops for dual land-use systems: a meta-analysis", *Agronomy for Sustainable Development*, vol. 42, no. 51, pp. 42–51, 2022, doi: 10.1007/s13593-022-00783-7/Published.
[29] M. Trommsdorff, J. Kang, C. Reise *et al.,* "Combining food and energy production: Design of an agrivoltaic system applied in arable and vegetable

farming in Germany," *Renewable and Sustainable Energy Reviews*, vol. 140, p. 110694, Apr. 2021, doi:10.1016/J.RSER.2020.110694.

[30] A. de Wit, H. Boogaard, D. Fumagalli *et al.*, "25 years of the WOFOST cropping systems model," *Agricultural Systems*, vol. 168, pp. 154–167, Jan. 01, 2019. doi:10.1016/j.agsy.2018.06.018.

[31] D. Wallach, "The chaos in calibrating crop models: Lessons learned from a multi-model calibration exercise," *Environmental Modelling and Software*, vol. 145, pp. 105206, Nov. 2021, doi:10.1016/j.envsoft.2021.105206.

[32] S. Amaducci, X. Yin, and M. Colauzzi, "Agrivoltaic systems to optimise land use for electric energy production," *Appl Energy*, vol. 220, pp. 545–561, Jun. 2018, doi:10.1016/j.apenergy.2018.03.081.

[33] N. Brisson, C. Gary, E. Justes *et al.*, "An overview of the crop model STICS," *in European Journal of Agronomy*, vol. 18, no. 3–4, pp. 309–332, Jan. 2003, doi:10.1016/S1161-0301(02)00110-7.

[34] J. Gu, X. Yin, C. Zhang, H. Wang, and P. C. Struik, "Linking ecophysiological modelling with quantitative genetics to support marker-assisted crop design for improved yields of rice (Oryza sativa) under drought stress," *Ann Bot*, vol. 114, no. 3, pp. 499–511, 2014, doi:10.1093/aob/mcu127.

[35] C. Pradal, S. Dufour-Kowalski, F. Boudon, C. Fournier, and C. Godin, "OpenAlea: A visual programming and component-based software platform for plant modelling," *Functional Plant Biology*, vol. 35, no. 10, pp. 751–760, 2008, doi:10.1071/FP08084.

[36] M. A. Ponce de León and B. N. Bailey, "Evaluating the use of Beer's law for estimating light interception in canopy architectures with varying heterogeneity and anisotropy," *Ecol Modell*, vol. 406, pp. 133–143, Aug. 2019, doi:10.1016/j.ecolmodel.2019.04.010.

[37] B. Willockx, B. Herteleer, and J. Cappelle, "TECHNO-ECONOMIC STUDY OF AGROVOLTAIC SYSTEMS FOCUSING ON ORCHARD CROPS.", 7th European Photovoltaic Solar Energy Conference and Exhibition, pp. 1761–1766, 2020, doi: 10.4229/EUPVSEC20202020-6DO.14.2

[38] S. Gorjian, E. Bousi, Ö. E. Özdemir *et al.*, "Progress and challenges of crop production and electricity generation in agrivoltaic systems using semi-transparent photovoltaic technology," *Renewable and Sustainable Energy Reviews*, vol. 158, pp. 112126, Apr. 01, 2022, doi:10.1016/j.rser.2022.112126.

[39] R. Meitzner, U. S. Schubert, and H. Hoppe, "Agrivoltaics—The Perfect Fit for the Future of Organic Photovoltaics," *Adv Energy Mater*, vol. 11, no. 1, pp. 2002551- 2002558, Jan. 2021, doi:10.1002/aenm.202002551.

[40] P. Thenkabai, R.B. Smith, and E. De Pauw, "Hyperspectral vegetation indices for determining agricultural crop characteristics," *Remote Sensing of Environment*, vol. 71, no. 2, pp.158–182, February 2000, doi:https://doi.org/10.1016/S0034-4257(99)00067-X

[41] L. A. Shepard, C. W. Higgins, and K. W. Proctor, "Agrivoltaics: Modeling the relative importance of longwave radiation from solar panels," *PLoS One*, vol. 17, no. 10 October, pp. 1–21, Oct. 2022, doi:10.1371/journal.pone.0273119.

[42] P. P. Aron Dobos, J. M. Freeman, and N. J. Blair, "Improvements to PVWatts for Fixed and One-Axis Tracking Systems", 2019 IEEE 46th Photovoltaic Specialists Conference (PVSC) , Chicago, IL, USA, pp. 1249–1254, 2019, doi:10.1109/PVSC40753.2019.8981312

[43] C. P. Cameron, W. E. Boyson and D. M. Riley, "Comparison of PV system performance-model predictions with measured PV system performance," 2008 33rd IEEE Photovoltaic Specialists Conference, San Diego, CA, USA, 2008, pp. 1–6, doi:10.1109/PVSC.2008.4922865.

[44] N. C. Park, W. W. Oh, and D. H. Kim, "Effect of temperature and humidity on the degradation rate of multicrystalline silicon photovoltaic module," *International Journal of Photoenergy*, vol. 2013, p.19, doi:10.1155/2013/925280.

[45] P. E. Campana, B. Stridh, S. Amaducci, and M. Colauzzi, "Optimisation of vertically mounted agrivoltaic systems," *J Clean Prod*, vol. 325, Nov. 2021, pp. 129091, doi:10.1016/j.jclepro.2021.129091.

[46] B. Willockx, B. Herteleer, and J. Cappelle, "Combining photovoltaic modules and food crops: First agrovoltaic prototype in Belgium," *Renewable Energy and Power Quality Journal*, vol. 18, no. August, pp. 266–271, 2020, doi:10.24084/repqj18.291.

[47] E. Mengi, O. A. Samara, and T. I. Zohdi, "Crop-driven optimization of agrivoltaics using a digital-replica framework," *Smart Agricultural Technology*, vol. 4, no. Aug. 2023, pp. 100168, Aug. 2023, doi:10.1016/j.atech.2022.100168.

[48] Tanino, T. "Sensitivity Analysis in Multiobjective Optimization," *Journal of Optimization Theory and Applications*, vol. 56, pp. 479–499, 1988, doi: https://doi.org/10.1007/BF00939554

[49] B. Willockx, C. Lavaert, and J. Cappelle, "Development of a quick-scan webtool to facilitate agrivoltaic system design," in 2022 International Conference on Renewable Energies and Smart Technologies, REST 2022, Tirana, Albania, pp. 1–5, 2022, doi:10.1109/REST54687.2022.10022628.

[50] N. Zhang, X. Zhou, M. Kang, B.-G. Hu, E. Heuvelink, and L. F. M. Marcelis, "Machine learning versus crop growth models: an ally, not a rival," *AoB Plants*, vol. 15, no. 2, pp. 1–7, Feb. 2023, doi:10.1093/aobpla/plac061.

[51] S. Moslehi, T. A. Reddy, and S. Katipamula, "Evaluation of data-driven models for predicting solar photovoltaics power output," *Energy*, vol. 142, pp. 1057–1065, Jan. 2018, doi:10.1016/j.energy.2017.09.042.

[52] S. J. Seidel, T. Palosuo, P. Thorburn, and D. Wallach, "Towards improved calibration of crop models – Where are we now and where should we go?," *European Journal of Agronomy*, vol. 94, pp. 25–35, Mar. 2018, doi:10.1016/j.eja.2018.01.006.

Chapter 7
Operations and maintenance considerations at agrivoltaic sites

James McCall[1], Brittany Staie[1], Kai Lepley[1] and Jordan Macknick[1]

The purpose of this chapter is to focus on changes to operations and maintenance (O&M) practices or design changes that are needed for Agri-PV projects. Adopting Agri-PV projects will require changes to site O&M practices from conventional PV projects to accommodate farming practices on-site that come with potential contracting, legal, and health and safety regulatory changes. The focus of this chapter is to outline O&M operations at Agri-PV sites, including a comparison with O&M practices at conventional utility-scale solar sites. These comparisons will be relevant, but different, to both new Agri-PV project developments and retrofitting existing conventional PV sites to Agri-PV. This chapter starts with the impacts of different PV design choices on crop management, grazing, and vegetation management practices coupled with other relevant agricultural considerations. Next, this chapter discusses differences in site design and O&M that arise from incorporating crop production and grazing, along with potential impacts of these agricultural activities on PV systems. Finally, non-technical considerations of Agri-PV site management are discussed to ensure that these pieces are accounted for throughout the project life cycle.

This chapter will discuss cropping operations based on insights from four Agri-PV sites in the United States: Jack's Solar Garden in Longmont, CO; the Bifacial Agrivoltaic Research at NREL (BARN) site in Golden, CO; a demonstration site at the University of Massachusetts Amherst; and a demonstration site at the University of Arizona's Biosphere 2 facility near Tucson. All projects listed here are hand-harvested but small equipment and tractors are used for tilling, ploughing, and other farming activities.

[1]National Renewable Energy Laboratory, 15013 Denver W Pkwy, Golden, CO 80401

Project name	Location	Crops cultivated	Cultivated land (acres)	System size (kW)	System design	Crop location
Jack's Solar Garden	Longmont, CO	Tomato, pepper, kale, radish, eggplant, 30+ crop types	3	1100	1-axis tracking 72″ panel height	Underneath, In-between Panels
BARN	Golden, CO	Tomato, pepper, kale, chard, carrot, basil, beans	0.1	80	1-axis tracking, 48″ panel height	In-between Panels
University of Arizona Bio-sphere 2 Site	Oracle, AZ	Tomato, pepper, 20+ crop types	0.4	20	Fixed, 96″ panel height	Underneath, In-between Panels
University of Massachusetts-Amherst	Amherst, MA	Pepper, kale, broccoli, beans, chard	0.5	26	Fixed, 84″ panel height, open space in between panels	Underneath, In-between Panels

7.1 PV design impacts on Agri-PV O&M

PV infrastructure (supports, modules, inverters, wiring, etc.) can provide many obstacles to farming and other Agri-PV activities on site. These obstacles can be mitigated during the planning and design process but these mitigations either come with a cost, energy, or agricultural yield penalty. This section will discuss some implications of PV design parameters (Figure 7.1) on O&M and farming practices on Agri-PV sites (Figure 7.1).

7.1.1 Wire management

Wire management and placement can impact Agri-PV O&M practices by limiting access to parts of the site, limiting where certain digging and trenching activities can occur, and presenting potential worker and animal safety issues.

7.1.1.1 Buried cabling

Agri-PV projects will often require more land disturbance than conventional PV projects due to the need to plough, till, and harvest crops. Because of this, there is a higher potential for workers to come into contact with, and/or damage, buried

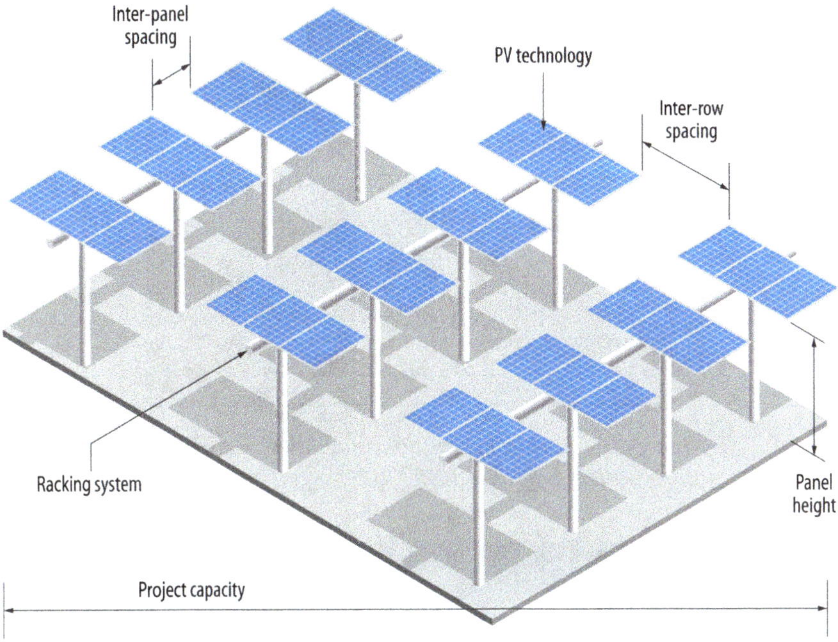

Figure 7.1 PV infrastructure design diagram that details the relevant parameters discussed in this section [1]

cabling on site. Also, after PV construction, some areas of land may be compacted, or topography changes may occur that may require ploughing, tilling, and other grading operations to make the land arable again. PV sites bury wiring, and according to the National Electric Code (NEC) the wires need to be buried 15–60 cm depending on installation practices. Burying will reduce the risk of workers on-site interacting with the wiring during normal operations, but it presents a higher cost of both installation and O&M to dig up wiring if problems occur.

The Agri-PV project owner wants to make sure that the cables trenches are deep enough such that after 30 years of operation the cables are still out of reach of the tiller and plough. The average tilling depth for crops is in the range of 20 cm, compared to the 45–60 cm needed for buried cabling in Figure 7.2 [3]. If more cables need to be put into one cable ditch, alternatively the ditch might be dug wider to avoid extra cost on digging deeper compared to conventional PV. Often, red warning tape is put into the ditch approximately 10 cm above the cable [4]. In such a way, the farmer recognizes danger if he/she is ploughing too deep and is about to impact buried cables.

Along with increased cost from trenching and securing the wires, the location of buried cables may impact where farming operations or decompaction of soils can occur if the wires are not buried to significant depth. For example, in Figure 7.3, trenching of the soil may result in compaction issues and remove the topsoil, which

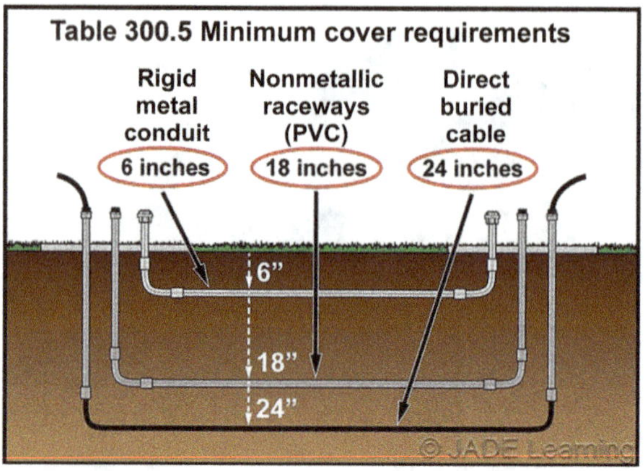

Figure 7.2 Wire depth requirements for solar facilities; each option comes with cost, time, and safety tradeoff [2]

Figure 7.3 Example of direct buried cable [5]

could require a portion of the farmable area to be remediated, or it could limit the depth of farming operations to avoid hitting wires.

Along with wiring considerations for Agri-PV sites, other PV electrical equipment (inverter boxes, meteorological monitoring, interconnection equipment, batteries, etc.) will need to be secured and demarcated for the type of Agri-PV operation. Ensuring that this electrical equipment is sited to allow for farm workers and equipment access needs to be included in the design process while still allowing for electrical maintenance activities to continue.

7.1.1.2 Above ground cabling

Above ground cabling practices have been deployed to reduce the risks and costs of digging to bury cables and to provide much easier access to electrical infrastructure [6]. This above ground cabling system. Figure 7.4 is quick and easy to install, but it limits access to the site for farming and other mowing equipment. If the above ground cabling system is deployed up the middle of the PV tracker array, the ability to turn equipment around is limited and can increase both time and risk for damaging PV equipment. Above ground cabling can also present a potential hazard to livestock on site, but fencing can limit this risk. An Agri-PV plan at sites with above ground cabling systems needs to account for this limitation and plan to use handheld or smaller tools in those areas which could add complexity and time. Wires will be more visible and easier to account for than buried, but the risk of damage is also higher. These considerations need to be balanced in the site operational plan.

7.1.1.3 Worker safety

Along with the wires from PV arrays to the inverters, there will be cabling on the back of the module for the array that runs to a junction box onsite. Proper wire management is also needed to prevent drooping wires and to minimize the risk of workers encountering those wires. Agricultural workers will be on-site and around

Figure 7.4 Above ground cabling system highlights ease of access to wire, but creates a barrier for agricultural and vegetation maintenance equipment [6], Credit: Solar Power World

electrical equipment more often than PV operators and this can create contact risks with electrical equipment. Tight wire management can reduce the risk, but installing a mesh or other impediment on the back of the array will further reduce risk to workers onsite.

7.1.1.4 Animal safety

Another issue of concern that will impact wire management is the presence of livestock on the site. Some animals, such as goats and cattle, may chew on or pull at the wiring so proper wire management safety will need to be implemented to ensure the wiring can withstand animal impacts. Above ground cabling systems also have the potential to limit animal movement and there is a small potential for animals to become stuck in the wires. Fencing and proper animal handling techniques are needed to prevent any impacts to the wires and animals in those systems.

7.1.2 Panel spacing and racking

Along with worker access to agricultural beds/rows in between the panels, different farm equipment will need access to perform ground preparation, tilling, seeding/planting, harvesting, and other farming operations.

7.1.2.1 Farming/O&M equipment access

Inter-row spacing, or the distance between the racking supports, is normally dictated by preventing shading of modules by other modules, but in an Agri-PV project further consideration is needed to allow for farm equipment access. A larger inter-row spacing distance will allow for easier access of larger farming equipment and more land area for farming, but will come at a reduced energy density, leading to more land being needed for the same amount of solar energy production. This trade off of spacing and energy production is often called ground coverage ratio (GCR) in the solar industry, and it is measured as the length of the panel divided by distance between the rows [7].

If a farmer has existing equipment that is used for their operations, then often the PV design will need to accommodate that (Figure 7.5). New farming equipment and/or tractors can be anywhere from $10,000–$75,000 that can often reduce the economic viability of Agri-PV projects [8]. Farm equipment can range in width from walk-behind tractors (1–1.5 m wheelbase) to 18+ m combine harvester for crops like wheat and corn.

Along with spacing, the racking system used in the project will impact the ability to get farm equipment in between the PV arrays. A fixed-tilt system will limit access of equipment, but the spacing needed to prevent shading will often accommodate tractors in more northern latitudes [10]. For a single-axis tracking system, increased panel height will allow for slightly easier access than fixed tilt. A tracking system allows the ability to operate equipment on one side of the plot when the array is tilted to extremes in the morning and afternoon times allows for planning of operations to mow or operate equipment at different times of the day for ease of access.

Figure 7.5 Farming equipment in operation at Jack's Solar Garden. PV row spacing is 5.5 m pole to pole, designed to incorporate the dimensions of needed farm equipment. Photo by Werner Slocum, NREL [9].

Figure 7.6 Mowing operations at a vertical bifacial site. PV row spacing, just over 30 ft, allows for larger equipment to be operated between panels [11].

Vertical bifacial systems are fixed-tilt systems that allow for easy farming access and more farming activities to occur due to their design (Figure 7.6). There is less risk of damaging equipment because of a slimmer profile and reduced time of operations due to the ability to use larger equipment. However, there is a higher risk of damage when using machinery that is prone to throwing rocks between rows of vertical panels. Other trackers will allow for the panels to be stowed vertically at the 90-degree position to allow for equipment to work under and around the panels and to reduce the potential for agricultural equipment hitting PV infrastructure.

7.1.2.2 Worker safety

Increased panel spacing will allow for more farmer access to the plots and reduce the risk of workers coming into contact with the PV array. The tracking system design will have a larger impact on worker safety issues. Single-axis trackers tilt throughout the day, so worker access will be impacted at different times of the day and present a shifting PV profile throughout the day.

7.1.3 Panel height

The height that PV panels are raised off the ground dictates what crops and animals can be incorporated into the Agri-PV project. The metric is often determined by the hub height, or the height off the ground for the main brace, for tracking systems. For fixed tilt system it is the ground clearance, or distance from the ground to the lowest edge of the panel.

7.1.3.1 Crop selection

There is no standard height for conventional utility scale solar projects, and the ground clearance can vary from 30 cm to 120 cm depending on the project and location. Systems can be raised or lowered depending on project economics and climatic conditions. Raising the panels will lead to more upfront installation cost but will allow for taller crops and vegetation underneath. Crop choices should be considered early in the development process because crop heights vary greatly. Failing to consider crop height early could lead to future conflicts between farming and photovoltaic activities on the site. Based on crop type, the project will need to consider rotation schedules. For example, if a cover crop is used to amend or strengthen the soil, the maximum height of the cover crop needs to be compatible with the final design height.

Revenue from PV operations is often higher on a per acre basis than agricultural products, so current systems are typically designed to maximize PV production. This maximization also includes reducing shading of the modules from vegetation which dictates crop selection and can impact mowing/grazing frequency dependent on growth rate of vegetation. This is also seen through use of herbicide to remove and maintain weed and invasive species that may shade the panels. PV production is often contractually mandated, and lost production due to crop shading on PV modules could result in expensive fines to system owners.

7.1.3.2 Equipment

Panel height will impact the type of equipment that can be used in the Agri-PV setting. If farming operations will be under or around the drip edge of the panels, then the height of the panels will need to be designed to incorporate farming equipment. With fixed tilt systems the minimum height of the panels will need to be adjusted to accommodate O&M equipment or specialized equipment may need to be purchased.

7.1.3.3 Worker safety

Depending on the type of crop, workers will need to be in and out of the crop beds to plant, maintain, and harvest crops throughout the growing season. Workers will have to navigate the PV equipment, and the lower the height of the panels the more likely workers are to encounter the panels. This could potentially result in electrical shocks and/or bodily injury. The edge of panels can be sharp so proper safety equipment and/or raising panels to avoid hitting the panels should be considered (Figures 7.7 and 7.8).

Figure 7.7 Low panel height of array impacts workers ability to access crop bed [12], Credit: Alison F. Takemura/Canary Media

Figure 7.8 Higher panel height allows for worker access to crops underneath panels with risk of hitting head, photo by Dennis Schroeder / NREL [9]

7.1.3.4 Animal safety

Like wire management, the panel design height will also be impacted by grazing animals on site. Sheep have already been integrated into traditional utility scale PV projects with low mounting heights with minimal impact, and there can be some benefits for sheep grazing vegetation underneath low panels where mowers cannot reach [13]. Other animals, may be strong enough to tear panels off the mount (cattle, bison) requiring increased panel height, may jump on the panels (goats) leading to damage, or need other animal management practices (such as wing clipping for chickens) to prevent impacts to panels.

7.2 Agricultural impacts on PV O&M

7.2.1 PV O&M

This section details the changes to PV site operations and timing from incorporating crops and grazing activities at PV sites. Planning for different Agri-PV activities will need to begin early in the PV development process to accommodate different activities on site. Lining up a project partner who has experience with the type of activity (crop production, native vegetation establishment, grazing, etc.) is important to ensure project success.

Heavy equipment to grade and clear the land is often used during the PV development process. The impact of these operations on the land can vary by activity and some developers are incorporating low-impact practices into site preparation to accommodate other land uses on solar sites. The following low-impact site development practices are adapted from the InSPIRE Low-Impact Solar Development Strategies Primer [14]:

Conventional site preparation	Low-impact site preparation
Clearing and grubbing of soil and roots	Existing vegetation is left intact for grazing or is replaced with crops
Topsoil stripping and stockpiling	Existing topsoil is left in place to allow for the successful growth of native vegetation or crops and to promote soil health post-decommissioning of the solar project
Land grading and leveling utilizing heavy machinery	Natural contours of land are worked into the design and configuration of the solar project, with minimal if any land grading required
Soil compaction utilizing heavy machinery	Soil and vegetation are left intact to facilitate the growth of native vegetation or crops, improve stormwater management through less runoff and erosion, and maintain soil health
Land footprint for the foundations of vertical support structures, often including concrete	Lower land footprint for foundations of vertical support structures, often driven piles
O&M activities include herbicide spraying, mowing of weeds and other vegetation	Minimal O&M activities due to low-growing native vegetation species, could involve livestock grazing

7.2.2 Agricultural inputs

This section details farming O&M steps to establish crops underneath solar panels. Many of these activities will be time sensitive and will need to be done before and at the start of the growing season for the first year. Missing any of these steps may delay operations by a year. Because some of these activities will be interdependent with PV construction, site preparation and crop planning should be incorporated into both the development and O&M schedules. All these details should be documented early on in the development cycle for a yearly cropping plan of all the activities on site.

7.2.2.1 Soil sampling

Soil sampling often occurs in the upfront PV development process but oftentimes these surveys are geared toward geotechnical considerations that would impact the PV site design (Figure 7.9). Different soil sample metrics will be needed to determine if soil amendments and modifications are needed for farming operations. Soil decompaction, before and after site operations, will be important to both the PV developer and O&M contractors to assist with establishment of vegetation and/or crops.

The following metrics should be sampled before cropping operations, especially early in the project, as it may impact site O&M activities:

- Organic matter—impacts soil structure, water holding capacity, and nutrient uptake.
- pH—impacts nutrient availability, ideal range varies by crop.

Figure 7.9 Pulling a soil sample core to test for farming operations, photo by Werner Slocum/NREL [9]

- Macro and micronutrients—important for plant growth and health.
- Soluble salts—high levels negatively impact plant growth and water uptake.
- Cation exchange capacity—assists with uptake of nutrients.
- Heavy metals—needed for food safety and plant health.

7.2.2.2 Ground and bed preparation

Compaction will be important to monitor in the cropping area and some soil preparation work may be needed at the start of the growing season. Correct equipment to perform this work needs to be considered to ensure that tractors and other equipment can operate safely within the array (Figure 7.10). Ploughing and/or tilling are often the first O&M activities at the start of the growing season. In small-scale crop production, farmers often prepare crop beds at this time. Bed number and creation will be dependent on the crops chosen for planting and the amount of space available in between the panel rows Some distance from the edge of the panels to the first bed will be needed for worker safety and to avoid shading concerns.

Once beds are established, worker access will be impacted, and care needs to be taken to avoid stepping in the areas where crops will be planted. After the growing season is complete, more O&M work is needed to prevent topsoil erosion and to provide nutrients to the crop beds for next season. Mulch (such as hay) and/or cover crops can be utilized for this.

Figure 7.10 O&M contractor tills soil to prepare for planting underneath a solar array—note the timing of operations with respect to the panel tilt, photo by Joe DelNero (NREL) [9]

7.2.2.3 Planting

Planting operations will have little impact on PV O&M but need to be accounted for during operations planning. The crops selected will determine timing, soil needs and planning operations. The cropping plan should include details for rotation of crops and different growing seasons. Depending on the equipment needed for planting, tracking systems could have their tracking algorithms manually adjusted to accommodate the machinery on-site for the duration of the planting.

7.2.2.4 Irrigation

Most (but not all) crops will need irrigation during the growing season, and obtaining water rights and access to water will need to be considered in the cropping plan. Along with water accessibility, irrigation piping needs to be planned onsite to prevent any tripping concerns and installation of piping needs to be planned around equipment access (Figure 7.11). If equipment access is needed often in each area, then a plan should be developed to prevent access concerns from irrigation piping.

On top of main irrigation piping, a plan for water access to crops is needed as conventional large scale irrigation equipment (e.g., center pivot irrigation) will likely be incompatible with solar operations. Some farmers are using lay flat drip tape to alleviate this concern and to increase watering efficiency of farming operations. For PV equipment and inverters around irrigation areas, a higher water-resistant rating may be needed to prevent equipment damage.

7.2.2.5 Weed management

Like conventional PV projects, vegetation management under and around the solar arrays will be a large O&M activity. This is often done through weeding, herbicide

Figure 7.11 Irrigation piping being installed at Jack's Solar Garden, photo by Werner Slocum (NREL) [9]

application, or hand pulling weeds [15]. Depending on the cropping system, weed management may require hand pulling, hand tools, spot spraying of herbicides, or use of a tractor in the crop production area. The PV design must allow for any equipment access needed to manage weeds. For the land directly under the panels, farmers may choose to plant crops, keep the native vegetation, plant a pollinator habitat for attraction of beneficial insects, or cover with landscaping fabric. Management of this area may increase the labor requirements for the Agri-PV farmer. It is important to have a vegetation plan for this area to discourage the growth of invasive species and weeds that matches the crop design.

7.2.2.6 Harvest and food storage

Specialized equipment may be needed to harvest crops, but in the United States, most Agri-PV projects are currently hand-harvested horticultural products. In other countries and PV designs, sites are harvested with traditional agricultural machinery. Worker and equipment access for harvesting will need to be incorporated into the Agri-PV design. After harvesting, additional infrastructure and plans for cleaning, sorting, packaging and transporting crops to the final marketplace are needed to ensure safe handling of food and reduce food spoilage. Finding a partner that has experience in this arena is important up front to determine infrastructure needs.

7.2.3 Grazing

Another agricultural activity is grazing of livestock in and around solar arrays, and sheep are already incorporated into many PV sites [16]. Current operations mainly incorporate sheep grazing (Figure 7.12) due to height, temperament, and cost, but some operations have looked at grazing a wide variety of animals including cows, chickens, rabbits, and aquaponic operations [13]. Costs for sheep grazing are often comparable to the cost of mowing operations at US solar sites [15]. Grazing animals can take advantage of shade environment of the panels reducing heat stress while providing a source of feed and forage for the animals. Many solar developers are considering sheep grazing to keep land in agricultural production while providing benefits to local graziers.

7.2.3.1 Forage

Suitable vegetation is needed to accommodate grazing operations at solar sites. Along with making sure that forage is of proper nutritional value for the animals, care is needed to select suitable crops that can withstand the grazing plan. Grazing operations can incorporate high intensity grazing operations that concentrate grazing over a short period of time and eat the vegetation mostly bare or low intensity operations that reduce the number of sheep per acre to slowly manage the vegetation. Either method needs to be incorporated into vegetation management plans and be planned around flowering seasons if pollinator habitat is included in the forage mix.

7.2.3.2 Water

One of the main costs of sheep grazing operations at solar sites, is providing access to a water source for the animals. Water is often stored onsite in tanks and/or

Figure 7.12 Sheep grazing at a fixed panel solar site in northeastern US, photo by Lexie Hain [17]

troughs for animal access but depending on the size of the storage container and the flock this may need to be filled up daily, requiring a grazier to transport water to the site. Some developers have started drilling water wells on site to reduce the cost of grazing and make their site more desirable to sheep graziers. PV sites have also shown potential to decrease water intake from sheep due to the shade provided from the panels [18]. There is also the potential to install gutters or other systems around the PV array to allow for collection of rainwater for both livestock and crop use.

7.2.3.3 Fencing

Including a plan for fencing early in the solar design process can reduce the cost of grazing operations. Dependent on the grazing plan and the intensity of grazing operations, fencing may be installed at the beginning of the season or may need to be moved often to incorporate rotational grazing practices (Figure 7.13). On top of inter-array fencing to manage where sheep can graze, external fencing needs to be considered. An external fence that is designed to keep out predators will be more desirable for grazing operations than fences that allow wildlife through. More monitoring may be required on non-suitable grazing sites that can increase the cost of grazing operations.

7.2.4 Impact on PV operations

7.2.4.1 Electrical O&M access

With crops in between panel rows, access for electrical O&M workers to repair wiring, tracking systems, and/or panels will be reduced (Figure 7.14). Often, PV operators will drive trucks through arrays to perform maintenance, but this option will likely be severely limited unless the beds are designed for truck access. This will cause delays in electrical O&M activities and increase the cost of these operations. Because these events are often infrequent, there should be little impact to the overall

Figure 7.13 Sheep fenced into an array to accommodate a high intensity grazing plan over one section of the PV array, photo by MNL [19]

project lifetime cost, but it needs to be accounted for during the O&M planning process. Legally, if O&M work is necessary to enable high electricity output, the O&M service for the PV equipment needs to compensate the farmer for any loss of crop yield. Often, the crop is not worth much compared to the lower electricity yield, but in land lease agreements these aspects shall be considered.

7.2.4.2 Soiling concerns

One potential concern for agricultural operations is the potential increase in soiling of the modules that can reduce electrical energy output. During tilling and other bed preparation activities, dust can be kicked up that may soil the panels. If the project is in an area without a lot of rainfall, further panel cleaning operations may be needed, but the frequency increase of these events is site dependent and currently unknown (Figure 7.15). To protect the coating of the PV module, cleaning should only be done with either (1) de-mineralized water or (2) tap-water that has an add-in that is certified to be eco-friendly and food safe.

Another potential concern is the application of farming inputs such as aerial herbicide and pesticide spraying that may soil the panels. Along with soiling concerns, it is currently unknown if these chemicals may impact wiring and other equipment on the PV site, so more study is needed to determine what agricultural input activities are compatible with Agri-PV operations. The spraying of pesticides and fungicide is often very precise but there is still potential for PV equipment soiling from chemical application. Yet, the distribution of fertilization, which are often pellets/granules, may lead to a similar effect as stone flicking and damage the PV module.

Figure 7.14 Workers accessing electrical junction box at the end of row for routine maintenance activities, photo by Werner Slocum (NREL) [9]

Figure 7.15 Demonstration of potential soiling impacts—more frequent panel cleaning may be required for Agri-PV operations [20]

7.2.4.3 Monitoring

Traditional monitoring systems for PV should not differ for an Agri-PV site. Ideally, inverter, string and module (when available) data should be available remotely using manufacturer or aftermarket data monitoring systems. Alerts should be properly configured, and staff should monitor these systems daily for potential production losses or maintenance issues. In addition to standard PV monitoring systems, installing a robust meteorological station on site to measure temperature, humidity, rainfall, wind speed and irradiance will provide benefits to PV operators and Agri-PV users alike. On-site meteorological data provides a first stop for PV troubleshooting and helps farmers make informed decisions on day-to-day operations. Depending on the type of Agri-PV operation, additional monitoring such as soil moisture, temperature and salinity sensors might be used to provide real-time data for precision agriculture and long-term trends for seasonal agricultural planning [21]. The costs of these monitoring systems are small compared to overall development costs for most PV installations, and the benefits to PV operators and agricultural producers can have large impacts on crop production Monitoring systems should be discussed between PV operator and agricultural producer early in the planning and development process to meet both party's needs.

One potential benefit of Agri-PV operations to solar O&M is a consistent presence of workers on site to monitor PV equipment (Figure 7.16). Dependent on the project, traditional solar operations may only have an operator onsite every week or so and a farmer or grazier can be trained to alert the operator when an equipment check may be needed. Cross training of farmers or graziers should be considered to provide another set of eyes to maximize PV performance.

Figure 7.16 Farmers onsite during harvest operations—cross training and identifying contact points can help identify potential equipment problems, photo by Werner Slocum (NREL) [9]

7.2.4.4 Shading and PV generation

There are potential impacts on PV generation from Agri-PV adoption. Some studies have shown an increase in PV generation from microclimate changes with vegetation underneath leading to a cooling effect of the panels [22]. This effect will not occur in all climates but should be relevant in hot and dry conditions. There is also the possibility of PV generation reductions from inadvertent shading from the farmer's presence every day on site and limiting sunlight to the modules (dependent on panel height) or crops growing taller than the lowest edge of the panels and impacting generation through the day (fixed systems) or in the morning or evening at high tilt angles (single-axis tracking).

7.3 Non-technical O&M considerations

Besides technical O&M specifications and how different operations will affect the site, there are non-technical aspects of Agri-PV projects to consider. As discussed before, a strong partnership between a solar developer, farmer, and O&M contractors is needed to ensure efficient and safe operations at Agri-PV sites. Many of the insights learned in this chapter are discussed in further depth in [1].

Often, the project developer is not the O&M service provider. Yet, the O&M service provider will work the longest time (30 years) with the farmer on the project site. To avoid misunderstandings, to allow knowledge sharing and information management, and to improve the project development phase and system layout, in Agri-PV it is important to include the O&M service in the early stage of project development and bring O&M in touch with the farmer. This will reduce the risk of site management.

7.3.1 Roles and responsibilities

At the start of project operations, there is a need to define the roles and responsibilities of different aspects of the project. This does not mean that every person onsite should have a specific and limited scope but should know who to inform before different activities begin and who should be consulted before changing operations on site. Some examples include, but are not limited to:

- Identify who is responsible for safety onsite for different operations.
- Identify who is responsible for vegetation O&M for different areas of the site, especially if there are combined crop, grazing, and vegetation operations on site.
- Establish communication pathways for those who need to be informed. An example of this is who to call if obvious PV equipment impacts are encountered or who needs to be consulted on farm operations.

7.3.2 Insurance and legal liability

Along with roles and responsibilities, liability insurance for the different entities needs to be obtained to ensure that workers and equipment are insurable for the given operations on site. All these responsibility and liability agreements should be incorporated into a management document to ensure proper documentation. There is also the potential impact of power purchase agreement stipulations that require a certain amount of electrical production to be sold by the developer that may lead to some conflict between the solar developer and farmer. Establishing these stipulations early on will help prevent miscommunication or breach of contract.

7.3.3 Land ownership and leasing

One of the more complicated agreements for Agri-PV projects may be the land leasing and ownership agreement. Traditional PV projects will often lease land from landowners and then the PV developer will be responsible for maintaining the land and vegetation. Under Agri-PV agreements, it can be unclear who will be responsible for vegetation O&M and if/how much the developer should pay the farmer to continue farmer operations. An equitable arrangement should be discussed at the beginning of the project to ensure that both partners are comfortable with the payment and responsibilities.

This can be complicated even further if the landowner is not currently farming the land and is instead leasing the land to a tenant farmer. Often, tenant farming agreements are going to be shorter in duration (roughly 5 years) than the overall project life (25 years). Changes in tenant famers during the project life may result in disruption to both farming and solar activities. Buy in from the tenant farmer, landowner, and developer would be needed if all parties would remain involved in the project.

7.3.4 Water rights

Unlike conventional PV projects, access to water is needed to provide irrigation to crops and water to grazing animals. Water rights and access to irrigation should be prioritized up front as this may cause a delay in farming operations. A change of land use from one farming activity to another may result in changes in water needs that may or may not be covered by the existing water agreements. Equipment may also be needed for storage of water onsite that is often not a part of the PV development process and can result in project delays.

7.3.5 Business models and contracting

Because Agri-PV projects are not currently of a standardized design or widespread in the PV market, innovations in the business models of crop and energy production need to be developed. One of the main questions is if there is a need for profit sharing agreements between the developer and farmer and to whom the revenue benefits of the different parties accrue. One idea is to create an independent business entity that can obtain financing and share profits equitability amongst the different parties onsite. These new business models will likely be needed to promote long term viability of Agri-PV operations.

In some projects, the farmer (and potentially landowner) is invited to become a minority investor of the equity-share of the Limited Liability Corporation that holds the project. In such a way, the farmer becomes a co-owner and has an additional income also from the PV site. This will also lead to the effect that the farmer is treating the PV hardware on site with more respect, as he is a co-owner, which diminishes Agri-PV project risks.

7.3.6 Decommissioning

One of the touted benefits of Agri-PV projects is either allowing for land recharge over the 25-year life of the PV project or to allow the continuation of agricultural activities over the project lifetime. At the end of the PV project, there are claims that the PV infrastructure can be removed, and farming operations can easily continue. Anecdotally, this is true, but many PV projects already developed are not at the end of the project lifetime and more research is needed to ascertain the truth of this. Decommissioning steps and O&M activities to maintain land quality should be incorporated into the project agreement to ensure this is accounted for and to assuage farmers and landowners.

Disclaimer
This work was authored by the National Renewable Energy Laboratory, operated by Alliance for Sustainable Energy, LLC, for the US Department of Energy (DOE) under Contract No. DE-AC36-08GO28308. Funding, in part, was provided by the InSPIRE project through the US DOE Office of Energy Efficiency and Renewable Energy (EERE) Solar Energy Technologies Office under award DE-EE00034165. The views expressed in the article do not necessarily represent the views of the DOE or the US Government. The US Government retains and the publisher, by accepting the article for publication, acknowledges that the US Government retains a nonexclusive, paid-up, irrevocable, worldwide license to publish or reproduce the published form of this work, or allow others to do so, for US Government purposes.

References

[1] Macknick J, Hartmann H, Barron-Gafford G, *et al. The 5 Cs of Agrivoltaic Success Factors in the United States: Lessons From the InSPIRE Research Study*. Golden, CO (United States): National Renewable Energy Lab. (NREL); 2022.

[2] Varga D. JADE Learning. 2021 [cited 2023 Sep 8]. Underground electrical installations. Available from https://www.jadelearning.com/blog/electrical-burial-depths-and-procedures/.

[3] Townsend TJ, Ramsden SJ and Wilson P. How do we cultivate in England? Tillage practices in crop production systems. *Soil Use and Management*. 2016;32(1):106–17.

[4] NFPA. *NFPA 70, National Electric Code* [Internet]. 2023. Available from https://www.nfpa.org/codes-and-standards/all-codes-and-standards/list-of-codes-and-standards/detail?code=70.

[5] Snake Tray. *Trenching* [Internet]. Snake Tray; 2020 [cited 2023 Sep 8]. Available from https://www.snaketray.com/solar-cable-management/trenching/.

[6] SPW. *Solar Power World.* 2023 [cited 2023 Sep 8]. The rise of above-ground solar cable management systems. Available from https://www.solarpowerworldonline.com/2023/03/the-rise-of-above-ground-solar-cable-management-systems/.

[7] Doubleday K, Choi B, Maksimovic D, Deline C and Olalla C. Recovery of inter-row shading losses using differential power-processing submodule DC–DC converters. *Solar Energy*. 2016;135:512–7.

[8] PriceItHere. *Compare Tractor Prices in* 2023 | Cost Guide | PriceItHere.com [Internet]. [cited 2023 Sep 8]. Available from https://priceithere.com/tractor-prices/.

[9] NREL. *NREL Image Gallery* [Internet]. 2023 [cited 2023 Mar 24]. Available from https://images.nrel.gov/mx/Profiles/en/landing/.

[10] Grana P. *Solar Power World.* 2018 [cited 2023 Sep 8]. The new rules for latitude and solar system design. Available from https://www.solarpowerworldonline.com/2018/08/new-rules-for-latitude-and-solar-system-design/.

[11] Kopecek R and Libal J. Bifacial photovoltaics 2021: Status, opportunities and challenges. *Energies*. 2021;14(8):2076.

[12] Takemura A. *Canary Media.* 2022 [cited 2023 Sep 8]. Can agriculture and solar farms coexist? It depends. Available from https://www.canarymedia.com/articles/food-and-farms/can-agriculture-and-solar-farms-co-exist-it-depends.

[13] SEIA, *ASGA. Solar & Multiuse Farming* [Internet]. 2019. Available from https://solargrazing.org/wp-content/uploads/2022/01/Solar-Multiuse-Farming-Practices-Fact-Sheet-2019-v3.pdf.

[14] National Renewable Energy Laboratory. *Low-Impact Solar Development Strategies Primer* [Internet]. 2022. Available from https://openei.org/wiki/InSPIRE/Primer#What_Is.

[15] McCall J, Macdonald J, Burton R and Macknick J. Vegetation management cost and maintenance implications of different ground covers at utility-scale solar sites. *Sustainability*. 2023;15(7):5895.

[16] National Renewable Energy Laboratory. *InSPIRE/Agrivoltaics Map | Open Energy Information* [Internet]. 2023 [cited 2023 Feb 10]. Available from https://openei.org/wiki/InSPIRE/Agrivoltaics_Map.

[17] Dreves H. *Growing Plants, Power, and Partnerships Through Agrivoltaics* [Internet]. [cited 2023 Sep 11]. Available from https://www.nrel.gov/news/program/2022/growing-plants-power-and-partnerships.html.

[18] Andrew AC, Higgins CW, Smallman MA, Graham M and Ates S. Herbage yield, lamb growth and foraging behavior in agrivoltaic production system. *Frontiers in Sustainable Food Systems*. 2021;126:6.

[19] Pickerel K. *Solar Power World.* 2022 [cited 2023 Sep 8]. Minnesota research finds sheep grazing at solar sites actually improves soil quality. Available from https://www.solarpowerworldonline.com/2022/01/minnesota-research-finds-sheep-grazing-at-solar-sites-actually-improves-soil-quality/.

[20] Cordero RR, Damiani A, Laroze D, *et al*. Effects of soiling on photovoltaic (PV) modules in the Atacama Desert. *Scientific Reports*. 2018;8(1):13943.

[21] Choi P, Macknick J, Li Y, Bloom D, McCall J and Ravi S. Environmental co-benefits of maintaining native vegetation with solar photovoltaic infrastructure. *Earth's Future*. 2023;11:e2023EF003542. https://doi.org/10.1029/2023EF003542.

[22] Barron-Gafford GA, Pavao-Zuckerman MA, Minor RL, *et al*. Agrivoltaics provide mutual benefits across the food–energy–water nexus in drylands. *Nature Sustainability*. 2019;2(9):848–55.

Chapter 8
Diverse perspectives and common goals: the importance of social aspects on Agri-PV diffusion

Alexis S. Pascaris[1]

Agriculturalists face mounting challenges as changes in environmental conditions combined with development pressures strain their farm's viability. Energy companies struggle to site and deliver cost-effective projects as public opposition to development intensifies. Regulators endeavor to meet multiple environmental objectives at once yet are strained to manage such cross-sector complexity. While oversimplified, these diverse perspectives demonstrate the compounding difficulties faced by a society striving for common goals: sustainable agriculture and energy systems.

Innovation in agrivoltaics provides a means to revolutionize how these complexities are managed; this integrated climate solution prompts the cross-sector collaboration needed to better address critical food-energy-water-biodiversity nexus challenges. By forging a synergistic relationship between agriculture and energy, agrivoltaics has the potential to drive socio-technical transition towards more regenerative practices and policies. Delivering this cross-cutting solution will therefore demand cross-cutting competencies, encompassing both techno-economic efficiency gains and broader social and institutional adaptations. To that end, a deeper consideration of the social aspects of agrivoltaics is warranted.

The purpose of this chapter is to draw attention to the importance of social aspects of technology diffusion and their implications on agrivoltaics. Aligned with the theme of this book—establishing agrivoltaics as a tool for climate resilience and sustainable development—this chapter demonstrates how creating an enabling social framework is essential to realize this potential. While current social science research on agrivoltaics is slim and dispersed, all scholars stress a common thesis: diffusion of the innovation requires effort beyond optimizing technical aspects—stakeholder acceptance and adoption, proper legal frameworks, and favorable market conditions are imperative. This chapter elaborates this thesis by presenting the current state of the social science, which at its core, is an investigation of agrivoltaics through the lens of innovation diffusion theory. Within this innovation diffusion approach, a few general streams of literature have emerged—one

[1]AgriSolar Consulting, LLC, United States of America

concerns micro-level technology adoption by investigating individual stakeholder perspectives; another looks at meso-level (community) aspects of social acceptance and multi-stakeholder interactions; and the other considers macro-scale dynamics related to economics and policy. These streams of literature are not isolated; rather, they often overlap, which is a testament to the complexity of the acceptance, adoption, and diffusion of agrivoltaics. The studies presented in this chapter leverage multiple social science research methods, including interviews and surveys, to understand this complexity and outline the enabling social framework needed to fully realize agrivoltaics as a tool for sustainable development.

High-level insights are organized thematically to set the conceptual foundation for discussion on the significance of social aspects on the diffusion of agrivoltaics. Section 8.1 highlights the multi-stakeholder perspectives on the drivers of agrivoltaics to demonstrate the valuable social co-benefits and priorities to be retained in development. Section 8.2 addresses challenges to adoption and discusses the technical, economic, cultural, and political factors that complicate the development landscape. Section 8.3 considers the interaction of these dynamic drivers and challenges, potential pathways forward, future research needs, and implications for the field.

8.1 An occasion for innovation

Agrivoltaics has emerged with promise at the confluence of many sustainable development challenges. Stakeholders across sectors are building constituency around this food-energy-water-biodiversity nexus solution. Early social science research on agrivoltaics has confirmed that farmers, solar developers, and citizens alike value the robust potential of the innovation; these studies maintain that upholding and harmonizing diverse values across the stakeholder groups who are involved and impacted is key to arriving at successful agrivoltaic outcomes. To that end, researchers have been mapping trends in perceptions on agrivoltaics to inform pathways to multi-stakeholder acceptance and adoption. The common interests identified center on resilience, sustainability, and justice—all of which are at the heart of global sustainable development goals (SDGs). Figure 8.1 demonstrates how the SDGs are embedded, rather than existing independently, and suggests that "Partnerships for the Goals" is the link that makes these common goals realizable.

8.1.1 Agrivoltaics for farm viability

Strengthening the role of the agriculture sector in the energy transition requires robust stakeholder engagement to prioritize farmer needs in system design and policy development. This understanding is reflected in researcher's exploration of farmer's values and interests regarding solar on farmland. Qualitative interviews [1–6] and quantitative surveys [1,7,8] across farm operations of varying scales and geographies have provided insight to the matter of farmer adoption of agrivoltaics. The common denominator driving farmers towards agrivoltaics is the potential to enhance farm viability, both in terms of climate resilience and income diversification. Therefore, positioning solar to satisfy both resilience and diversification

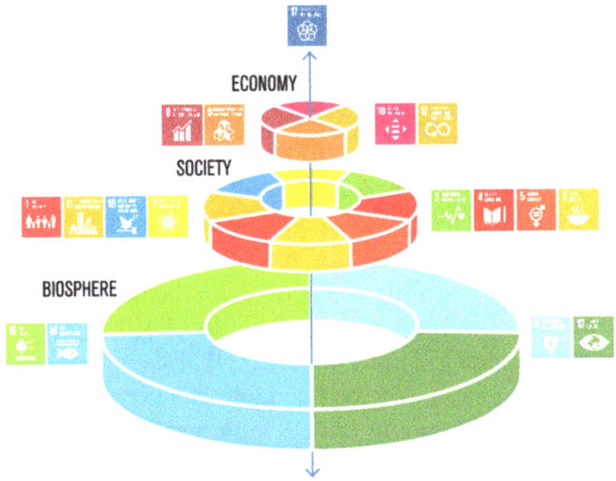

Figure 8.1 *"The Sustainable Development Goals Wedding Cake" reprinted from the Azote for Stockholm Resilience Centre, Stockholm University CC BY-ND 3.0. https://www.stockholmresilience.org/research/research-news/2016-06-14-the-sdgs-wedding-cake.html*

needs will be key to increasing the long-term demand for agrivoltaics from the agricultural sector.

Climate resilience and adaptation are at the core of agrivoltaic innovation. Farmers in Japan [1], Austria [3], Turkey [5], the Netherlands [6], China [7], and the United States [2,9] are actively assessing benefits to production made possible by combining solar with their operation. Producers in arid regions who face chronic water-related issues have exhibited the most positive attitude towards agrivoltaics [5]. Through in-depth interview of 16 farmers in Turkey, Ağir *et al.* demonstrate a strong biophysical and social fit for agrivoltaics in arid climates. Study participants focused on the potential for agrivoltaics to address local agricultural challenges associated with drought and heat stress—envisioning how their native crops would work well in an agrivoltaic environment, they noted the likelihood of increased productivity, reduced water and herbicide use, and sustained operation despite climatic changes [5]. The solar canopy is also perceived as an asset by crop producers in Austria [3] and livestock farmers in the United States [2], indicating that the unique microclimatic conditions and aerial protection provided by solar are valuable to a diversity of farm operators. This perceived usefulness, for various technical and environmental reasons, demonstrates how climate resilience is a driving force of farmer adoption of agrivoltaics [6,7].

Underlying the pursuit for climate resilience is economic viability. Agrivoltaics is perceived by famers as an opportunity to achieve both technical and economic co-benefits for their operation—the prospect of energy independence through net-metering or income diversification through lease payments is paramount [1,5,6,9]. Research has demonstrated how farmers are primarily concerned with "supplementing meager returns" by leasing land for solar or reducing on-farm energy costs [9]; while these two potential goals require different technical and regulatory configurations, agrivoltaics can be designed to meet a range of farmer priorities related to income diversification and energy independence. Other appealing income diversification aspects of agrivoltaics are opportunities for value-add labeling schemes and associated marketing advantages [2,6]. Overall, much of the discourse on solar among farmers centers on farm viability and is grounded in values related to diversification, independence, pride, land stewardship, and private propriety rights [4]. Some farmers in the United States have discussed an acceptable trade-off between energy and agriculture, being willing to convert some land to solar production to realize economic benefits [4,10], which is perceived as a farmland preservation strategy if done properly [9]. Taken together, these research findings emphasize the prospect of compounding benefits, including climate resilience and income diversification, as critical to farmer adoption of agrivoltaics.

These initial farmer engagement efforts showcase how the decision to adopt solar on farmland involves multi-layered criterion across economic, environmental, cultural, and socio-political factors [11]. From the northeast Unites States to the Mediterranean and beyond, there is consistency in farmer's values in agrivoltaics—climate resilience, energy independence, economic security, and land stewardship. This growing body of research demonstrates how driving agriculture sector adoption of agrivoltaics will require farmer-informed systems that enhance farm viability.

8.1.2 A strategy for the solar industry

Delivering robust agrivoltaic solutions is largely the responsibility of the solar industry. To understand the agrivoltaic development landscape and identify the key motivations driving industry adoption, researchers have been engaging solar professionals through in-depth interviews. These targeted industry adoption studies are few, and are concentrated in the United States [9,12,13]. Multi-stakeholder interview studies that include insights from industry actors have also occurred in Japan [1], Austria [3], the United States [4], the Netherlands [6], and Germany [14]. Industry actors across geographies and operational scales share common motivations regarding agrivoltaic adoption, including siting conflict mitigation, enlarged area potential for PV deployment, enhanced community relations and reputation, continued access to rural markets, and delivery of "triple-bottom-line" benefits. These initial industry perspective studies help elaborate the social, environmental, and economic motivations behind the emerging agrivoltaic market segment.

Industry actors are growing keen on the diverse co-benefits associated with agrivoltaics [13]. In two separate industry perspective studies, Pascaris *et al.* have catalogued the key reasons US solar developers choose to pursue agrivoltaic projects. They found that developers associate agrivoltaics with better community relationships and ecosystem services, and that they consider such co-benefits integral to company success. These co-benefits are part of the broader package of agrivoltaics, which industry actors anticipate they will be able to leverage in the development process to win projects [12]. In fact, the potential to increase community acceptance of solar was discussed among study participants as the greatest opportunity for agrivoltaics from an industry perspective—developers often face opposition to solar on farmland, but some believe that agrivoltaics offers a unique strategy to mitigate this siting conflict, enhance community relations, and ultimately smooth the permitting process, all at once [12,13]. Solar professionals in Austria also raised the potential for agrivoltaics to provide a strategic advantage compared to traditional solar, anticipating that implementing these projects would boost company competitiveness and long-term reputation [3]. Across industry perspective studies, participants alluded to how delivering socio-environmental co-benefits with solar in a cost-effective manner translates to market success. These studies demonstrate how community relations and land stewardship have implications for company prosperity, which poises agrivoltaics as uniquely valuable. Additionally, agrivoltaics is considered among some developers a strategy to reduce operations and maintenance (O&M) costs in the long-term [13], which contribute to cost-optimization objectives. In alignment with social and economic values of solar companies, agrivoltaics is gaining recognition across the industry as a means to capture multiple, diverse benefits at once.

While limited in geographic scope, these solar industry engagement studies provide the initial insight needed to understand the emerging agrivoltaic development landscape. The findings strengthen the thesis that the industry values the social, environmental, and economic potential of agrivoltaics and that the innovation satisfies their many objectives in a unique way. Having the potential to increase community acceptance and enhance company reputation through delivery of multi-functional projects that prioritize local economies and ecosystems, solar developers are coming to appreciate agrivoltaics as critical to the long-term success of the entire industry [13]. Emphasizing the role of solar developers in agrivoltaics, who operate at a critical nexus between landowners, farmers, community members, and regulators, is an essential aspect of broader innovation diffusion efforts.

8.1.3 Local food, local power

Both farmers and solar developers operate within communities, therefore, understanding the "*acceptance*" of agrivoltaics from a community (meso-level) perspective is essential. Studying community perspectives entails a different unit of analysis than individual (micro-level) "*adoption*" perspectives and thus provides more dynamic to the discourse on agrivoltaic diffusion. To learn about community acceptance factors relative to agrivoltaics, researchers have been cataloging meso-level perspectives through citizen workshops [11,15] and public survey [1,16]. These studies reveal how community members perceive agrivoltaics, how such projects relate to their values and interests, and how projects can embody those values to increase acceptance of solar development in the long-term. Core values related to participatory planning and local resilience have emerged as critical to community acceptance of agrivoltaics, which inform the key factors of importance to be prioritized in development [1,6,11,15,16].

Community-engaged research efforts have found favorable perspectives towards agrivoltaics compared to traditional solar development and have identified opportunities to create meaningful local benefits [6,11,15,16]. Community stakeholder involvement is critical to realizing these benefits, as demonstrated by the pioneering work conducted by Ketzer *et al.* [11,15] in Germany, which emphasized the value of participatory approaches in agrivoltaics. Through a series of workshops pre- and post-construction of an agrivoltaic system, researchers determined higher community acceptance as a result of public participation in project planning. The prioritization of local practical knowledge, including agricultural practices, economic interests, and site selection, created the basis for a "locally legitimate planning framework" [11].

The significance of a locally driven planning framework is further validated by research in the Netherlands [6], Germany [14,15], and the United States [16], which have all found consistency in the perceived opportunity for and importance of local economic and agricultural benefits. Interestingly, these core community values related to local resilience are reflected in the results of a survey conducted in two US counties; Pascaris *et al.* found that 81.8% of respondents would be more likely to accept a solar project in their community if the project supported agricultural production. Additionally, survey respondents stressed economic benefits to farmers and their community as key priorities and noted the potential of agrivoltaics to support these ideals [16].

These findings indicate that incorporating local criteria and interest in the development of agrivoltaic projects can result in positive community sentiment, tangible economic benefits, and increased acceptance—which are all critical to diffusion efforts. From a community perspective, participatory planning and local resilience are central to acceptable agrivoltaic outcomes. Understanding the dynamics between these meso-level "*acceptance*" factors and micro-level "*adoption*" criteria is an important next step as researchers seek to outline the proper framework conditions for agrivoltaics.

8.1.4 More value for more people

The growing body of social science scholarship on agrivoltaics emphasizes how this form of solar development creates diverse value for multiple stakeholders. Farm viability, sustainability and resilience, and economic prosperity have been elevated across all sets of actors—these trends in farmer, solar developer, and community member values underscore the key driving forces of agrivoltaic acceptance and adoption. The potential to create manifold social, environmental, and economic benefits is recognized by many, and this perception of value creation is evinced in the growing willingness to adopt agrivoltaics across stakeholders and countries. By prioritizing these diverse stakeholder perspectives and values, the insights from this first generation of social science research on agrivoltaics can help outline a pathway for the long-term acceptance, adoption, and diffusion of the innovation.

8.2 When two complex systems collide

The agriculture and energy sectors are independently complex—forging their union through agrivoltaics requires understanding how these systems and their

Figure 8.2 Integrated photovoltaics—areas for the energy transformation. Fraunhofer Institute for Solar Energy Systems ISE.

stakeholders are engrained in existing paradigms and how those paradigms impact innovation diffusion. Although mutual learning across sectors to create integrated photovoltaic solutions is not extraordinary (Figure 8.2), developing common goals and standards for combined agriculture and solar systems presents unique multi-sector, multi-level complexities. By uncovering these complexities within and across key sectors, researchers have outlined the micro-, meso-, and macro-level dynamics that challenge agrivoltaic adoption. Findings across sectors and stakeholders point towards various technical, economic, and institutional challenges—perceived threats to long-term agricultural viability and heritage, cost–benefit uncertainties, misaligned regulatory regimes, and distributional justice issues have been identified as central roadblocks to upscaling agrivoltaics. Researchers concerned with the multi-level criterion for agrivoltaic adoption stress that proactive attention to these sectoral and stakeholder challenges is vital.

8.2.1 Deep roots in existing practice

Multi-layered technical, economic, political, and cultural factors challenge the uptake of agrivoltaics by the agricultural sector. To understand the magnitude and complexity of these challenges from a farmer perspective, researchers have leveraged qualitative interview methods [1–6] and quantitative surveys [1,7,9]. These studies provide insight into micro-level adoption concerns, meso-level socio-cultural challenges, and macro-level institutional dynamics that all have implications on agrivoltaics. This research suggests that agriculture sector engagement with agrivoltaics does not merely involve farm integration of solar, but instead requires a willingness to alter existing practices, engage in risk, navigate bureaucratic hurdles, and transcend paradigms around the cultural heritage of farmland.

Farmers in Japan [1], Austria [3], Turkey [5], the Netherlands [6], China [7], and the United States [2,9] have disclosed their hesitations about the feasibility and viability of agrivoltaics. Both farmers who are keen on the concept and those who are skeptical have stressed a need for agrivoltaics to be optimized in practice through proof of concept before they seriously consider adoption [3,6]. Study participants across countries and farm operation types have flagged gaps in knowledge as concerning, especially unknowns related to impacts on production [1,2,5,9]. A range of potential adverse effects on agricultural production have been raised by farmers, including questions about end-of-life impacts to soil, effects on crop quality and yield, and changes in wind patterns on crop fertilization success, which are all important for long-term farm viability [1,5,6]. Farmers also perceive many operational challenges associated with agrivoltaics and have expressed concern about incompatibility of solar with their existing cultivation practice and its implications on future farming flexibility [1,2]. Study participants across the Netherlands, Turkey, Japan, and the United States perceive navigating fixed infrastructure as a short-term burden and a long-term impediment to crop rotation and diversification [1,2,5,6]. As these technical factors are all inherently related to economic viability, farmers have also emphasized concerns about the cost and profitability of agrivoltaics [1,5,7], constraints on business planning to accommodate market changes [2,5], and hesitations around fair agreements with solar companies [2,6]. This diversity of technical and economic challenges perceived by farmers evaluating agrivoltaic adoption are critical for informing future research, system designs, technical assistance, business models, and more.

These micro-level adoption challenges are entangled within meso- and macro-level dynamics of agricultural heritage and regulatory regimes. Farming is both an occupation and an identity, which complicates agrivoltaic adoption, as these stakeholders operate within communities and systems that often regard farmland as a public good rather than private property [4]. Solar siting on farmland is thus not only an individual decision, but a matter that involves broader political and community considerations. In a study on multi-stakeholder interactions around farmland solar in the United States, Moore *et al.* emphasized how issues regarding the rights of agricultural landowners will have implications on the diffusion of agrivoltaics [4]. Researchers in Germany [17], the Netherlands [6], and the United States [9] also found concern about adverse impacts on tenant farmers whose landowners engage in solar leasing but do not properly design business models and systems to accommodate existing actors. These studies demonstrate that socio-economic and cultural issues are connected to broader institutional challenges that perpetuate the segregation of agriculture and energy; Ağir *et al.* call for attention to the institutional setting in which farmers operate, stressing that low political feasibility and predictability challenge farmer investment in agrivoltaics and undermines a supportive legal framework [5]. Complex tax issues, burdensome bureaucratic procedures, and conflicting stakeholder opinions present formidable barriers to farmer adoption of agrivoltaics [1,2,4,6].

As traditional farming is both a matter of livelihood and culture, a full spectrum of technical, economic, social, and political factors must be assimilated into agrivoltaic introduction efforts. Social scientists are working to better understand how the interplay of agricultural heritage and disparate regulatory regimes has implications on diffusion and what adaptations are needed to enable greater adoption among farmers. These early research insights demonstrate that the broad uptake of agrivoltaics by the agriculture sector will require reconciling farmer concerns about production and operation, as well as socio-cultural and political challenges around solar siting on farmland.

8.2.2 The liability of newness

Agrivoltaic projects are brilliant in theory but challenging in practice. This is made clear by research concerned with understanding barriers to industry adoption, which has found that solar professionals perceive agrivoltaic development in terms of uncertainty, complexity, and cost. These micro-level adoption studies [12,13], multi-stakeholder assessments [1,3,6], and expert engagements [18] provide key insight into the agrivoltaic development landscape and market conditions. An underlying "liability of newness" impedes agrivoltaic development; industry actors acknowledge uncertainty regarding the cost–benefit balance, design challenges, community acceptance, and political feasibility. This small but growing collection of research on factors impacting industry adoption of agrivoltaics all concur that a complex web of techno-economic and socio-political factors are challenging the expansion of this market segment.

Techno-economic unknowns and uncertainties have been stressed by solar developers as critical barriers to their adoption of agrivoltaics. In a multi-stakeholder interview study, Torma and Aschemann-Witzel found a stark contrast in perceptions about the viability of agrivoltaics across their industry participants; while highly regarded among some, many others expressed extreme skepticism based on a wide range of techno-economic unknowns and uncertainties [6]. The essence of these unknowns and uncertainties is elaborated by Pascaris *et al.* in discussion about the "liability of newness" of agrivoltaics and the compounding effect it has on industry adoption [13]. Regarding agrivoltaic projects as complicated, expensive, and risky relative to conventional solar, many industry actors are averse to development practices beyond their core scope of work [6,12]. Uncertainties about optimal designs, political feasibility, and agricultural management directly impact the ability to perform cost–benefit analysis, which is central to decision-making and investment confidence [13]. These challenges result in hesitancy among the industry, who are asking for systematized research, data validation, market assessments, knowledge dissemination, and curation of best practices to inform their adoption [3,6,13,18,19]. In a study devoted to deducing major challenges to agrivoltaic development, Vorast found lack of crop data validation, high capital costs, and unclear risk evaluation all disabling industry adoption of agrivoltaics [18]. These interrelated techno-economic challenges are interfering with market diffusion.

Socio-political factors further complicate the agrivoltaic development landscape. Solar on farmland is not only subject to organized public opposition [12], but also to complex taxation, zoning, permitting, and siting [4,6,19–21]. Developers pursuing agrivoltaics are burdened with bureaucratic complexity, which is a direct result of insufficient cohesion and standardization across regulatory agencies [13], unclear approval processes [18], and prolonged planning and implementation [6]. Political feasibility thus presents a barrier to industry adoption; not only does it belabor project delivery but results in higher soft costs, which reduces the perceived value of agrivoltaics. Further, in the United States, restrictive land use policy can preclude solar installation on designated farmland, which is a distinct barrier to agrivoltaic development [19–21]. Yet a lack of supportive policy is not exclusive to the United States—researchers have found missing incentives and complicated regulation pervading the agrivoltaic development landscape in Mali [22], Austria [3], the Netherlands [6], and beyond [18]. As a cross-sector innovation, regulators and developers are strained to manage agrivoltaic permitting and programs. Absent clear development standards, a cohesive legal framework, and supportive market conditions, industry actors are overburdened in attempts to advance agrivoltaic projects.

While solar developers strive to deliver cost-optimized projects with speed and ease, research suggests that the perceived complexities and uncertainties associated with agrivoltaics does not exactly align with their ideals of efficiency. As a nascent development practice, many developers are skeptical about the economic, technical, and socio-political feasibility of agrivoltaics—the liability of newness, unclear economics, and lack of a supportive legal framework is invoking hesitancy among the majority of industry actors, which is perpetuating business-as-usual. This research demonstrates how both techno-economic and socio-political challenges are thwarting broad industry adoption of agrivoltaics.

8.2.3 Cultural landscapes matter

Agrivoltaic development efforts are subject to the values and concerns held by the broader community in which the project is sited. Therefore, researchers have sought to understand what potential impacts are most concerning to community members and to identify mitigation strategies through workshops [11,15], interviews [1,4,6,22], and surveys [16]. Community engagement has revealed a range of concerns with the potential environmental, economic, and cultural impacts of agrivoltaics, as well as unease with perceived justice implications.

A dominant conviction among agricultural constituents is that farmland is a public good that underpins food security, rural identity, and agrotourism [4]. Agricultural community members value farmland as more than a parcel of private property, culturally, environmentally, and economically. In a series of workshops with members of a community hosting an agrivoltaic pilot project, Ketzer *et al.* found concern that agrivoltaics would disrespect local land characteristics and change cultural landscapes [11,15]. A full range of concerns have been elevated across various facilitated discussions; not only do study participants care about discrete aesthetic factors around project size and proximity, as well as crop production standards and monitoring, but potential effects on broader system impacts related to recreation, tourism, and the functional landscape have been flagged (see Figure 8.3, example of landscape integration) [1,6,11,15]. Findings from workshops

Figure 8.3 Integration of solar with agricultural landscape—"Agri-PV Obstbau" project in Gelsdorf, Rhineland-Palatinate, Germany by BayWa r.e. and the Fraunhofer Institute for Solar Energy Systems ISE. https://www.ise.fraunhofer.de/en/press-media/press-releases/2021/first-agrivoltaic-system-for-carbon-neutral-orcharding-being-tested.html

indicate that perceived negative impacts on local interests and landscapes can reduce the social acceptability of agrivoltaics; many study participants perceived risks and cascading effects on other sectors [11,15]. Concern about effects on other sectors and aspects of agricultural communities was further emphasized by Pascaris *et al.* in a survey of two rural US communities that catalogued key concerns related to social acceptance of agrivoltaics [16]. Among a range of agrivoltaic development factors, economic opportunities for farmers and the host community, fair distribution of economic benefits, and minimal threat to local interests were raised by community members as most critical. The importance of broader agricultural stakeholder participation in agrivoltaic development that Ketzer *et al.* and Pascaris *et al.* suggested was additionally supported by Torma and Aschemann-Witzel who similarly found that distributional justice and socio-economic impacts are central to community-level concerns [6,11,15,16]. These seminal studies on community perspective conclude that preserving cultural landscapes, ensuring just processes and agreements, and avoiding the outcome of "winners and losers" are all critical to agricultural communities' acceptance of agrivoltaics.

Because there is an intrinsic, inexplicable value to farmland, perceived threats to rural character, agricultural economies, and farming heritage are met with aversion. Research interested in how agrivoltaics fits or conflicts with these community values demonstrates the importance of public participation in planning to ensure outcomes that are aligned with local interests and do not negatively impact cultural landscapes. As many rural communities thrive and depend on agriculture, both economically and culturally, distributional justice and farmland preservation must be key tenets of successful agrivoltaic developments.

8.2.4 Compromise

Harmonizing agriculture and energy through agrivoltaics can be perceived either as a collision of two complex systems or as an opportunity for mutually beneficial compromise. Social science research focused on identifying and mitigating innovation diffusion challenges recognizes a complex interplay across sectors and stakeholders. These early studies clarify the interests, values, and concerns of farmers, solar developers, and community members, and illustrate how micro-level adoption and meso-level acceptance are embedded within macro-level dynamics of legal frameworks and existing paradigms. Taken together, this body of research offers an outline of key barriers to agrivoltaic acceptance and adoption across stakeholders—strengthening agricultural viability, optimizing cost-benefit trade-offs, developing coherent and supportive policy, and ensuring distributional justice will be critical to ongoing diffusion efforts.

The realization of agrivoltaics at scale will require an element of compromise across sectors and stakeholders. Farmer adoption is most contingent on techno-economic viability and compatibility; solar industry adoption hinges on risk mitigation and cost-optimization; and community acceptance rests in participatory planning that maintains cultural and economic interests. Yet underlying these diverse perspectives are common SDGs. The research that has identified these key factors has also suggested that arriving at an optimal outcome for all will entail structured processes and incentives to work across domains to make agrivoltaics operational [23]. The interrelated technical, economic, and socio-political barriers to agrivoltaic acceptance and adoption identified by this body of research warrants further investigation.

8.3 Toward "socio-agrivoltaic" systems

Innovation in agrivoltaics prompts the cross-cutting collaborations needed to deliver resilient and equitable solutions for global food and energy challenges. As research and demonstration further validate the food-energy-water-biodiversity nexus benefits provided by agrivoltaics, it also reveals how the technology interacts with agricultural heritage and economies, cultural landscapes, existing markets, and regulatory regimes—all of which have major implications on innovation success. The cases presented in this chapter provide the basis for understanding the socio-political and techno-economic complexities associated with agrivoltaic acceptance, adoption, and diffusion. This body of scholarship stresses that advancing agrivoltaic solutions will largely require reconciliation of social challenges across sectors and stakeholders; researchers are calling for an improved theoretical approach that acknowledges "socio-agrivoltaic" systems as embedded within communities, businesses, economies, and agricultural landscapes [4,13,24]. This approach honors the importance of social aspects of technology diffusion and better positions agrivoltaics to serve as a tool for sustainable development.

Recognizing the necessity for an improved approach, social scientists have prioritized stakeholder perception studies to contextualize agrivoltaic diffusion.

These researchers are keen on the processes and pathways that could move the innovation beyond niche-application and have therefore focused on identifying key elements of the enabling social framework for widescale deployment. Of their own accord, scholars across geographies have concluded a common thesis: ensuring success of agrivoltaics requires effort beyond optimizing technical aspects—stakeholder acceptance and adoption, proper legal frameworks, and favorable market conditions are imperative. They have arrived at this thesis by investigating agrivoltaics from a variety of theoretical lenses, including the diffusion of innovations theory [25], social acceptance of renewable energy [26], and socio-technical transitions theory [27]. Future research could draw insight from "the ambiguity-conflict model of policy implementation" [28] or "the joint-decision trap" [29] to properly approach the difficulties of cross-sectoral, multi-level policy making for agrivoltaics. These social theories help us better understand the interplay of stakeholders, sectors, economics, culture, and policy, and how these dynamic forces impact technology diffusion. Addressing the broader social setting in which agrivoltaic diffusion will occur is thus a priority for future research and development efforts, which underscores a need to move towards "socio-agrivoltaic" systems.

8.3.1 Emerging data patterns and future needs

This chapter showcases how a diversity of perspectives about technical, economic, social, and political factors shape the agrivoltaic development landscape. By systematizing and interpreting multi-stakeholder perspectives on agrivoltaics, this body of research provides a roadmap that flags key opportunities, values, interests, concerns, and barriers that need be accounted for in ongoing research and development. Farmers are driven towards agrivoltaics to enhance farm viability through climate resilience and income diversification but are challenged to adapt their practices and manage unknowns across environmental and market impacts. Solar developers are pursuing agrivoltaics to better deliver triple-bottom-line solutions that increase community acceptance and company competitiveness but are hindered by risk assessment uncertainties and complicated bureaucratic processes. Community members value the food, energy, and economic resilience potential of agrivoltaics, but are concerned about fair outcomes and broader impacts on landscapes and agricultural heritage. These findings demonstrate how multi-stakeholder acceptance and adoption is contingent on optimizing trade-offs across these core drivers and challenges. Advancing agrivoltaics in a way that capitalizes on common goals while minimizing differences is thus central to the future research agenda.

Interpretation of the key social science findings presented in this chapter informs recommendations for enhanced agrivoltaic development practice and policy, as well as future research needs. This work explains how arriving at agrivoltaic outcomes where benefits are maximized and concerns are mitigated across all stakeholders will require participatory planning, a supportive legal framework, and cost-benefit optimization. Farmer- and community-driven projects can nurture greater acceptance, ensure just distribution of benefits, and reduce conflict with existing local interests and values. It is clear that positioning agrivoltaics to boost farm viability and community resilience is key to long-term industry success; this farmer-first and community-engaged approach can soften public resistance to development early on through inclusive planning and sensitivity to culture of agrarian communities. Further, farmer- and community-driven projects can reduce the "liability of newness" challenging the solar industry's ability to deliver agrivoltaic solutions. By informing the appropriate location, design, and business model, participatory planning for agrivoltaics can boost the developer's ability to achieve cost-benefit optimization. This virtuous cycle of inclusion and information sharing can compound benefits for all stakeholders, and eventually drive the needed adaptations in practice and policy to provide the stable environment for innovation diffusion. By pointing towards the interactions across technology, practice, and regulation, social science research on agrivoltaics provides the basis for developing an enabling framework and specifies the next generation of research priorities.

More social science research is needed to fully capture how complex drivers and challenges interact and impact the diffusion of agrivoltaics—characterizing these dynamics could support efforts to increase stakeholder acceptance and adoption, develop proper legal frameworks, and establish favorable market conditions—all of which have been stressed as critical to success of the innovation. At present, the body of scholarship presented in this chapter provides a general typology of drivers and challenges across stakeholders; key technical, economic, and political factors have been identified in isolation, yet more attention is needed to the interaction of these factors and actor groups, and the resulting implications on agrivoltaic development. Torma and Aschemann-Witzel propose the "Pentagon Tower of Social Innovation Diffusion" as an improved theoretical approach to understand the complex drivers across micro-, meso-, and macro-levels—future social science research on agrivoltaics could aim to elaborate this emerging theory [6] (Figure 8.4).

Ultimately, the current focus on multi-stakeholder interactions must be deepened as well as expanded to include the perspectives of regulators, urban planners, community organizers, farm bureaus, and land trusts. There are gaps in knowledge pertaining to their interests and concerns in relation to agrivoltaics, yet these actors are vital to strengthening cross-sector collaboration and developing robust policy outcomes. Progressing the social science research priorities identified by the studies collated in this chapter will be beneficial for the agrivoltaic sector more broadly. Table 8.1 provides a simplification of the findings, recommendations, and future research needs.

8.3.2 Conclusions

Sustainability transition demands both technical and social change. Advancing research, development, and policy for agrivoltaics from a socio-technical perspective can better position the innovation to achieve high acceptance and successful market diffusion. Framing agrivoltaics as a "techno-fix" for global food and energy

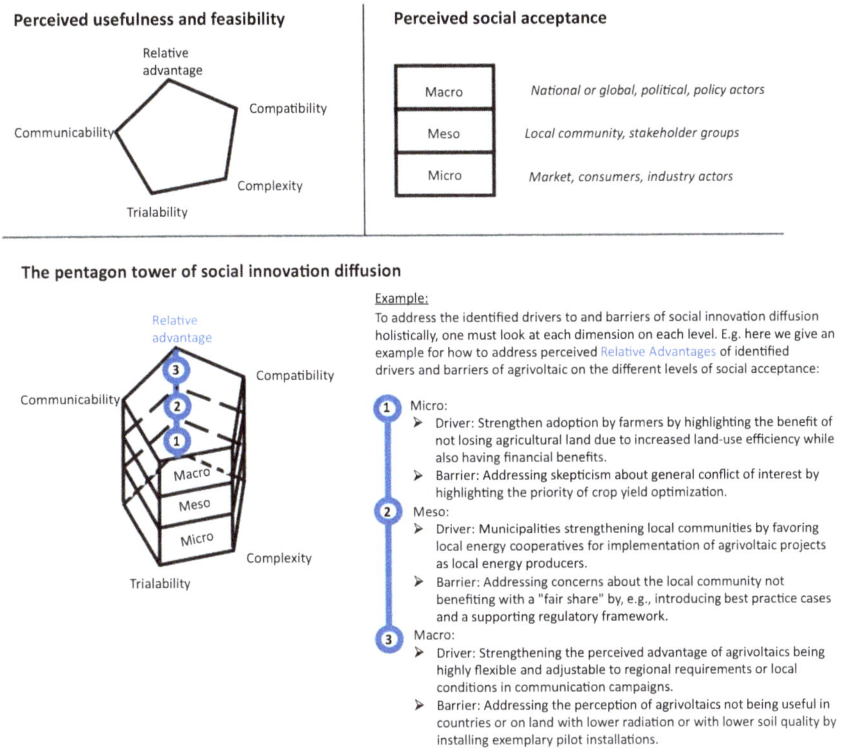

Figure 8.4 "The Pentagon Tower of Social Innovation Diffusion" reprinted from "Social acceptance of dual land use approaches: Stakeholders' perceptions of the drivers and barriers confronting agrivoltaics diffusion," by G. Torma and J. Aschemann-Witzel, 2023, Journal of Rural Studies (97), p. 622

challenges neglects the relational, cultural, and political realities of solar integration with agrarian landscapes. This framing risks foregoing prioritization of societal values in technology diffusion, which could curtail efforts to realize agrivoltaics as a multi-functional solution to diverse sustainable development challenges.

To that end, this chapter grounds the discussion on agrivoltaics in the reality of stakeholder adoption, legal frameworks, and market conditions to demonstrate that both techno-economic efficiency gains and broader social and institutional adaptations are vital to innovation success. The research presented herein suggests that social aspects, including acceptance, economics, and political feasibility are the core determinants of agrivoltaic diffusion. By highlighting importance beyond technical dimensions, this chapter on social aspects contributes a valuable perspective that is intended to enhance research and policy agendas for agrivoltaics. As sustainability challenges intensify, building robust partnerships and policies around common goals is our surest path forward.

Table 8.1 Simplified overview of key drivers and challenges of agrivoltaics across stakeholder groups, relevant recommendations, and future research needs

Stakeholder	Drivers	Challenges	Recommendation	Future research
Farmer	– Climate resilience – Income diversification	– Impacts on production and operation – Lack of political and financial support	"Need for research and demonstration for farmers, but also attention to institutional setting in which farmers operate" [5]	Workshops to identify key system design modifications to enable agricultural production long-term and identify technical assistance needs
Solar developer	– Deliver multi-functional projects – Increase community acceptance and enhance company reputation	– Cost–benefit and risk evaluation uncertainty – Bureaucratic challenges	"Synchronistic improvements in technical performance, price, social acceptance, user practice, and regulation are necessary to improve industry adoption of agrivoltaics" [13]	Longitudinal study of changes in economic viability and the effect on industry adoption rate as market mechanisms are introduced
Community member	– Local energy and food resilience – Rural economic development	– Impacts on cultural landscapes and agricultural economies – Distributional justice	"Research and innovation should be designed as a transparent, interactive process by which societal actors and innovators become mutually responsive to each other with a view on the acceptability, sustainability, and societal desirability of the innovation process" [15]	Surveys to assess differences in perspectives across rural and urban communities relative to acceptance of solar versus agrivoltaics
Policy makers and regulators	– Integrated solution that addresses multiple sustainable development goals – Satisfy many diverse constituents' interests at once	– Cross-sector policy integration and program administration – Curation and communication of benefits and costs	"Regional working groups and task forces appointed to facilitate information exchange, curate best practices, define standards, and provide directionality for regulators could alleviate current interagency struggles to manage agrivoltaic permitting and program administration" [13]	Interviews and workshops to identify key challenges faced by regulators interested in agrivoltaic policy development and program administration

References

[1] Irie, N., Kawahara, N., and Esteves, A. M. (2019). Sector-wide social impact scoping of agrivoltaic systems: A case study in Japan. *Renewable Energy*, *139*, 1463–1476.

[2] Pascaris, A. S., Schelly, C., and Pearce, J. M. (2020). A first investigation of agriculture sector perspectives on the opportunities and barriers for agrivoltaics. *Agronomy*, *10*(12), 1885.

[3] Ressar, K., Muhar, A., and Schauppenlehner, T. (2021). Agrivoltaics in Austria: A stakeholder perspective on the opportunities and constraints of synergetic land use. In *AIP Conference Proceedings* (vol. 2361, no. 1, p. 020001). AIP Publishing LLC.

[4] Moore, S., Graff, H., Ouellet, C., Leslie, S., and Olweean, D. (2022). Can we have clean energy and grow our crops too? Solar siting on agricultural land in the United States. *Energy Research and Social Science*, *91*, 102731.

[5] Agir, S., Derin-Gure, P., and Senturk, B. (2023). Farmers' perspectives on challenges and opportunities of agrivoltaics in Turkiye: An institutional perspective. *Renewable Energy*.

[6] Torma, G. and Aschemann-Witzel, J. (2023). Social acceptance of dual land use approaches: Stakeholders' perceptions of the drivers and barriers confronting agrivoltaics diffusion. *Journal of Rural Studies*, *97*, 610–625.

[7] Li, B., Ding, J., Wang, J., Zhang, B., and Zhang, L. (2021). Key factors affecting the adoption willingness, behavior, and willingness-behavior consistency of farmers regarding photovoltaic agriculture in China. *Energy Policy*, *149*, 112101.

[8] Neupane Bhandari, S., Schlüter, S., Kuckshinrichs, W., Schlör, H., Adamou, R., and Bhandari, R. (2021). Economic feasibility of agrivoltaic systems in food-energy nexus context: Modelling and a case study in Niger. *Agronomy*, *11*(10), 1906.

[9] Pascaris, A.S., Winter, E., and Gazillo, C. (2023). *Smart Solar in Connecticut: Survey Findings and Initial Recommendations*. Northampton, MA: American Farmland Trust.

[10] Goldberg, Z. A. (2023). Solar energy development on farmland: Three prevalent perspectives of conflict, synergy and compromise in the United States. *Energy Research and Social Science*, *101*, 103145.

[11] Ketzer, D., Schlyter, P., Weinberger, N., and Rösch, C. (2020a). Driving and restraining forces for the implementation of the agrophotovoltaics system technology–A system dynamics analysis. *Journal of Environmental Management*, *270*, 110864.

[12] Pascaris, A. S., Schelly, C., Burnham, L., and Pearce, J. M. (2021). Integrating solar energy with agriculture: Industry perspectives on the market, community, and socio-political dimensions of agrivoltaics. *Energy Research and Social Science*, *75*, 102023.

[13] Pascaris, A. S., Gerlak, A. K., and Barron-Gafford, G. A. (2023). From niche-innovation to mainstream markets: Drivers and challenges of industry adoption of agrivoltaics in the US. *Energy Policy*, *181*, 113694.

[14] Schindele, S. (2023). *Nachhaltige landnutzung dank doppelernte eine mehrdimensionale politikanalyse der agri-photovoltaik-diffusion in deutschland* (Doctoral dissertation). Institute of Political Science, Faculty of Economics and Social Sciences, Eberhard Karls University Tübingen.

[15] Ketzer, D., Weinberger, N., Rösch, C., and Seitz, S. B. (2020b). Land use conflicts between biomass and power production–citizens' participation in the technology development of agrophotovoltaics. *Journal of Responsible Innovation*, *7*(2), 193–216.

[16] Pascaris, A. S., Schelly, C., Rouleau, M., and Pearce, J. M. (2022). Do agrivoltaics improve public support for solar? A survey on perceptions, preferences, and priorities. *Green Technology, Resilience, and Sustainability*, *2*(1), 1–17.

[17] Trommsdorff, M., Vorast, M., Durga, N., and Padwardhan, S. M. (2021). Potential of agrivoltaics to contribute to socio-economic sustainability: A case study in Maharashtra/India. In *AIP Conference Proceedings* (vol. 2361, no. 1, p. 040001). AIP Publishing LLC.

[18] Vorast, M. (2022). *Challenges for Agrivoltaics in the International Context*. [Master's thesis, University of Graz].

[19] Brunswick, S. and Marzillier, D. (2023). The new solar farms: Growing a fertile policy environment for agrivoltaics. *Minnesota Journal of Law, Science and Technology*, *24*(1), 123.

[20] Pascaris, A. S. (2021). Examining existing policy to inform a comprehensive legal framework for agrivoltaics in the US. *Energy Policy*, *159*, 112620.

[21] Guarino, J. and Swanson, T. (2022). Emerging agrivoltaic regulatory systems: A review of solar grazing. *Chicago-Kent Journal of Environmental and Energy Law*, *12*(1).

[22] Cheo, A. E., Adelhardt, N., Krieger, T., *et al.* (2022). Agrivoltaics across the water-energy-food-nexus in Africa: Opportunities and challenges for rural communities in Mali, Diskussionsbeiträge, No. 2022-03, Albert-Ludwigs-Universität Freiburg, Wilfried-Guth- Stiftungsprofessur für Ordnungs- und Wettbewerbspolitik, Freiburg i. Br.

[23] Pascaris, A. S., Schelly, C., and Pearce, J. M. (2022). Advancing agrivoltaics within the U.S. legal framework: A multidimensional assessment of barriers and opportunities. In *AIP Conference Proceedings* (vol. 2635, no. 1, p. 050002). AIP Publishing LLC.

[24] Choudhury, M. R. (2022). Transition to agriphotovoltaics applying a systems level approach. In *AIP Conference Proceedings* (vol. 2635, no. 1, p. 070003). AIP Publishing LLC.

[25] Rogers, E. *Diffusion of Innovations*. 1st edn. New York, NY, USA: Free Press; 1963.

[26] Wüstenhagen, R., Wolsink, M., and Bürer, M. J. (2007). Social acceptance of renewable energy innovation: An introduction to the concept. *Energy Policy, 35*(5), 2683–2691.
[27] Geels, F. W. (2018). Socio-technical transitions to sustainability. In *Oxford Research Encyclopedia of Environmental Science*. Oxford University Press, Oxford, United Kingdom.
[28] Matland, R. E. (1995). Synthesizing the implementation literature: The ambiguity-conflict model of policy implementation. *Journal of Public Administration Research and Theory*, *5*(2), 145–174.
[29] Scharpf, F. W. (1988). The joint-decision trap: Lessons from German federalism and European integration. *Public Administration*, *66*(3), 239–278.

Chapter 9

Economic performance of agrivoltaic systems: a comprehensive analysis

Max Trommsdorff[1,2], *Jakob Nekolla*[1], *Niklas Schwendemann*[1], *Nauman Zafar Butt*[3] and *Arndt Feuerbacher*[4]

This chapter presents a comprehensive analysis of the economic performance of agrivoltaic systems, examining their potential benefits, drawbacks, and the factors influencing their feasibility. By evaluating case studies, economic models, and policy implications, we aim to provide a clear understanding of the economic viability of agrivoltaics and its contribution to a more resilient and sustainable future.

9.1 Introduction

The increasing global demand for both renewable energy and food has spurred interest in innovative solutions that can simultaneously address energy and food production. Agrivoltaics offers a promising approach to harvest clean energy while maintaining agricultural production of the land. This chapter delves into the economic dimensions of agrivoltaics, analyzing the factors contributing to its economic viability, potential risks, and the role of policy frameworks. For this, the following central questions guide through this chapter:

- What is the motivation behind agrivoltaics from an economic perspective?
- Which key parameters help to describe the performance of agrivoltaic systems?
- How does cost differ to those of conventional ground-mounted PV (GMPV)?
- What are decision criteria for stakeholders to implement agrivoltaics?
- What differences in business models do exist and what are their challenges?
- Which implication do the economic findings have on policy making?

[1]Department Analysis Modules and Power Plants, Fraunhofer Institute for Solar Energy Systems ISE, Germany
[2]Wilfried Guth Chair, Department of Economics, University of Freiburg, Germany
[3]Syed Babar Ali School of Science and Engineering, Lahore University of Management Sciences, Pakistan
[4]Ecological-Economic Policy Modelling Research Group, University of Hohenheim, Germany

To answer these questions, we introduce key parameters and indicators and shed light on the current state of technology and research. Responses differ significantly based on the specific agricultural application domain, system configuration, and local circumstances, mirroring the variation in benefits, motivations, and obstacles within these distinct contexts. Limitations of this analysis are set by difficulties to quantify the ecologic impact of agrivoltaics in monetary terms and the general early stage of the technology that, still, implies a thin data basis.

Main findings:

- While agricultural revenues in arable farming and permanent grassland applications are neglectable compared to revenues from power generation, in horticulture and vegetable applications, agricultural revenues can be even higher compared to the electrical production.
- Depending on the system design, the CAPEX of agrivoltaics exceeds those of GMPV by up to +50%, with lowest cost for extensive forms of interspace agrivoltaics in permanent grassland and highest cost for systems with a high degree of PV integration like overhead systems.
- Regarding challenges to comply to minimum agricultural productivity for maintaining the primary agricultural use of the land as set by several governments, competition forces market players to focus on PV production and to neglect agricultural yield particularly in application areas. This holds particularly in domains with low agricultural revenue streams like arable farming and permanent grassland.
- Economics of scale play an important role for reducing the cost of agrivoltaics which is expected to lead to large scale agrivoltaics mainly on arable farming and permanent grassland.
- An effective and sustainable market introduction of agrivoltaics depends on an appropriate and differentiated regulatory framework that considers the externalities and challenges of each application area.

9.2 Economic motivation for agrivoltaics

A fast and comprehensive ecological transformation is without a doubt one of the most pressing tasks of this century—also *and particularly* from an economic point of view. This includes, amongst others, ecosystem preservation, decarbonizing the economies, and adapting to the rapidly changing climatic conditions in a sustainable way. David Pearce and Dominic Moran revealed in their book "The Economic Value of Biodiversity" [1] already back in 1994 the importance of biodiversity protection and how inappropriate economic tools fail to assess the value of species and their variety. Similarly, in his 2007 review "The Economics of Climate Change," Nicholas Stern [2] effectively demonstrated that taking ambitious actions against global warming is not only justifiable but also critically imperative when considering cost–benefit factors. While a sufficiently high price for greenhouse gas emissions is key to reduce our carbon emissions [3], however, carbon taxation that also addresses land use impacts could lead to increasing water depletion in the farming sector because of increased cost of converting land for agricultural use. This effect was recently shown by Flannery Dolan *et al.* who employed an integrated global hydrologic model that incorporates economic parameters to analyze the increasing threats of water scarcity [4].

In the context of agrivoltaics, a pivotal query arises: is it more effective to tackle the pressing environmental challenges individually or to pursue integrated approaches? Stated differently, should a multi-dimensional crisis truly be met with multi-dimensional solutions? Upon closer examination of the challenges inherent in agrivoltaics implementation, one-dimensional solutions seem appealing at first sight. Higher system cost, complex stakeholder structures, and awkward policy making reduce the attractiveness of agrivoltaics compared to already existing green energy sources like GMPV that are cheap, well-established, and immediately available.

However, there is a set of sound economic arguments justifying efforts to better explore possible advantages of agrivoltaics, among them a diversification of income streams, an increased land productivity, reduced operating costs, energy costs reductions through own electricity consumption, and increasing social acceptance toward PV projects.

Regarding diversification of income streams, integrating PV modules with agriculture allows farmers to generate income not only from traditional farming activities but also from selling the electricity generated by PV modules, the provision of a service on the project site, such as green maintenance, or by additional land rent payments [5,6]. This diversification can provide a more stable and predictable income, reducing the dependency on the volatile agricultural market.

Regarding land productivity, co-locating agriculture and PV significantly increases land use efficiency which is particularly relevant in regions with scarce land resources [7–9]. From an economic view, agrivoltaics capitalizes on land as a valuable input factor for agricultural production and renewable energy generation with potential lower land cost for both parties.

Reduced operating costs: PV modules can provide shade for crops, reducing evapotranspiration rates and, hence, the need for excessive irrigation and potentially lowering water costs [10]. Additionally, the PV modules can protect crops from extreme weather conditions, minimizing the need for expensive protective measures [11]. For PV operators, there are less operating cost for land maintenance due to the cultivation of agricultural land and the inclusion of the farmer as a service provider to the project [12].

Reduced energy costs by own consumption: By generating on-site renewable energy, farmers can offset their electricity consumption from the grid, leading to lower energy bills and potentially eliminating energy costs altogether. While the use of generated electricity on-site is typically only relevant for smaller systems, generally, own consumption can significantly contribute to cost savings over the long term.

Increasing social acceptance: Agrivoltaic projects can enhance both a farmer's and a PV developer's reputation within the community and among consumers. Demonstrating a commitment to sustainable practices and renewable energy can lead to positive public relations and potentially attract environmentally conscious consumers [13]. For PV developers, access to land for PV projects might be easier when original agricultural cultivation can continue, or horticulture farmers benefit from crop protection effects from the agrivoltaic system leading to less resistance of local stakeholders [14].

The following sections present parameters and concepts to assess whether and under which conditions the effects of the mentioned arguments are sufficient to offset the disadvantages of agrivoltaics.

9.3 Key parameters and concepts for assessing economic performance

9.3.1 The land equivalent ratio

The land equivalent ratio (LER) is a concept in agronomy and agriculture that quantifies the efficiency of land use when multiple crops or activities are combined in the same area presented first by Mead and Willey [15,16]. In the context of agrivoltaics, the LER assesses how effectively both agricultural crops and PV modules can coexist on the same parcel of land. This can be written as the sum of the respective yield ratios of dual land use to mono land use:

$$LER = \frac{Yield_a(dual)}{Yield_a(mono)} + \frac{Yield_b(dual)}{Yield_b(mono)} \tag{9.1}$$

9.3.2 Capital expenditure

Capital expenditure (CAPEX) refers to the upfront costs required to build a facility. For agrivoltaics, CAPEX varies based on factors like installed capacity and the PV technology chosen which again depends on the kind of agricultural activity and the local climate. These costs are typically higher than conventional GMPV due to the more complex mounting structure and factors like larger row distances or transparent PV modules. Also, the substructure's cost is influenced by the clearance height and post spacing. Accordingly, interspace systems generally show lower CAPEX compared to overhead agrivoltaics. Within overhead systems, smaller farming machinery and more manual operations can improve cost-effectiveness as requirements for height and width clearance are lower. Similarly, applications with perennial row crops offer cost benefits by integrating substructure posts without losing acreage. Generally, overhead agrivoltaics can eliminate the need for fencing which also leads to cost savings. Figure 9.1 shows comparative CAPEX estimates from an agrivoltaics guideline of the Fraunhofer Institute for Solar Energy Systems ISE (Fraunhofer ISE) [17].

The estimates originate from 2020 data collections based on pilot facility costs and expert interviews comparing four different systems: a vertical interspace agrivoltaics on permanent grassland, two overhead agrivoltaics—one on horticulture and one on arable farming—, and a traditional GMPV system. The horticulture

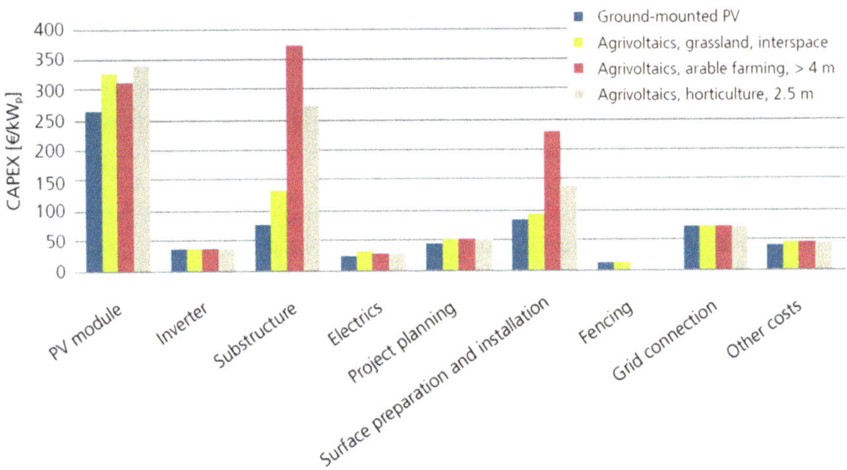

Figure 9.1 Comparison of CAPEX for ground-mounted PV and three different agrivoltaic system classes. © Fraunhofer ISEA LER greater than 1 implies that the combined land use is more productive than the sum of its individual components, indicating a more efficient utilization of the land. Typical values for agrivoltaics range between 1.3 and 1.8 and differ depending on crop choice and local climate [7–9].

agrivoltaics represents average values from both perennial crops like fruits and wine, as well as specialty crops. The calculation is performed on facilities with 2 ha size only considering PV components.

Main differences can be seen for the cost items "PV modules," "Substructure," and "Surface preparation and installation." While PV module cost are higher for bifacial and semi-transparent PV modules as assumed to be used for the vertical interspace agrivoltaics and the overhead horticulture agrivoltaics, respectively, the costs on mounting structure and site preparation and installation increase with vertical clearance being highest for the overhead agrivoltaics in arable farming and horticulture agrivoltaics (assumed vertical clearances of >4 m and 2.5 m, respectively).

Within the EU-funded project "HyPErFarm," a pilot installed in 2022 in Straßkirchen, Bavaria, demonstrated that, with EUR165 per kWp, tensile structures enable high vertical clearances at significantly lower cost compared to the Fraunhofer ISE guideline [18]. Another effect that is not considered in the cost analyses of Figure 9.1 is the effect of up-scaling projects leading to lower per-unit cost. This effect is particularly relevant for agrivoltaics in permanent grassland or arable farming where typically large land areas are available. Scaling effects are discussed in more detail in Section 9.3.7.

9.3.3 Operating expenditure

Operating expenditure (OPEX) refers to ongoing costs to operate and maintain a facility over its lifetime. When calculating the OPEX, the residual value either for disposal cost or the remaining value of the facility after the lifespan of the facility and replacement cost, e.g., for inverters also must be considered. In contrast to CAPEX, the OPEX of agrivoltaics tend to not differ much from those of GMPV even allowing for potential savings compared to GMPV systems. These savings can stem from reduced land provision costs, with costs decreasing in arable farming or permanent grassland, and a slight reduction in horticulture, based on agricultural lease rates shared between the farm and the agrivoltaics operator. Savings might be greater in arable farming due to commonly lower land lease rates compared to horticulture. Further, the agricultural activity decreases or fully eliminates land management costs traditionally borne by PV system operators. However, maintaining overhead agrivoltaics, especially cleaning or repairs, can be costlier when conducted at greater heights.

In the calculation of the Fraunhofer ISE guideline, the overall effects result in OPEX of agrivoltaics ranging between 4% and 9% lower compared to GMPV. Regarding potential risks of increased OPEX, long-term effects of fertilizers and crop protection products on the system's components are not yet well-understood.

9.3.4 Cost of capital

To account for the high CAPEX and to better understand the financing cost of agrivoltaics projects, economic analyses of agrivoltaics often employ the concept of weighted average cost of capital (WACC). The first foundations of the concept trace back to works of Modigliani and Miller and were developed further through many other theoretical advancements in corporate finance in the following decades [19]. Generally, the WACC helps firms to understand the overall cost of financing their operations and can aid in making decisions about capital structure, such as the balance between equity and debt. The WACC represents the average rate of return a company is expected to pay to its investors for using their capital, with both debt and equity capital being appropriately weighted.

According to the standard textbook of corporate finance by Brealey, Myers, and Allen, components of the WACC are the cost of equity, C_e, the cost of debt, C_d, the respective weights for equity, W_e, and dept, W_d, and the tax rate T [20]. The cost of equity, C_e, is the return required by equity shareholders for investing their capital in the business. It is often estimated using the capital asset pricing model (CAPM) which considers the risk-free investment, the volatility relative to the market, and the expected market return [21,22]. The cost of debt, C_d, is the effective interest rate that a company must pay on its debts. It is adjusted for tax because interest expenses are tax-deductible, reducing the effective cost of debt. The two weights W_e and W_d stand for the proportion of the company's financing that comes from equity and debt, respectively. The WACC can be calculated using the following equation:

$$WACC = (W_e * C_e) + [W_d * C_d * (1 - T)] \tag{9.2}$$

In analyses of agrivoltaics, business models with more stakeholders might include different WACC to account for varying parameters in the agricultural and energy sectors and specific price developments. When evaluating potential projects, farmers or PV developers might use the WACC as a discount rate to determine the net present value (NPV) of an agrivoltaics investment.

9.3.5 Levelized cost of electricity

The levelized cost of electricity (LCOE) is an economic measure used to compare the cost-effectiveness of different energy technologies on a consistent basis. Essentially, it represents the per-unit cost of installing and operating a generating power facility over an assumed financial life and duty cycle. The LCOE can be calculated using the following formula:

$$LCOE = \frac{\sum \text{Costs over life of the asset}}{\sum \text{Electricity produced over life of the asset}} \tag{9.3}$$

The numerator (costs over life of the asset) typically includes the CAPEX and OPEX as well as the cost of capital as calculated by the WACC. In case there are any cost expected for decommissioning or waste disposal not included in the OPEX, this must be considered here, too. The denominator (electricity produced over life of the asset) is the total amount of electricity that the facility produces over its operational lifetime. To account for the time value of money, LCOE calculations typically use a discounted cash flow approach. This means that future costs and electricity outputs are discounted back to present value terms based on the capital cost. Equation (9.4) shows the LCOE calculation with discounted OPEX and the total generated electricity M_t.

$$LCOE = \frac{\text{CAPEX} + \sum \frac{\text{OPEX}}{(1+WACC)t} - \text{R}}{\sum \frac{M_t}{(1+WACC)t}} \tag{9.4}$$

Note that, for consistency in calculation, both the numerator *and* the denominator must be discounted. While it might appear counterintuitive to discount electricity, an exemplary calculation with just two years and constant cost for generating electricity illustrates well that not discounting future electricity generation would lead to wrong results. Let us imagine a situation in which the cost of generating 1 MWh within year 1 and 2 is EUR100. The discount factor between year 1 and year 2 amounts to 5%. Obviously, the LCOE in both years amounts to 100 EUR/MWh regardless of the discount factor and regardless of whether we look at it from todays or a future perspective, since, in both years, the electricity is generated in the same year in which the costs are incurred. Accordingly, also the LCOE over both years must be 100 EUR/MWh. Putting these figures in (9.5) leads to the correct LCOE:

$$LCOE = \frac{\text{EUR}100 + \frac{EUR100}{(1+0.05)}}{1\,MWh + \frac{1\,MWh}{(1+0.05)}} = \frac{EUR100 + EUR95.2}{1\,MWh + 0.952\,MWh} = \frac{EUR195.2}{1.952\,MWh}$$
$$= 100\,EUR/MWh \tag{9.5}$$

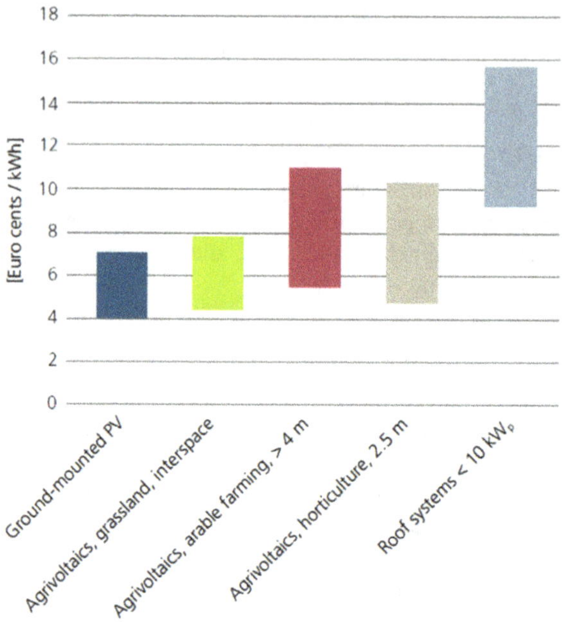

Figure 9.2 Comparison of LCOE for GMPV, agrivoltaics, and small rooftop PV systems

If, however, we would not discount the 1 MWh in the second year, the total LCOE would be slightly too low with 97.6 EUR/MWh.

The accuracy of an LCOE calculation depends heavily on the assumptions made, such as the expected lifetime of a facility and future operation cost. It is also worth noting that the LCOE does not consider external costs of generation methods, such as environmental and health impacts. The LCOE is frequently used by policymakers, utility operators, and researchers to make informed decisions about which energy technologies to pursue or subsidize. It can be particularly useful in understanding how competitive renewable technologies are relative to more traditional forms of generation. The LCOE of agrivoltaics varies according to differences in CAPEX, OPEX, solar irradiation, and the system's efficiency. In the calculations of the Fraunhofer ISE guideline, the LCOE of agrivoltaics are between 10% and 50% higher compared to GMPV with interspace agrivoltaics in grassland being most competitive. Compared to small rooftop PV systems, the LCOE of all three agrivoltaics are significantly lower (see Figure 9.2).

9.3.6 Levelized cost of agricultural products

To compare the cost of electricity generation with those of agricultural production, in agrivoltaics analyses, the concept of LCOE can be applied to cost of respective farming activities and investments [11]. This facilitates a direct performance comparison of PV and agriculture and their contribution to overall economic feasibility. Different to PV power generation, however, distinct challenges can arise due to the varied quality of agricultural products and their respective market prices. This can be addressed by considering a weighted average price that considers marketed quantities and prices. While calculating levelized cost of agricultural products is not a common concept in agriculture, it is a new method from energy sector applied to food sector to integrate the food cost into LCOE. Section 9.5.3 presents more details regarding how levelized cost of agricultural products can be calculated for an example of apple farming in Germany.

9.3.7 Economies of scale

Another useful concept for assessing the economic performance of agrivoltaics is the economies of scale (EoS). Generally, EoS refer to the cost advantages that enterprises obtain due to their scale of operation, with cost per unit of output generally decreasing with increasing scale. Essentially, as a company produces more, the average cost of each produced output goes down. This phenomenon is a result of spreading fixed costs over a larger volume of products and achieving efficiencies in production. For GMPV systems as well as for agrivoltaics, EoS arise from various sources such as bargaining power, spread of fixed costs, and centralized operation and maintenance.

Regarding bargaining power, developers often obtain discounts when purchasing PV modules, inverters, and other equipment in large quantities. Similarly, larger projects can give developers better negotiation power with suppliers, potentially resulting in more attractive prices for electricity uptake or land acquisition. With respect to spread of fixed costs, expenditures on feasibility studies, permits, grid connection, or project management can distribute over a larger capacity, reducing the cost per unit of electricity. In terms of centralized operation and maintenance, larger systems might allow for more centralized and efficient maintenance operations, and cost on monitoring and management systems can decrease per-unit as one system can often handle a larger installation without a linear increase in costs.

To analyze EoS of agrivoltaics, we evaluated 21 pilot facilities in Germany, Italy, and Chile with different system heights. Additionally, we employed data from Agostini, Colauzzi and Amaducci [23], Chalgynbayeva, Mizik and Bai [24], Schindele *et al.* [25], Vaccaro *et al.* [26], and Jung and Salmon [27], and conducted interviews with PV developers.

The logarithmic regression model used for estimating the relation between system size and CAPEX follows the form

$$y = a + b * \ln(x) \tag{9.6}$$

with y = CAPEX in Euro per kWp and x = system size in hectare. Figure 9.3 depicts three resulting EoS functions and data points for agrivoltaics with vertical clearances of more than 4 m, between 2.1 m and 4 m, and below 2.1 m.

Average CAPEX of agrivoltaics with a system size of 1 ha range between EUR845 for systems with a vertical clearance (VC) below 2.1 m up to EUR1,914 for systems with a VC above 4 m (see Table 9.1). When increasing the system size

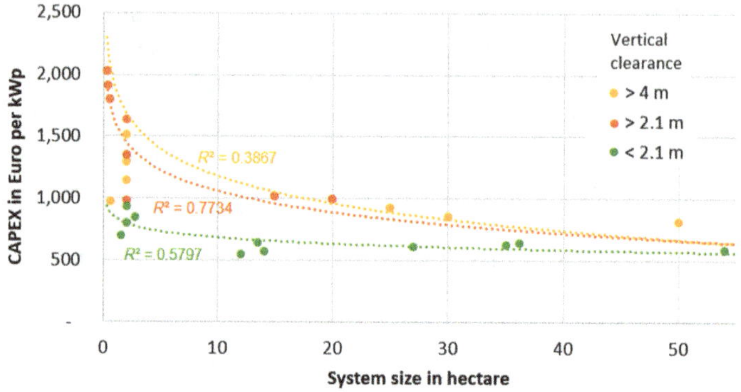

Figure 9.3 Economies of scale of agrivoltaics with a vertical clearance above 4 m (yellow), between 2.1 m and 4 m (orange), and below 2.1 (green)

Table 9.1 Economies of scale for three different agrivoltaic system classes and three different system sizes in Euro per hectare Note that the lower capacity density of the interspace system (<2.1 m) does not allow a direct comparison in term of cost per installed kWp

Vertical clearance	Unit	System area		
		1 ha	5 ha	20 ha
>4 m	[EUR]	1,914	1,401	959
	Change* in %		−27	−50
>2.1 m	[EUR]	1,621	1,227	887
	Change* in %		−24	−45
<2.1 m	[EUR]	845	732	635
	Change* in %		−13	−25

*Compared to 1 ha

Note that the lower capacity density of the interspace system (<2.1 m) does not allow a direct comparison in term of cost per installed *kWp*

to 5 ha, average CAPEX reduces between 13% for low mounted systems and 27% for overhead systems above 4 m. Further increasing the system size to 20 ha yields 25% for low mounted systems and 50% for overhead systems above 4 m. The coefficients of determination R^2 for the three regressions amount to 0.387 (VC > 4 m), 0.773 (2.1 m < VC < 4 m), and 0.387 (VC < 2.1 m). EoS seem to be larger for overhead systems compared to low mounted interspace systems. Reasons for this observation could be higher fixed planning cost of overhead systems, their higher installed capacity per hectare compared to interspace systems, and a bias because of the high number of funded overhead research facilities with a small system size in the data set which are likely to be more expensive compared to systems financed in the free marked.

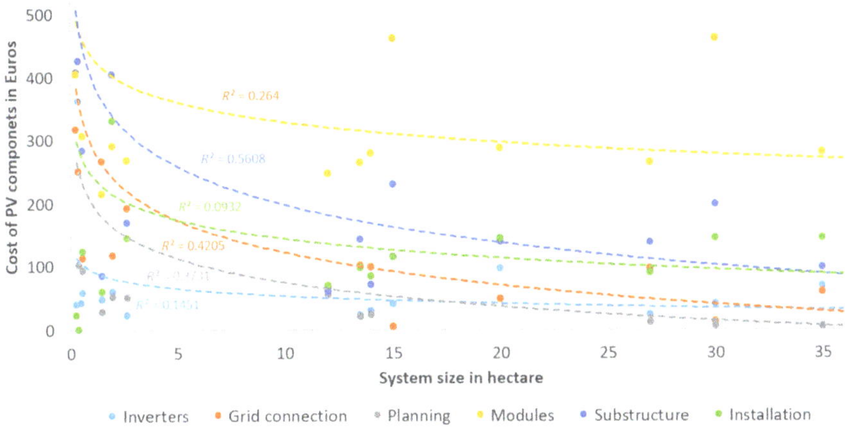

Figure 9.4 Economies of scale for the main PV components of three different agrivoltaic systems classes. 10 data points with cost >EUR500 and systems sizes >40 ha are missing for the sake of a better visualization

The same logarithmic regression model of (9.6) was used for estimating the relation between system size and the level of the six main cost items of the CAPEX inverters, grid connection, planning, PV modules, substructure, and installation (see Figure 9.4). As shown by the coefficients of determination R^2 for the six regressions, the regression predictions approximate the real data points less well than those of the above presented regressions of the three agrivoltaics categories with different vertical clearances. Only the R^2 values of the regressions on grid connection ($R^2 = 0.420$) and substructure ($R^2 = 0.561$) are higher than the lowest R^2 value of the three vertical clearances (VC > 4 m, $R^2 = 0.387$).

Being mainly fixed costs, as expected, the highest EoS result for the two cost items grid connection and planning with cost reductions of 82% and 76%, respectively, when increasing the system size from 1 ha to 20 ha (Table 9.2). In contrast, PV modules show the lowest EoS with cost reductions of 32% when increasing the system size from 1 ha to 20 ha.

Main shortcomings of the analysis are the small data set and the bias due to the above-mentioned high number of small research facilities that are expected to lead to higher EoS than in reality.

It's worth noting that while EoS offer cost reductions, there are practical limits to scaling. For instance, a too-large PV system may face challenges with land acquisition, environmental concerns, or grid integration complexities. The balance between gaining from EOS and facing increasing challenges is crucial for optimal project development.

9.3.8 Revenues

Revenues are defined as the total income or monetary gains generated from a particular activity, venture, or source. In the field of agrivoltaics, revenues are

Table 9.2 Economics of scale of the main PV components in Euro per hectare

Cost item	Unit	System area		
		1 ha	5 ha	20 ha
Inverters	[EUR]	94	66	41
	Change* in %		−30	−56
Grid connection	[EUR]	293	174	70
	Change* in %		−41	−76
Planning	[EUR]	201	113	37
	Change* in %		−44	−82
PV-Modules	[EUR]	434	360	297
	Change* in %		−17	−32
Mounting structure	[EUR]	400	259	137
	Change* in %		−35	−66
Installation	[EUR]	246	175	113
	Change* in %		−29	−54

*Compared to 1 ha

typically generated through sales of electricity and agricultural products. Regarding PV, various mechanisms facilitate the sales of electricity such as feed-in tariffs (FiT), power purchase agreements (PPA), or net metering. FiT represent fixed payments for the generated electricity typically offered by governments [28,29]. This guarantees facility operators a steady revenue stream over a specified period. In contrast, PPA are usually made individually between electricity producers and buyers,—usually utilities or businesses—to sell the generated electricity at a pre-determined price over a defined time frame [30].

Net metering is a billing arrangement and incentive program that allows owners of typically smaller renewable energy systems to receive credit for the excess electricity they generate and feed back into the grid [31]. This process involves installing a special meter that tracks both the electricity consumed from the grid and the surplus electricity that the renewable energy system exports to the grid. Other sources of revenue might be generated by tradable certificates or indirectly by own consumption of electricity. Certificates that represent the environmental attributes of a certain amount of renewable energy generated can be received by producers and sold to entities seeking to offset their carbon footprint or meet renewable energy targets. Own consumption of the generated electricity can reduce the electricity off-take from the grid and, hence, reduce the expenses on power purchases of an entity similarly to net metering. In agrivoltaics, own consumption of electricity for farms can only contribute to indirect earnings in a significant way when the installed capacity of the facilities is relatively small, and electricity purchases from the grid are high because of high loads for processing food or cooling. In the following, we only consider direct revenues from electricity sales as this is usually the main source of revenue for the PV operation. As (9.7) shows, the revenues of period t are typically calculated by the quantity of generated electricity M_t and the price.

$$Revenue_t = M_t * price_t \tag{9.7}$$

To calculate M_t, the solar irradiation S, the PV system's efficiency μ, and the annual degradation of the system d have to be considered as given by the following equation:

$$M_t = \sum_t^n S * \mu * (1 - d)^t \tag{9.8}$$

Details of how to calculate the electrical yield of PV systems can be found in the standard textbook of Arno Smets *et al.* (2016) [32]. Properly assessing and estimating potential revenues is a crucial aspect of conducting feasibility studies and cost-benefit analyses for agrivoltaics.

9.3.9 Efficiency criteria

To evaluate whether an investment in agrivoltaics does make sense from an economic point of view, as for other investments, a cost-benefit analysis can help to identify and to quantify both the costs associated and the benefits it can yield. Beneath CAPEX, OPEX, and revenue, on a macro-economic level, also impact on the public welfare should be considered like externalities attributed to the respective carbon footprint, impact on the landscape, or effects on biodiversity. On micro-economic level, though, a cost-benefit analysis of agrivoltaics can be boiled down to the question whether revenues from the agricultural activity are enough to off-set the higher cost of agrivoltaics compared to GMPV. A methodology to answer this question was provided by Schindele *et al.* (2020) by developing the concept of the price-performance ratio (ppr) [25]. Elements of the *ppr* are the land preservation cost, or price p, and the performance benefit pb.

$$ppr = \frac{\mathrm{p}}{pb} \tag{9.9}$$

According to Schindele *et al.*, p is calculated by the differences of costs per hectare stemming from agrivoltaics and GMPV electricity generation, respectively.

$$p = LCOE_{APV} * M_{APV} \quad - LCOE_{GMPV} * M_{GMPV} \tag{9.10}$$

The costs of agrivoltaics and GMPV are defined by their respective LCOE and quantities of generated electricity. Note that the unit per hectare implies that agrivoltaics with an installed capacity per hectare lower than those of GMPV can partially or completely off-set higher LCOE. The denominator pb represents the gain from preserving the land, which is the agricultural revenue generated on the area. Accordingly, cash crops as typically cultivated in horticulture agrivoltaics imply a high pb, while agrivoltaics in arable farming and on grassland rather lead to low values of pb and, ceteris paribus, a higher absolute value of ppr. Theoretically, the ppr can take any value from $-\infty$ until ∞.

For agrivoltaics projects with a *ppr* value higher than 1, from a micro-economic perspective, investing in a GMPV project represents a superior option as the additional cost for agrivoltaics cannot be justified by its agricultural benefits. In contrast, a *ppr* value lower than 1 indicates that the land preservation cost can be offset by the agricultural activity. While Schindele *et al.* only consider systems with positive land preservation cost $p>0$, indeed, *ppr* can also become negative if the regarded agrivoltaic system has low LCOE and a low installed PV capacity per area. To illustrate possible *ppr* values, we assume three agrivoltaic systems over a lifetime of 20 years with the following properties:

1. An overhead horticulture system with an electricity generation of 11 GWh/ha, LCOE of EUR0.08, and agricultural revenues of EUR80k/ha.
2. An overhead arable system with an electricity generation of 10 GWh/ha, LCOE of EUR0.07, and agricultural revenues of EUR10k/ha.
3. A vertical interspace system on permanent grassland with an electricity generation of 9 GWh/ha, LCOE of EUR0.06, and agricultural revenues of EUR10k/ha.

As a reference GMPV system, we assume an electricity generation of 12 GWh/ha with LCOE of EUR0.05.

Table 9.3 shows the assumptions and the results for the three agrivoltaic facilities. The results indicate that, for the horticulture system, with a *ppr* of $0.35<1$ agricultural revenues are high enough to justify the higher cost compared to the GMPV reference system. The opposite holds for the arable farming agrivoltaics: here, the *ppr* of $1.25>1$ indicates that earnings of the farming activity cannot compensate the higher cost of the agrivoltaics. In the permanent grassland system, the negative value of $ppr = -0.75$ indicates that—despite of the higher LCOE compared to the GMPV system, the cost per ha is still lower due to the lower electricity generation per ha.

Table 9.3 Assumptions for three different agrivoltaic systems in horticulture, arable farming, and permanent grassland with exemplary ppr calculations over a lifetime of 20 years

System	**Electricity generation**	**LCOE**	**Agriculture revenue**	**Land preservation cost**	**Price-performance ratio**
Unit	M_{APV} **(GWh/ha)**	**(EUR)**	*pb* **(EUR/ha)**	*p* **(EUR/ha)**	*ppr*
Horticulture	11	0.08	80,000	28,000	0.35
Arable farming	10	0.07	8,000	10,000	1.25
Permanent grassland	9	0.06	8,000	−6,000	−0.75
GMPV[1]	12	0.05	–	–	–

[1]Reference system

However, the negative value $ppr = -0.75$ of the agrivoltaics in permanent grassland does not necessarily mean that the system is preferable to the agrivoltaics in horticulture with a *ppr* of 0.35. In fact, the net benefit of the horticulture system is higher (80.000−28.000 = 52.000) compared to the agrivoltaics in permanent grassland (8.000− (−6.000) = 14.000). This illustrates the limits of the *ppr* concept.

9.4 Economic challenges and risks

9.4.1 Expected behavior of rent-seeking market players

As shown in the previous section, for $ppr < 1$, an investment in agrivoltaics is beneficial from an economic point of view since the benefits from preserving the land for agriculture offset the additional cost of agrivoltaics compared to GMPV. Though, in most countries, projects with a *ppr* lower than 1 are usually still not competitive with GMPV because of legal barriers that do not allow or do not sufficiently consider a dual land use. Legal barriers can occur in several legal areas such as energy law, agricultural law, permitting law, or tax law. Examples for barriers are FiT that are excluded for PV systems on agricultural land, agricultural subsidies that are not granted for areas also used for PV power production, permitting regulations that do not allow PV installations on agricultural land, or the loss of tax advantages when agricultural areas lose their tax privileges once PV power generation enters the land [33]. While such legal barriers represent economic challenges due to the higher cost they induce, other economic challenges come into play once the legal framework for agrivoltaics is set.

Legal privileges for agrivoltaics incentivize rent-seeking behavior of market players. While definitions and criteria for agrivoltaics can help to differentiate from GMPV, from an agrivoltaics investor's perspective, considering agricultural activities typically raises cost and reduces profitability of their projects. This is mainly due to the usually much lower added value of agriculture activity compared to electricity production. Accordingly, rent-seeking market players will probably reduce agricultural activities to a minimum for the sake of increasing electrical yield while still benefitting from agrivoltaics privileges. Findings of Feuerbacher *et al.* [34] regarding negative contribution margins for most agrivoltaics arable farms also support this argument (see Section 9.5.2). For horticulture applications or projects with a $ppr < 1$, rent-seeking behavior seems less critical because the financial relevance of cash crops incentivize a sufficient consideration of agricultural performance.

9.4.2 Economic relevance of agriculture and revenue split

Indeed, the relevance of agricultural revenues varies widely from crop to crop with least revenues in permanent grassland and arable farming and high revenues in horticultural applications (see Figure 9.5). A comparison between revenues from nine different crops and their respective revenues from PV power generation per hectare and year in six different countries shows average agricultural revenue splits of less than 4% in sheep grazing for meat production on permanent grasslands and

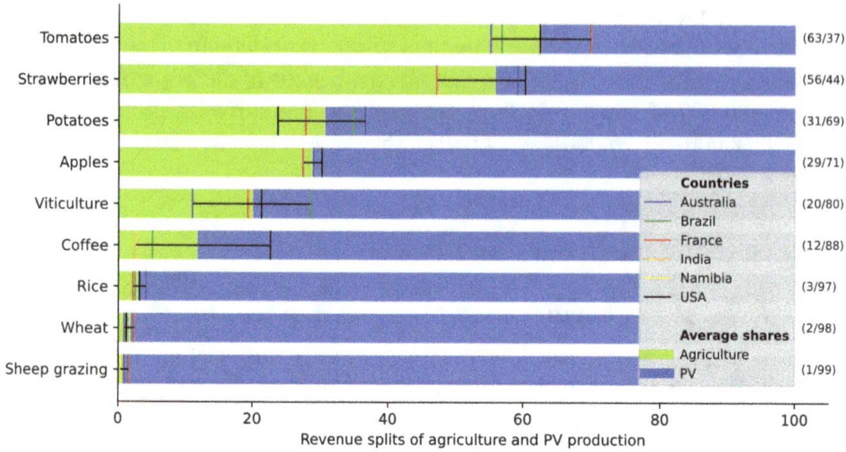

Figure 9.5 Revenue splits of agriculture and PV production. The colored lines indicate the variations of the different countries considered. Note that, for arable crops, crop-rotation should be considered. Agriculture data: FAO Stat. PV data: Solar GIS and own calculations.

rice and wheat production. In contrast, average agricultural revenue splits for tomato and strawberry production amount to more than 55% exceeding annual revenues from PV. The highest agricultural revenue split was found for tomato production in France with about 70% while grazing in permanent grasslands in Namibia reveals the lowest agricultural revenue split with less than 1%.

In the case of animal husbandry on permanent grasslands, drawing the line between GMPV and agrivoltaics seems difficult since an integration of sheep grazing between PV components is easy and comes at low cost [35]. For this reason, sheep grazing in GMPV systems for weed control is already well-established in many countries but was not considered as agrivoltaics in the past. One strategy to deal with rent-seeking behavior is setting clear criteria for agrivoltaics with animal husbandry, e.g., in terms of density and duration of stockage.

However, controlling for definitions and criteria for agrivoltaics being met and appropriately sanctioning not complying projects during the operation of the system can be challenging, too. This is particularly difficult in the case of criteria that set minimum agricultural yields as implemented in some pioneer countries of agrivoltaics [36].

While, with some efforts, it might be possible to control the actual agricultural yield, there is typically a lack of reference figures indicating the expected yield without agrivoltaics. This is mainly due to the large range of parameters influencing plant growth that differ from region to region, year to year, and crop to crop.

Considering such uncertainties, an implementation of too strict sanctioning mechanisms might endanger the bankability of agrivoltaics projects. On the other hand, too weak sanctioning probably encourages rent-seekers to abuse privileges for agrivoltaics.

Against this background, tasks for future economic research and policy makers are analyzing different business models and their effect on incentives for rent-seeking behavior and finding parameters, mechanisms, and sanctions for an appropriated control scheme to avoid excessive rent-seeking.

9.5 Economic models and case studies

9.5.1 *Micro-economic optimization approach by Alam* et al.

The model by Alam *et al.* [37] breaks down the *LCOE* into the hardware and soft costs to evaluate their respective impacts on the price of agrivoltaics. Typically, the hardware customization (i.e., the elevated mounting and stronger foundations) is the main contributor to the price of agrivoltaics while the soft costs including EPC, taxes, and land lease may have a relatively small difference as compared to the standard GMPV. While the hardware costs (C_M) are usually modulated by global economics, soft costs (C_L) depend more on the country specific policies and can depend on the type of land and business models.

With the bifurcation of *LCOE* into hardware and soft costs, (9.4) can be re-written as:

$$LCOE = \frac{C_M + C_L}{YY_T.\chi} = \frac{c_M \quad A_M + c_L A_L}{YY_T.\chi} = \frac{M_L + A_L/A_M}{YY.\chi/c_L} \tag{9.11}$$

where A_M and A_L are the total PV module and land areas for agrivoltaics, respectively, and YY_T and YY are the total energy and energy production per PV module area, respectively. $\chi \equiv \sum_{k=1} (1-d)^k (1+r)^{-k}$, where d and r are rates for depreciation and discount rates, respectively. ($M_L = c_M/c_L$) is the hardware to soft cost ratio, where c_M and c_L are the hardware costs per unit module area and soft cost per unit land area, respectively. It is an important quantity which can influence the relative price for the agrivoltaic system. For GMPV, M_L is typically close to 10 in the United States and varies between 5 and 35 worldwide [38].

Using (9.10), the price for the agrivoltaic system as given in (9.11) can be re-written as:

$$p = \left[\left(\frac{M_L + \frac{A_L}{A_M}}{\frac{YY_T}{A_{M_{APV}}} \cdot \frac{\chi}{c_L}} \right)_{APV} - \left(\frac{M_L + \frac{A_L}{A_M}}{\frac{YY_T}{A_{M_{GMPV}}} \cdot \frac{\chi}{c_L}} \right)_{GMPV} \right] \times YY_T \tag{9.12}$$

Since the difference in the hardware cost between agrivoltaics and GMPV usually plays a dominant role in the price of agrivoltaics, it is convenient to normalize the price with the hardware cost of the standard GMPV . The normalized price (p') is given by:

$$p' = \frac{p}{C_{M\,GMPV}/\chi} \tag{9.13}$$

which after simplifications can be written as [37]:

$$p' = \left[\left(\frac{c_{M_{AV}}}{c_{M_{GMPV}}} \right) + \left(\frac{A_{L_{AV}}}{A_{M_{AV}}} \frac{c_{L_{AV}}}{c_{L_{GMPV}}} \frac{1}{M_L} \right) - \left(\frac{A_{L_{GMPV}}}{A_{M_{GMPV}}} \frac{1}{M_L} + 1 \right) Y_{PV} \right] \tag{9.14}$$

where $Y_{PV} = A_{M_{PV}} / A_{M_{GMPV}}$ is the ratio of total PV module area for agrivoltaics to that for *GMPV*.

The first and second terms represent the difference in hardware and soft cost for agrivoltaic relative to *GMPV*. The third term represents the impact of relative energy generation per module area for agrivoltaics as compared to that for *GMPV*.

For the ideal limit, when the energy generation per module area is the same and there is no difference in soft and hardware costs for agrivoltaics vs. *GMPV*, p' approaches to zero. For practical cases, p' is typically between 0.4 and 0.8.

To compute *ppr*, we divide the performance benefit with the same normalization factor as we have used for the price. The normalized performance benefit (pb') can be written as:

$$pb' = \frac{pb}{A_{M_{APV}} \cdot c_{M\,GMPV} / \chi} = \frac{A_{L_{APV}}}{A_{M_{APV}}} \left(\frac{pb / A_{L_{APV}}}{c_{M\,GMPV} / \chi} \right) \tag{9.15}$$

where the ratio in the brackets represents the crop profit earned from a unit area of land divided by the hardware cost of installing the same unit area of *GMPV* module. pb' is typically smaller than p' and can vary across a wide range for low vs. high value crop.

Equation (9.16) signifies that the pb' is proportional to the land area in the agrivoltaic system available for crop cultivation. The performance can therefore be enhanced by using larger land for a given PV module area (i.e., increase in the row to row spacing between the PV module rows). Figure 9.6 highlights this trend for the case of a high value crop in a single axis tracking module configuration [39]. The figure also implies that higher values for M_L indicating a relative decrease in hardware to soft cost ratio lower the price of agrivoltaics.

When government incentives such as FIT are available, their economic impact has to be included in the performance term [39].

$$pb' = \frac{A_{L_{AV}}}{A_{M_{AV}}} \left(\frac{pb / A_{L_{AV}}}{c_{M\,GMPV} / \chi} \right) + \Delta FIT \left(\frac{YY_T / A_{M_{AV}}}{c_{M\,GMPV} / \chi} \right) \tag{9.16}$$

where ΔFIT is the difference in *FIT* for agrivoltaics and *GMPV* and is assumed to be a positive number. For a given agrivoltaic system, a threshold ΔFIT can be computed to enhance pb' so that *ppr* becomes close to one.

In summary, this modeling framework provides a convenient way for assessing and predicting the economic feasibility of agrivoltaics compared to *GMPV* as a function of PV module configurations, crop value, hardware vs. soft module costs, and energy tariffs. At the design stage, it can be a valuable tool for optimizing the technical and economic parameters of the system for maximum cost-performance benefits.

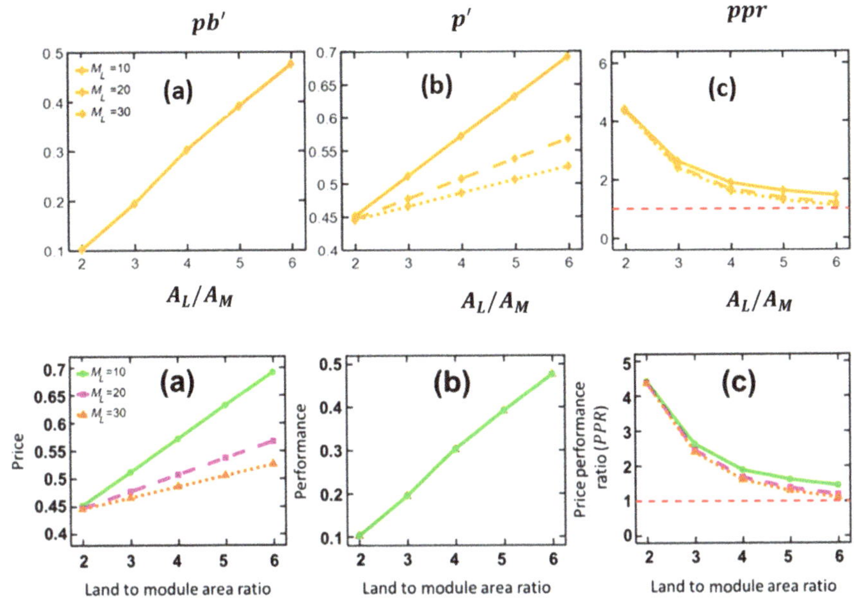

Figure 9.6 (a) Performance benefit, (b) price, and (c) price-performance ratio

9.5.2 Modelling macro-economics using the FEADPLUS framework to estimate the agrivoltaics adoption potential for arable farming in Germany

The FEADPLUS framework stands for "Framework to assess the economic benefits and the adoption potential of dual land-use systems" [34]. It allows to determine whether the investment in agrivoltaics is economically viable in the context of arable farming under given levels of solar irradiation, electricity tariff, agrivoltaics CAPEX and OPEX, and the changes in the agricultural contribution margin. The decision of a farmer or investor to invest in agrivoltaics is described as:

$$\beta(1-\varepsilon)\left(\sum_{i=1}^{I} U_i\delta_i - \sum_{v=1}^{V}\gamma_v C_v\right) - \beta\varepsilon E_{Base}^{Agri} + cap_{AV}(Haeta - CRFINV - M) > 0 \tag{9.17}$$

where β is the share of owned arable land used for agrivoltaics, ε the share of area lost within agrivoltaic system, i index for crops, U_i annual crop revenue from arable farming on that land, δ_i crop class specific yield change (in % yr^{-1}), v index for inputs, γ_v annual changes in input cost (in % yr^{-1}), C_v annual input costs, E_{Base}^{Agri} agricultural contribution margin in the base, cap_{AV} installed capacity of the agrivoltaics (in kWp), H annual full load hours, ae average lifetime efficiency (%), ta electricity tariff (EUR-cents per kWh), CRF capital recovery factor (in %), INV

investment costs in agrivoltaics (in EUR per kWp) and M is the agrivoltaics maintenance cost (in EUR per kWp and year).

The first two terms of the equation's left-hand side describe how the adoption of agrivoltaics change the agricultural contribution margin because of yield or input cost changes, and losses in arable land within the agrivoltaics. The third term describes the annual contribution margin due to the power production. Put simply, any possible economic loss in arable farming has to be offset by corresponding profits from the production of electricity, or vice versa. Simple rearrangements of this framework allow to determine the break-even electricity tariff:

$$ta > \frac{-\beta(1-\varepsilon)\sum_{i=1}^{I}\left(U_i\delta_i - \sum_{v=1}^{V} C_{i,v}\gamma_{i,v}\right) + \beta\varepsilon E_{Base}^{Agri} + cap_{AV}(CRF \quad INV + M)}{cap_{AV} \quad H \quad ae} \tag{9.18}$$

which can be expanded to calculate the LCOE of the agrivoltaics (see Appendix B in [34]). The static nature of the framework requires various simplifications. For example, no adjustments in crop rotations are assumed neither does it capture time varying levels of agricultural yields or solar radiation. Accounting for crop rotation adjustments requires a land-use optimization approach. This is rarely considered in the agrivoltaics literature, with Sponagel *et al.* (2022) being a notable exception [40]. However, its comprehensive but still parsimonious form makes it suitable to apply it to large datasets containing basic information about farmers revenue and input cost structure, e.g., for the bottom-up analysis of agrivoltaics adoption potential. This was demonstrated in Feuerbacher *et al.* using *inter alia* nationally representative farm-level data with the objective to understand what farm types and regions are most likely to adopt agrivoltaics, how EoS and levels of annual solar radiation influence the spatial pattern of adoption potential, and how agronomic effects such as crop shading and land losses affect the farm-level adoption potential [41].

Figure 9.7 shows one of the key results of this study, the average cost in EUR-cents per kWh for NUTS-3 level districts ("Landkreise") in Germany assuming that either all or only the 10% most cost-efficient farmers adopt agrivoltaics in either the absence or presence of EoS.

The main findings of the Feuerbacher *et al.*'s study using the FEADPLUS framework were manifold [41]. It revealed that certain farm types like specialist field crop farmers are most likely to be among the early adopters, which are predominantly found in Southern and Eastern Germany. It was found that the 10% most cost-efficient agrivoltaics adopters could supply about 9% of Germany's total annual electricity demand requiring only 1% of arable land area. Yet, this would require an annual subsidy of 1.18 billion EUR. Lastly, it is highlighted that adequate policies are needed to avoid farmers to abandon farming within the agrivoltaics given that agricultural contribution margins within the agrivoltaics turn negative for about 62% of the early adopters in the presence of EoS.

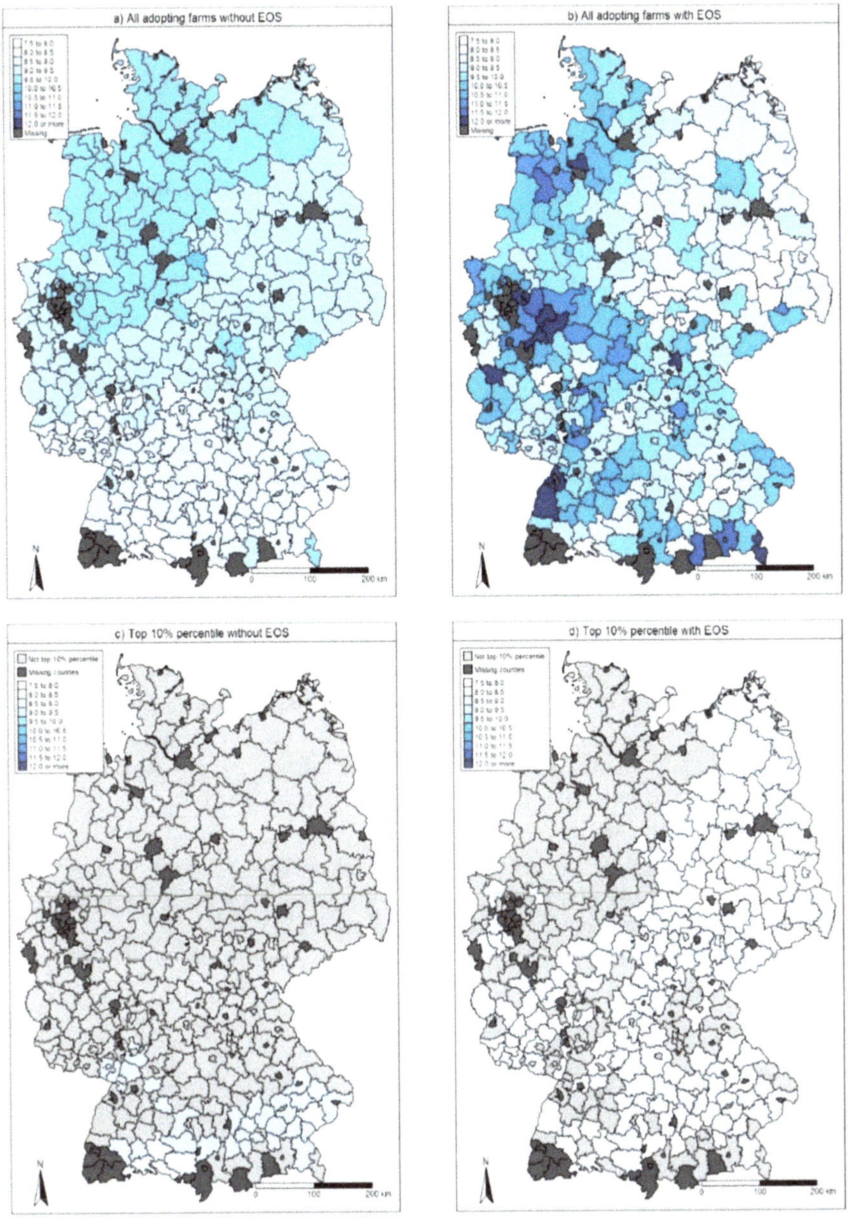

Figure 9.7 Average cost in EUR-cents per kWh for NUTS-3 level districts in Germany

9.5.3 *Case study on apple farming in Germany: can synergies in agriculture reduce the cost of agrivoltaics?*

Despite the advantages of enhanced land use efficiency and possible synergies between agriculture and power generation provided by agrivoltaics, higher investment costs, compared to traditional ground-mounted photovoltaics, pose a significant hurdle to its widespread market introduction in numerous countries [25]. Notable synergies are mainly expected for horticulture applications such as orchards and berry farming [42].

The study of Trommsdorff *et al.* [11] examines the economic feasibility of agrivoltaics in apple farming, focusing on potential synergies in agriculture and how they translate into possible cost reduction of LCOE. Based on literature, expert interviews, and data from three pilot projects in German apple farms, the authors analyze synergies and challenges related to investment costs and operational expenses. Main findings indicate that integrating agrivoltaics in apple farming could lead to a 26% reduction in average investment costs for the farming system. This reduction is primarily attributed to the partial replacement of hail protection structures with photovoltaic modules. Additionally, annual operating costs could decrease by up to 9%, primarily due to lower expenses for land use and maintenance. However, the study also highlights potential challenges. The integration of agrivoltaics may lead to a decrease in annual agricultural revenues by about 9% due to reduced light availability, resulting in an expected reduction in high-quality apple yield. Despite this decrease in revenues, the overall cost of apple production could be reduced by approximately 5%. In terms of the overall expense of agrivoltaics, though, the potential reduction of LCOE contributed by savings in the farming sector is less than 1%.

To calculate the effects of agrivoltaics on the net present cost of apple production, the authors modify the LCOE formula (see (9.4)) to apply it to apple farming, introducing the term levelized cost of apples (LCOA).

$$LCOA = \frac{CAPEX_{0,agri} + \sum_{t=1}^{n} \frac{OPEX_{t,agri}}{\left(1+WACC_{agri}\right)^{t}}}{\sum_{t=1}^{n} \frac{M_{t,agri,m}}{\left(1+WACC_{agri}\right)^{t}}} \tag{9.19}$$

with

$WACC_{agri}$ = Weighted average cost of capital for apples

$M_{t,agri,m}$ = Annual marketable agricultural yield (kg/ha/a)

In contrast to the produced kWh of electricity, produced apples vary in terms of qualities. To consider this, (9.20) includes the annual marketable agricultural yield differing between two apple qualities at three different market prices: Low-quality apples typically sold to lower market prices and high-quality apples that can either be directly marketed or sold to the wholesale trade.

To determine the annual marketable agricultural yield $M_{t,agri,m}$, the authors weight the two produced apple qualities against their respective relative prices to the average of prices in the reference scenario with hail nets (Scenario HN) as follows:

$$M_{t,agri,m} = M_{t,agri} * S_{low} * \frac{p_{low}}{p_{HN,a}} + M_{t,agri} * S_{high} * S_{dm} * \frac{p_{dm}}{p_{HN,a}} + M_{t,agri} * S_{high} * (1 - S_{dm}) * \frac{p_{wm}}{p_{HN,a}} \tag{9.20}$$

with

S_{low} = Share of low-quality apples (%)
S_{high} = Share of high-quality apples (%)
p_{low} = Price of low-quality apple (€/kg)
$p_{HN,a}$ = Average price of apples in Scenario HN (€/kg)
S_{dm} = Share of direct marketing (%)
p_{dm} = Price of direct marketing (€/kg)
p_{wm} = Price of wholesale market (€/kg)

The results of the LCOA calculations are shown by the bars HN and agrivoltaics_agri in Figure 9.8. The LCOA Scenario agrivoltaics (agrivoltaics_agri) are approx. 5% lower than the LCOA of the Scenario HN. Whereas agrivoltaics_agri presents the results in Scenario agrivoltaics, considering solely agricultural activities, the bar AVS_total also includes both the CAPEX and OPEX from the energy sector. In this case, the LCOA would be more than three times higher as in Scenario HN with EUR3.48 per kg instead of EUR0.91 per kg.

A sensitivity analysis of LCOA in Scenario agrivoltaics highlights the focal role of labor cost and agricultural yield. For instance, if apple yield quality remains stable, without the anticipated 10% reduction due to diminished light availability, LCOA would decrease by 7%. Labor cost constitutes 41% of LCOA, with harvest work making up the largest segment at 34% of total labor cost. Most of labor work

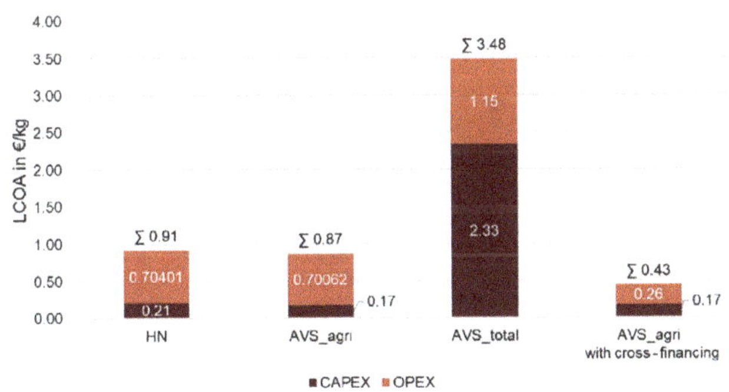

Figure 9.8 LCOA in €/kg for Scenarios HN and AVS. The final bar illustrates the potential influence of earnings from the photovoltaic (PV) sector to decrease the LCOA (AVS_agri with cross-financing).

is unaffected by PV integration and thus remains stable. If a more comprehensive PV shelter of the agrivoltaics systems allows for a complete replacement of hail nets, the LCOA might decrease by an additional 3.5%. Similarly, eliminating costs for freeze protection and fungicide use through rain shelter could further decrease LCOA by 5% and 1.5%, respectively.

The study also underscores the significance of regulatory support, such as FiT or other subsidies, for making agrivoltaics economically viable in orcharding. The authors estimate that the theoretical potential of agrivoltaics in apple farming in Germany could reach 23.8 GWp, contributing to around 13% of the photovoltaic development required to meet Germany's climate goals by 2030.

9.5.4 Case study 2: the economic potential of agrivoltaics on grape farms in India

Several studies on agrivoltaics on vineyards were conducted mainly pointing out the high potential for synergies that can address rising challenges for viticulture due to climate change [43–45]. A techno-economic analysis of Malu *et al.* [46] examines the feasibility of deploying agrivoltaics on existing grape farms in India determining the per unit area electrical energy generation potential and quantifying the economic benefits for cultivators across various design options. In the study, the energy aspects and anticipated output per area of an agrivoltaic system are analyzed for the Nashik district of Maharashtra, India (Lat: 19 °59′0″ Long: 73 °48′0″ E Alt: 700 m). Utilizing the National Renewable Energy Lab System Advisory Model (SAM) version 2014.1.14 [47], the performance of the PV system is simulated. Detailed system design inputs include the system array size, DC to AC ratio, azimuth angle, required type of axis, and the tilt angle, which was optimized through a sensitivity analysis with SAM. Based on the simulations and data on historical grape cultivation in India, the authors quantify revenues from both the electricity produced and agriculture activity.

Table 9.4 provides a summary of the revenue produced. The total annual gross revenue from 1 ha of grape farming amounts to USD3.511, calculated using an exchange rate of approximately 66 Indian rupees per US dollar. Conversely, the revenue from solar energy is derived by multiplying the annual energy generated with the cost of electricity, summing up to USD51.364. The results indicate that the economic value of agrivoltaics on grape farms could potentially increase over

Table 9.4 Expected generated revenue of agrivoltaics for 1 ha of farm in India

Unit	Annual revenue (1rs year) USD/hectare	Share of revenue %
Grape production	3.511	6.3
PV production	51.364	93.6
Total	54.876	100

15 times in comparison to traditional farming, while preserving roughly equivalent grape production.

Implementing agrivoltaics on grape farming nationwide in India has the potential to create a substantial impact by generating over 16,000 GWh of electricity, thereby meeting the energy needs of over 15 million people.

9.6 Business models and policy implications

9.6.1 Business models

Business models of GMPV are shaped by government incentives, declining cost of PV installations, rapid technological developments, and the global urgency to transition toward renewable energy sources. Various business models pervade the sector, including direct ownership models where entities invest in and own the solar installations, and third-party ownership models, such as PPAs, where a developer owns and operates the facility, selling the electricity at a negotiated rate. Community solar projects also present an inclusive model, enabling multiple stakeholders to share the benefits derived from a common solar facility. In scenarios where land acquisition is a potential barrier, partnerships with landowners, governmental bodies, or private entities come into play, establishing mutually beneficial arrangements.

Business models for agrivoltaics projects are typically more complex compared to conventional GMPV. The project execution can frequently involve various stakeholders or responsibility sectors, each carrying distinct functions, contingent upon the entities participating in the development. According to Schindele *et al.* [23], agrivoltaic business models typically include the following four functions or roles of the involved stakeholder groups:

1. Provision of land (ownership)
2. Agricultural cultivation of the land
3. Supply of the PV system (ownership/investment)
4. Operation of the PV system

In the case of a single entity business case, e.g., a farming business, one stakeholder might oversee all four areas. This approach tends to be utilized more for smaller agrivoltaic systems situated in proximity to farms, largely because farmers who own the agrivoltaic system, might find the capital expenditure to be within manageable limits. This business model presents several benefits: It minimizes project planning costs and simplifies contract negotiation due to its reduced complexity and allows for a more straightforward assessment of the agrivoltaic system's synergies since the revenue from both the agricultural and photovoltaic activities accrue to the same economic unit. This is notably pertinent in agrivoltaic systems where potential interactions between farming and power generation activities exist. Additionally, the capability of the producer to utilize the generated energy, coupled with the fact that many farms might have pre-existing experience with PV systems via roof installations, further underscores the merits of this business model (see "Base case" of Figure 9.9).

Business model	Function			
	Providing land	Agricultural management	Providing the PV system	Operating the PV system
1. Base case	Farm			
2. External land ownership	Land owners	Farm		
3. External PV investment	Farm		PV investors	Farm
4. Cultivation and operation only	Land owners	Farm	PV investors	Farm
5. Cultivation only	Land owners	Farm	PV investors	PV operators

Figure 9.9 Possible stakeholder arrangement for agrivoltaics business models. Source: Fraunhofer ISE (2022) based on [23].

Frequently, though, the land might be leased and not be directly owned by the farming business. This constellation is represented by the business model "external land ownership" in Figure 9.9. Similarly, in the context of larger agrivoltaic systems, it becomes increasingly atypical for the farming business to retain ownership of the PV system, thereby elevating the probability of external investment. Establishing a part-ownership structure could foster an incentive framework for synergistic dual-use of the land. However, as the portion of leveraged capital enlarges, maintaining a comprehensive perspective on the utilization of both production levels during continuous operations becomes more challenging. On the other side, this business model harbors the potential for achieving EoS and optimization through an enhanced labor division.

9.6.2 Policy implications

To analyze implications on policy making, it seems helpful to consider the different political motivations behind implementing agrivoltaics. As elaborated in preceding discussions, the multitude of business models are also determined by system sizes. Observing global disparities, stark contrasts emerge in system sizes from nation to nation. For example, China predominantly exhibits large-scale setups, while in contrast, 97% of Japan's system are installed on land parcels with less than 1 ha [48]. This divergence indicates that the politic motivation behind agrivoltaics varies from country to country, dictated by a spectrum of factors emanating from agricultural, energy, and land use sectors. Such motivations envelop considerations like farmers' income diversification, energy cost reductions for farmers, diminished operational expenses for PV companies, higher social acceptance of PV projects, enhanced land productivity, and positive external effects such as biodiversity enhancement. Depending on the constellation and the weighting of these motivations behind implementing agrivoltaics, the policy implications can differ widely with more farmer centered approaches if the policy pursuits rather goals of the agricultural sector and rather PV oriented large-scale applications if the policy targets at the energy sector.

Independent from the different political motivations, though, implementing a robust, clear, and measurable definition of agrivoltaics seems paramount to facilitate a successful market introduction. A central aim of most existing definitions of agrivoltaics is to ensure that the agricultural activity on the land is not negatively affected by the PV integration. With this, a definition shall draw a clear line between agrivoltaics and GMPV and might also enable politicians to distinguish between different agrivoltaic approaches in terms of the areas of applications, the system cost, and the synergy content of the different approaches. In case on a wide definition of agrivoltaics, this might also include extensive forms of agriculture activities like beekeeping or biodiversity measures. Indeed, the need to address the ongoing vast extinction of species together with the relatively low additional cost of including biodiversity measures clearly advocates ambitious political actions toward harnessing the high potential of extensive agrivoltaic applications. When it comes to those forms of agrivoltaics, though, a transparent and differentiated definition and regulative treatment of agrivoltaics seems imperative to avoid loss of social acceptance that might result when the intensity of the agricultural activity appears too low for being considered as agrivoltaics in the perspective off some stakeholders.

Challenges of ensuring a sufficient level of agricultural activity and extensive forms of agrivoltaics can be addressed by a strong farmers' involvement and a comprehensive embedment of agrivoltaics in the agricultural legislation.

Another elemental step toward harnessing the potential of agrivoltaics is to identify and to eliminate or at least mitigate legal barriers for agrivoltaics. Barriers typically exist in the following legal areas:

- Energy law
- Agriculture regulation
- Construction law
- Environmental law
- Safety regulation
- Tax law

To overcome legal barriers in these areas, a well-working communication between the respective political departments seems imperative to align sectoral interests. Details regarding how to address challenges within these legal areas are presented in Chapter 10.

Further challenges for policy making are managing the extensive heterogeneity of agrivoltaic systems in terms of costs, synergies, and social acceptance. Also here, the dialogue between the different sectors is of great importance. Similarly, identifying which variant of agrivoltaic systems mandates additional financial backing necessitates critical examination. This task is not limited to the respective cost of the systems but should also consider the expected synergies the different systems imply for agriculture, the environment, and society. This can be particularly challenging due to changing relevance of synergies depending on the local current and future climate.

To avoid over-regulation and windfall profits, once agrivoltaics is established in the market, additional financial backings should only consider synergies and externalities that cannot be internalized through market mechanisms. For a successful market launch, though, the lack of practical experiences and information might justify an initial financial support to overcome particularly barriers of risk averse market players in the agricultural sector.

Closely related to financial supporting schemes, formulating appropriate control and sanctioning mechanisms to prevent rent-seekers to exploit agrivoltaic privileges demands strategic policy innovation.

In conclusion, agrivoltaic systems, operating at the nexus of agricultural and photovoltaic energy production, beckon for policies and regulations that are not merely reflexive of economic and environmental considerations but are also intricately woven with societal, legal, and geographical threads.

9.7 Outlook and conclusion

The economic viability of agrivoltaics is influenced by a complex interplay of factors. As technology continues to advance, solar panel costs decrease, and policy frameworks evolve, the economic benefits of agrivoltaics are likely to become more pronounced. However, addressing challenges related to crop selection, shading optimization, and market fluctuations remains essential. To better understand the interplay of relevant factors necessary for a techno-economic optimization and an appropriate legal framing, investigating existing small- and large-scale facilities and governmental supporting schemes represents a crucial task for academics, enterprises, and policy makers in the agriculture and energy sector.

Agrivoltaics presents a novel approach to simultaneously address energy and food production challenges. The economic benefits, though dependent on various factors, are promising. By analyzing case studies, economic models, and policy implications, this book chapter underscores the importance of well-informed decision-making for implementing agrivoltaic systems. As societies strive for sustainable solutions, agrivoltaics emerges as a noteworthy contributor to a resilient and environmentally friendly future.

Acknowledgments

Parts of this book chapter were prepared within the framework of APV-Obstbau (funded by the Federal Ministry of Food and Agriculture, BMEL, grant number 2819101A19; and the Ministry for Climate Protection, Environment, Energy and Mobility of Rhineland-Palatinate, MKUEM, grant number 81077102), Modellregion Agri-Photovoltaik Baden-Württemberg (funded by the State of Baden-Württemberg, grant number L7522114), HyPErFarm (grant number LC-FNR-06-2020), and SynAgri-PV (funded by BMBF grant number 033L244A). We thank the companies BayWa r.e., BELECTRIC, Krinner Carport, Next2Sun Technology, REM TEC, and Solverde Bürgerkraftwerke Energiegenossenschaft for their support on data provision for the estimation of EoS. ChatGPT was used for research on general background information and on specific background information in Section 9.3.2 Capital expenditure, Section 9.3.3 Operating expenditure, Section 9.3.4 Cost of capital, Section 9.3.5 Levelized cost of electricity, Section 9.3.7 Economies of scale, and Section 9.3.8 Revenues. The language and orthography were checked and improved with DeepL.

References

[1] Pearce D and Moran D. *The Economic Value of Biodiversity*. London: Routledge; 1994.

[2] Stern N. *The Economics of Climate Change: The Stern Review*. Cambridge: Cambridge University Press; 2007.

[3] Nordhaus WD. Reflections on the economics of climate change. *Journal of Economic Perspectives* 1993;7(4):11–25.

[4] Dolan F, Lamontagne J, Link R, Hejazi M, Reed P and Edmonds J. Evaluating the economic impact of water scarcity in a changing world. *Nature Communications* 2021;12(1):1915.

[5] Pascaris AS, Schelly C and Pearce JM. Advancing agrivoltaics within the U. S. legal framework: A multidimensional assessment of barriers & opportunities. In: *AGRIVOLTAICS2021 Conference: Connecting Agrivoltaics Worldwide*. AIP Publishing; 2022. p. 50002 (AIP Conference Proceedings).

[6] Moore S, Graff H, Ouellet C, Leslie S and Olweean D. Can we have clean energy and grow our crops too? Solar siting on agricultural land in the United States. *Energy Research & Social Science* 2022;91:102731.

[7] Dupraz C, Marrou H, Talbot G, Dufour L, Nogier A and Ferard Y. Combining solar photovoltaic panels and food crops for optimising land use: Towards new agrivoltaic schemes. *Renewable Energy* 2011;36(10):2725–32.

[8] Amaducci S, Yin X and Colauzzi M. Agrivoltaic systems to optimise land use for electric energy production. *Applied Energy* 2018;220:545–61.

[9] Trommsdorff M, Kang J, Reise C, *et al.* Combining food and energy production: Design of an agrivoltaic system applied in arable and vegetable farming in Germany. *Renewable and Sustainable Energy Reviews* 2021;140:110694.

[10] Hassanpour Adeh E, Selker JS and Higgins CW. Remarkable agrivoltaic influence on soil moisture, micrometeorology and water-use efficiency. *PLoS ONE* 2018 [cited 2020 Jun 10];13(11):e0203256. Available from: URL: https://journals.plos.org/plosone/article/file?id=10.1371/journal.pone.0203256&type=printable.

[11] Trommsdorff M, Hopf M, Hörnle O, Berwind M, Schindele S and Wydra K. Can synergies in agriculture through an integration of solar energy reduce the cost of agrivoltaics? An economic analysis in apple farming. *Applied Energy* 2023;350:121619.

[12] Trommsdorff M. *An Economic Analysis of Agrophotovoltaics: Opportunities, Risks and Strategies Towards a More Efficient Land Use [Working Paper]*. University of Freiburg; 2016.

[13] Torma G and Aschemann-Witzel J. Social acceptance of dual land use approaches: Stakeholders' perceptions of the drivers and barriers confronting agrivoltaics diffusion. *Journal of Rural Studies* 2023;97:610–25.

[14] Ketzer D, Weinberger N, Rösch C and Seitz SB. Land use conflicts between biomass and power production – citizens' participation in the technology development of agrophotovoltaics. *Journal of Responsible Innovation* 2019; 7(2):193–216.

[15] Mead R and Willey RW. The concept of a 'Land Equivalent Ratio' and advantages in yields from intercropping. *Experimental Agriculture* 1980;16 (3):217–28.

[16] Willey RW. Evaluation and presentation of intercropping advantages. *Experimental Agriculture* 1985;21(2):119–33.

[17] Fraunhofer ISE. *Agrivoltaics: Opportunities for Agriculture and the Energy Transition: A Guideline for Germany*. 2022.

[18] Schnaiker M-A. *Sustainability Assessment of Agrivoltaics Systems and Design Optimization of Mounting Structures [Master thesis]*. University of Freiburg; 2021.

[19] Modigliani F and Miller MH. The cost of capital, corporation finance and the theory of investment. *The American Economic Review* 1958;48(3):261–97.

[20] Brealey RA, Myers SC and Allen F. *Principles of Corporate Finance*. New York NY: McGraw-Hill Education; 2011.

[21] Sharpe WF. Capital asset prices: A theory of market equilibrium under conditions of risk. *The Journal of Finance* 1964;19(3):425–42.

[22] Lintner J. The valuation of risk assets and the selection of risky investments in stock portfolios and capital budgets. *The Review of Economics and Statistics* 1965;47(1):13–37.

[23] Agostini A, Colauzzi M and Amaducci S. Innovative agrivoltaic systems to produce sustainable energy: An economic and environmental assessment. *Applied Energy* 2021;281:116102.

[24] Chalgynbayeva A, Mizik T and Bai A. Cost–benefit analysis of Kaposvár solar photovoltaic park considering agrivoltaic systems. *Clean Technology* 2022;4(4):1054–70.

[25] Schindele S, Trommsdorff M, Schlaak A, *et al.* Implementation of agrophotovoltaics: Techno-economic analysis of the price-performance ratio and its policy implications. *Applied Energy* 2020;265:1–19.

[26] Vaccaro C and Jäger M. *Agrivoltaic Pretrial: Experiment Report*. Wädenswil: ZHAW Zurich University of Applied Sciences; 2022.

[27] Jung D, Salmon A and Gese P. Agrivoltaics for farmers with shadow and electricity demand: Results of a pre-feasibility study under net billing in central Chile. In: *Agrivoltaics 2020 Conference: Launching Agrivoltaics World-wide*. AIP Publishing; 2021. p. 30001 (AIP Conference Proceedings).

[28] Ackermann T, Andersson G and Söder L. Overview of government and market driven programs for the promotion of renewable power generation. *Renewable Energy* 2001;22(1–3):197–204.

[29] Sijm J. *The Performance of Feed-in Tariffs to Promote Renewable Electricity in European Countries*. Petten: Energy Research Centre of the Netherlands ECN 2002; p. 7.

[30] Cory K, Canavan B and Koenig R. *Power Purchase Agreement Checklist for State and Local Governments*. Golden, CO: National Renewable Energy Laboratory (NREL); 2009.

[31] Eid C, Reneses Guillén J, Frías Marín P and Hakvoort R. The economic effect of electricity net-metering with solar PV: Consequences for network cost recovery, cross subsidies and policy objectives. *Energy Policy* 2014; 75:244–54.

[32] Smets AH, Jäger K-D, Isabella O, van Swaaij RA and Zeman M. *Solar Energy: The Physics and Engineering of Photovoltaic Conversion, Technologies and Systems*. Cambridge: UIT Cambridge Limited; 2016;462.

[33] Vollprecht J, Trommsdorff M and Hermann C. Legal framework of agrivoltaics in Germany. In: *Agrivoltaics 2020 Conference: Launching Agrivoltaics World-wide*. AIP Publishing; 2021. p. 20002 (AIP Conference Proceedings).

[34] Feuerbacher A, Laub M, Högy P, *et al.* An analytical framework to estimate the economics and adoption potential of dual land-use systems: The case of agrivoltaics. *Agricultural Systems* 2021;192:103193.

[35] Horowitz K, Ramasamy V, Macknick J and Margolis R. *Capital Costs for Dual-Use Photovoltaic Installations: 2020 Benchmark for Ground-Mounted PV Systems with Pollinator-Friendly Vegetation, Grazing, and Crops*. 2020.

[36] DIN SPEC 91434:2021-05, *Agri-Photovoltaik-Anlagen: – Anforderungen an die landwirtschaftliche Hauptnutzung*. Berlin: Beuth Verlag GmbH; 2021.

[37] Alam H, Alam MA and Butt NZ. Techno economic modeling for agrivoltaics: Can agrivoltaics be more profitable than ground mounted PV? *IEEE Journal of Photovoltaics* 2022;13(1):174–186.

[38] IRENA, ed. *Renewable Power Generation Costs in 2019*. Abu Dhabi; 2020.

[39] Alam H and Butt NZ. How does module tracking for agrivoltaics differ from standard photovoltaics? *Performance & Technoeconomic Implications*; 2023 Oct 25. Available from http://arxiv.org/pdf/2310.16946v1.

[40] Sponagel C, Feuerbacher A, Bendel D, Weber TKD and Bahrs E. Economic and agronomic impacts of agrivoltaics on arable land use at the example of the Stuttgart region. *German Journal of Agricultural Economics (GJAE)* 2023 May 11;23(2):101–116.

[41] Feuerbacher A, Herrmann T, Neuenfeldt S, Laub M and Gocht A. Estimating the economics and adoption potential of agrivoltaics in Germany using a farm-level bottom-up approach. *Renewable and Sustainable Energy Reviews* 2022;168:112784.

[42] Laub M, Pataczek L, Feuerbacher A, Zikeli S and Högy P. Contrasting yield responses at varying levels of shade suggest different suitability of crops for dual land-use systems: A meta-analysis. *Agronomy for Sustainable Development* 2022;42.

[43] Toledo C and Scognamiglio A. Agrivoltaic systems design and assessment: A critical review, and a descriptive model towards a sustainable landscape vision (three-dimensional agrivoltaic patterns). *Sustainability* 2021;13 (12):6871.

[44] Ahn SY, Lee DB, Lee H in, *et al.* Grapevine growth and berry development under the agrivoltaic solar panels in the vineyards. *Journal of Bio-Environment Control* 2022;31(4):356–65.

[45] Ferrara G, Boselli M, Palasciano M and Mazzeo A. Effect of shading determined by photovoltaic panels installed above the vines on the performance of cv. Corvina (Vitis vinifera L.). *Scientia Horticulturae* 2023; 308:111595.

[46] Malu PR, Sharma US and Pearce JM. Agrivoltaic potential on grape farms in India. *Sustainable Energy Technologies and Assessments* 2017;23:104–10.

[47] National Renewable Energy Lab. *(NREL).* System Advisor Model (SAM); 2023 [cited 2023 Oct 13]. Available from https://sam.nrel.gov/.

[48] Doedt C. *The Socio-Technical Dynamics of Agrivoltaics in Japan*. Daegu, South Korea; 2023. (*AgriVoltaics2023 Conference*).

Chapter 10
Legal frameworks in agrivoltaics

Jens Vollprecht[1], Laurent Battoue[2], Antoine Bois-Minot[2], Amandine Collard[2], Christian Doedt[3], Tienan Fan[4], Jessica Guarino[5], Jijiang He[5], David Moser[6] and Yuanyuan Xiao[5]

10.1 Introduction

This chapter presents the legal frameworks in agrivoltaics in China, France, Germany, Italy, Japan and the United States.

The introductory part will explain the role of the legal framework in the implementation of Agri-PV projects in general and the challenges that may arise in practice. At the same time, it will be shown how legal barriers can be identified and how legal advice can contribute to the successful implementation of a project. Figuratively speaking, this is a brief look at the 'workbench' of a lawyer. In combination with the comments on the individual countries, the reader is thus provided with legal assistance for the implementation of Agri-PV projects.

10.1.1 Importance of the legal framework

In addition to the technical and commercial aspects, the legal framework in particular plays an important role in the successful implementation of Agri-PV projects.

This is because the legal framework is of decisive importance for many issues: Can the project be implemented on the planned area in the desired design? Where can the plant be connected to the grid? What subsidies are available for the electricity generated and can the respective requirements be met? What are the costs for the implementation of the project? What are the conditions for financing the project? Which steps must be taken when and how long do the individual steps take?

[1]Becker Büttner Held Rechtsanwälte Steuerberater Unternehmensberater, Berlin
[2]Watson Farley & Williams LLP, Paris
[3]Institute for Sustainable Energy Policies, Tokyo, Japan
[4]Group Agrivoltaics, Fraunhofer Institute for Solar Energy Systems ISE, Germany
[5]Research Center for Energy Transition and Social Development, School of Social Sciences, Tsinghua University, China
[6]Institute for Renewable Energy, Eurac Research, Viale Druso 1, 39100 Bolzano, Italy

This is only a selection of questions that can arise from the legal framework and which – exaggeratedly formulated – can be extended (almost) at will.

In addition, the legal framework gives rise to rights and obligations. These can be defined by the legislator and anchored in a law. Negotiating conditions is not necessary here. The if and how is already determined by the law, thus ensuring a high degree of planning security. In this case, it is only necessary to know 'where it stands'. However, rights and obligations can also be established by a contract. The prerequisite for this is that the parties involved agree on the relevant provisions. This means that rights and obligations can only be predicted to a very limited extent. However, the legal framework cannot always be disregarded here either. This is because it is specified at certain points whether and within what framework contracts can be concluded. This shows that legal rights and obligations ensure a higher legal and therefore planning security compared to contractual rights and obligations. Therefore, legal rights and obligations are often the basis for promoting the expansion of renewable energies.

Thus, the legal framework is of high importance for all actors that are directly or indirectly involved in the project. For Agri-PV projects authorities, such as licensing or tax authorities, the grid operators must also deal with the legal framework.

10.1.2 Policy analysis: identification of the relevant legislation

The first step is to identify the relevant legislation for the project. Since there is no 'Agri-PV law' that lists all the relevant regulations for the project, the relevant regulations must be compiled from quite different laws. Ultimately, therefore, a customised 'patchwork law' must be created for each project. If a relevant regulation is overlooked, there is a risk that this will result in costs that jeopardise the economic viability of the project. Or regulations that are beneficial to the project may be overlooked. Under certain circumstances, the project is then abandoned, even though the framework conditions would have aided.

To obtain a better overview of the relevant regulations, it is useful to organise the various laws, e.g. into domains of law (civil, public, criminal law), policy-levels (e.g. EU, national, local) and legal areas (energy, agriculture, tax law) impacted by the project.

As a rule, there is a roughly comparable 'consonance' of the legal set-up across countries. Thus, a distinction can be made between civil law, public law and criminal law. The civil law determines the relations of natural and legal persons among each other and beyond that their relations to things as well as the relations of the things to each other. A land lease agreement between a farmland owner and the Agri-PV project owner (no matter if legal or private legal entity) is an example of civil law. Public law is the part of the legal system that regulates the relationship between holders of public power (the state) and citizens or other holders of public power. For example, is there a building permit necessary for the approval of the construction and operation of an Agri-PV project or is there subsidy available for the farming entity within the Agri-PV project area supporting financially the project's success? Criminal law comprises those legal norms by which certain conduct is prohibited and sanctioned with a penalty as 'criminal offenses'. For example, if an operator of an installation demands financial support for the electricity produced in the Agri-PV even though he knows that he does not fulfil the requirements for financial support.

Another categorisation ties in with the different 'levels of legislation'. As a rule, there is not only the legislative body that is responsible for the entire country. Rather, there are, for example, legislative bodies at the federal state (e.g. Baden-Württemberg) or municipal level (e.g. Lake Constance County) in addition to the national (e.g. Germany) and supranational (e.g. Europe) level. Therefore, it is possible that for an Agri-PV project at site A other legal regulations must be considered than at site B.

Once the 'patchwork law' has been put together, it must be examined whether and to what extent the project meets the respective legal requirements. It should also be considered whether and to what extent optimisation is possible. However, interactions between the various laws must always be considered. In some cases, optimisation in one law leads to serious disadvantages from another law. Optimisation must therefore be successful across all laws. In practice, this is often a great challenge. This can be vividly illustrated by the so-called Liebig's minimum law. This law states that the growth of plants is limited by the relatively scarcest resource (resources such as carbon dioxide, water and light). Applied to the legal framework, this means: The success of an Agri-PV project is measured by the 'strictest' law. If the requirements of the 'strictest' law cannot be met, the project fails. If the requirements of the 'strictest' law can only be met in part, the implementation succeeds only to the extent that the 'strictest' law allows. This also shows how important it is to identify all relevant regulations – and thus also the 'strictest' law.

10.1.3 Legal uncertainties and their sources

Once all relevant regulations have been found, it must be clarified whether and to what extent their requirements can be met. In practice, this cannot always be answered with legal certainty. What are the reasons for these legal uncertainties?

In some cases, the laws contain loopholes, i.e. they do not address the question raised at all. In some cases, the regulations themselves are not sufficiently clear. This may be because the wording allows for various interpretations. This is vividly illustrated by the term 'installation' in the prerequisite 'the surface area of the installation must not exceed 25,000 square meters'. Is it only the modules or does the elevation also belong to the installation. What about the line through which the generated electricity is transported to the grid, or the battery storage in which the electricity is temporarily stored? Do these technical facilities also belong to the 'installation' in the mentioned example? In addition, the result found may not 'harmonise' with the rest of the law: If, for example, 'installation' is mentioned in various places in the law, the same technical equipment must actually be meant in all these places. In some circumstances, this understanding makes sense in one place, but not in another. An example: If the term 'module' is used in one place for an installation, but the installation must have a capacity of more than 1 MW in another place, then the term 'module' cannot be used in the other place, even though the same word is used. Because there is no module with a capacity of more than 1 MW.

Uncertainties may also arise from the fact that the considerations of the legislator regarding the provision under consideration point in a different direction than the wording of the law suggests: According to the wording of a provision, for example, three requirements must be met cumulatively, but according to the considerations of the legislator, it is sufficient that only one of the three requirements is met. In addition, it is sometimes apparent that the result found does not correspond to the objective of the law. For example, the wording of a regulation suggests that in the event of grid bottlenecks, electricity from fossil fuels is to be given priority over electricity from renewables, but the law is intended to promote the faster expansion of renewables.

The probability that such legal uncertainties will arise is particularly high in the case of innovative concepts. This is because if the legislator is not aware of such a concept when drafting the legal provision, it cannot take into account any special features that may exist and cannot draft the provision accordingly. Figuratively speaking, this means that the 'piece of the puzzle' does not fit into the legal framework – with the consequences described above. In practice, this became apparent in Germany, for example, when battery storage systems were introduced. These are characterised by the fact that electricity is consumed during the storage process and generated during the withdrawal process. Since the legal framework for the same issue contained regulations for both consumption and generation, the question arose as to which of the regulations should be applied. Similar challenges can arise with Agri-PV. As explained, this technology is characterised by the fact that the land is used simultaneously for electricity generation and agricultural production. Thus, this technology differs from the 'classic' ground-mounted systems that the legislator had in mind when designing the legal framework for ground-mounted systems. In addition, due to the cross-sectoral nature of Agri-PV, as mentioned before, many different policy areas (e.g. agricultural law, energy law, tax law) are effected. The interdisciplinary regulations that need to be taken into account are therefore particularly important. Especially at the beginning of the Agri-PV market launch, a regulatory learning process takes place. The balance between restrictive legislation to prevent subsidy abuse and liberal legislation to enable investment into various Agri-PV innovations must therefore be considered in the legislation.

10.1.4 Possibilities for eliminating or minimising legal uncertainties

If legal uncertainties arise, they must be eliminated or at least minimised as far as possible. Failure to do so can have serious consequences for financing and thus for the project's chances of realisation.

10.1.4.1 Evaluation of case law

It goes without saying that case law must also be evaluated during the legal review of a regulation. This can help to reduce or even eliminate legal uncertainties.

If a judgement is available, it must be clarified in particular whether and to what extent the facts on which the judgement is based correspond to the facts to be assessed. If the facts correspond, the decision can in principle be transferred '1:1'. This means that a statement can be made as to how the court would decide if it had to decide on the facts to be assessed. If the facts of the case are different, it must be considered whether and to what extent the judgement can be transferred to the facts to be judged. If, for example, the court had to decide on the EU payments for sheep keeping in a solar project area, it must be considered whether and to what extent there are differences between sheep keeping and cereal and vegetable farming in a solar project area.

In this context, the changing legal framework must always be considered. If the decision was made on the provision that is also applicable to the project to be assessed, the court's considerations can be transferred. Otherwise, it must be clarified where the differences lie between the regulation referred to by the court and the regulation to be applied to the project. Under certain circumstances, for example, the regulation has only been changed editorially, so that a '1:1' transfer is possible. Otherwise, it must also be considered here whether and to what extent the decision can be used.

Furthermore, it must be considered that the jurisdiction is usually hierarchical, i.e. made up of several instances. It is thus possible to have decisions of a court reviewed. Judgements of the court of last instance can only be subject to review in exceptional cases.

What does this mean with regard to the legal uncertainties described above? If there is a judgement by the court of last instance, it is to be expected that the other instances will follow the view of the court of last instance. Because if the other instances deviate from it, they must expect that the judgement will be overturned. Thus, if a court of last instance has ruled on the provision under consideration, the legal uncertainties described above can usually be eliminated.

However, since it sometimes takes several years for the court of last resort to deal with a case, there is often not yet sufficient 'robust' case law for innovative technologies – such as Agri-PV. The same applies with regard to regulations that are newly introduced into the legal framework. Thus, when the legal framework is amended to remove barriers to Agri-PV installations, either no case law is yet available or at least case law of last resort is still lacking.

It should be noted that supranational jurisdiction may also have to be observed. This applies, for example, if the project is to be implemented in a member state of the European Union (EU) such as France, Italy and Germany. This is because at the EU level there are legal acts that must be considered by the member states. This applies, for example, to directives that must be transposed into national law in accordance with European law. Also worth mentioning in this context are the so-called 'state aid guidelines' of the European Commission, which play an important role in particular in connection with the financial support of renewable energies. Finally, European regulations should be mentioned, which apply directly in the member states without any act of transposition. If there are corresponding legal acts, the rulings of the European Court of Justice or the European Court of Justice may therefore also have to be taken into account.

10.1.4.2 Further options

Legal advice regularly raises the question of how remaining legal uncertainties can be eliminated or minimised.

During the development phase of a project, it may be possible to modify the concept so that the identified uncertainty no longer arises. For example, the installation can be dimensioned smaller in case of possible exceeding of performance limits and the associated disadvantages (e.g. loss of financial support for the generated electricity).

Or the concept is modified in such a way that the legal uncertainty does not affect the entire project, but only a part of it. Sometimes the legal uncertainty only affects an additional payment. If the project is only 'profitable' with the legally secure part or only with the basic remuneration, the legal uncertainties do not represent a 'knock-out criterion'. On the contrary, the profitability of the project improves if the legal uncertainties do not materialise.

In addition, as far as possible, discussions should be sought with the 'other side', e.g. the responsible authority or the network operator, at an early stage – of course, only after appropriate, well-founded preparation. This is because in some cases the 'other side' does not classify the situation as legally uncertain. Or it is possible to develop a common understanding with the other side.

In addition, acceptance plays an important role: If acceptance can be created, the risk that the uncertainties will be clarified in court and the risks associated with the uncertainties will be realised is reduced. In some cases, the legal framework explicitly opens the possibility for measures that promote acceptance, such as allowing affected communities to share in the revenues from electricity production or privileges for citizen energy projects. Acceptance can also be created by informing citizens 'on site' at an early stage, by offering favourable green power products, by financial participation or by structuring land use contracts (so-called pool contracts) accordingly.

Finally, there is of course the possibility of having the open question clarified in court. This path will be chosen if, among other things, the prospects of success are considered positive. If the facts relevant for the assessment have not been established, evidence may have to be taken. This means, for example, that expert opinions must be prepared or 'on-site meetings' must be held. This often not only takes a lot of time, but the additional costs increase the financial risks. This is because the party that loses usually must bear the costs associated with the judicial proceedings. As explained, if a party loses, it is generally possible to have the judgement reviewed. Therefore, several years may pass before a final judgement is reached. As a rule, the legal system provides for the possibility of resolving legal issues 'quickly' by means of so-called interim relief. However, in some cases very high hurdles have to be overcome. In addition, the decisions are often only provisional. This means that the 'normal' procedure must still be carried out for final clarification.

10.1.4.3 Impulses to policy-makers and the authorities

If uncertainties remain in the analysis of the legal framework that cannot be eliminated or sufficiently minimised in time, consideration should be given to whether and to what extent the legal framework needs to be changed. In view of the advancing climate change, the rapid expansion of renewable energies – and thus also of Agri-PV – is of considerable importance.

Organisations in the respective sectors can often contribute practical experience through their members and develop proposals for the political decision-makers for the design or improvement of the legal framework. For example, agricultural associations, together with solar and environmental organisations, can work towards a specific support scheme for Agri-PV and present the results in public hearings on legislative procedures. Public hearings can therefore also help to reduce or eliminate legal uncertainty. New laws must also be implemented by the authorities. In practice, however, they are often confronted with a large number of new regulations, so that the necessary expertise may be lacking. Therefore, targeted opportunities for exchange and networking should also be created for authorities. This is also in the interest of all other actors directly or indirectly involved in Agri-PV projects.

10.1.5 Outlook

The explanations show that the legal framework plays an important role for the implementation of Agri-PV projects. Whether and to what extent the legal framework in China, France, Italy, Germany, Japan and the United States is designed in such a way that Agri-PV projects can be implemented with legal certainty will be examined in the following chapters.

10.2 Legal framework of agrivoltaics in China

10.2.1 Background information

In China, agrivoltaics (Agri-PV) is a 'photovoltaic+' project. According to official statistics, by the end of 2021, China has more than 64 GW of photovoltaic facilities installed on agricultural land, ranking first in the world in terms of quantity [1]. The policy factor is the primary force behind the growth of Agri-PV in China. The national photovoltaic leader base plan and the photovoltaic poverty alleviation policy, which is a part of the exact poverty alleviation strategy, have encouraged the rapid growth of Agri-PV in China [2].

2014 marks the beginning of Agri-PV significant expansion in China. The National Energy Administration of China and the State Council Poverty Alleviation Office jointly issued the photovoltaic poverty alleviation policy. Established in

2015, the National Strategic Alliance for Collaborative Innovation of Modern Agricultural Photovoltaic Industry (AMAPV) released the first Agri-PV group standard (non-national standard) in China in July 2016. Agri-PV in China developed quickly in 2016–2017 as a result of subsidies for photovoltaic power installations. The development of the Agri-PV industry has suffered as a result of the new policy, which mandated that the amount of photovoltaic subsidies had been significantly cut since 31 May 2018. In the next few years, nine provinces and cities in China have formulated the technical specifications of Agri-PV, which have standardised the minimum height of photovoltaic panels and the spacing of photovoltaic arrays, but there is no clear regulation on agricultural operation and maintenance management. In September 2020, China released the goal of carbon neutrality in 2060, and the Agri-PV industry resumed its upward trend. Many hundreds of megawatts Agri-PV projects were approved.

At present, China has entered the era of photovoltaic parity, and there is almost no subsidy for photovoltaic agricultural projects, except for a few cities such as Huzhou City, Zhejiang Province. In addition, agricultural photovoltaic has very strict land use norms and land-related taxes and fees. Overall, the policy of Agri-PV industry in China tends to tighten, which is obviously different from that in Germany, Japan and other countries.

10.2.2 Agricultural regulations

10.2.2.1 Cropland Act

The development of Agri-PV in China is highly related to the land management policy, and the land management department has imposed strict restrictions on the installation of photovoltaic in farmland. And there are many types of Agri-PV in China, including the combination of photovoltaic with agriculture, forestry, animal husbandry and fishery. Therefore, the land management policies involved in agricultural photovoltaic projects in China are more complicated.

1. In March 2023, the General Office of the Ministry of Natural Resources, the Office of the State Forestry and Grassland Bureau and the General Department of the National Energy Administration issued the *Notice on Supporting the Development of Photovoltaic Power Generation Industry and Standardizing Land Management*, encouraging the development of photovoltaic power generation industry by using unused land and existing construction land [3].
   (a) Under the premise of strict ecological protection, encourage the construction of large-scale photovoltaic bases in desert, Gobi, and desert areas.
   (b) New and expanded photovoltaic power generation projects shall not occupy permanent basic farmland, basic grassland, Grade I protected forest land and key state-owned forest areas in Northeast Inner Mongolia. Photovoltaic square land shall not occupy cultivated land or other agricultural land, to avoid the impact on ecology and agricultural production.
   (c) Where the land for photovoltaic square array involves the use of forest land, the complementary mode of solar PV and forestry shall be adopted,

and the shrub land with annual precipitation below 400 mm and other areas with coverage below 50% can be used; it is not allowed to destroy the original vegetation or set up photovoltaic panels after transforming the original woodland into shrub land; the lowest point of photovoltaic mounting bracket should be more than 1 m higher than the height of shrub, and the net spacing should be set reasonably in the north–south direction of each column of photovoltaic panels; it is encouraged to take effective soil and water conservation measures to ensure that shrub cover and other growth conditions do not deteriorate from pre-forestry-photovoltaic levels; after the expiration of the project service, the forest land should be restored to its original state.

   (d) If the land for photovoltaic square array involves occupying the grassland outside the basic grassland, the mode of Grassland-Photovoltaic-Complementary is encouraged. The local forestry and grassland authorities should scientifically assess the grassland resources and ecological conditions in the region, and reasonably determine the appropriate areas, construction modes and construction requirements for the project.

2. According to the Current Land Use Classification (GB/T 21010-2017), in addition to cultivated land, woodland and grassland, other agricultural land also includes garden, reservoir water surface, pit water surface, facility agricultural land, rural roads, ditches and fields. It is particularly emphasised that the construction of photovoltaic power generation projects in the branches of reservoirs should meet the relevant requirements in the Guiding Opinions of the Ministry of Water Resources on Strengthening the Spatial Control of Shoreline in Rivers and Lakes (Shuihehu [2022] No. 216).

China has strict land property management for agricultural photovoltaic project construction, but flexible operation and maintenance management. This stems from the fact that there is no special legal system and supervision mechanism in China to restrict the operation and management measures of Agri-PV, and it is difficult to form a standardised and legalised operation and management of Agri-PV only by national land management policies, local standards and construction guidelines.

10.2.3 Directives on agrivoltaics

In recent years, many provinces have formulated technical regulations for agricultural photovoltaic projects according to local land conditions. Take China's eastern provinces of Zhejiang, Shandong, and central and western provinces of Shaanxi as examples (Figure 10.1), and it is generally stipulated that:

1. The lowest edge of photovoltaic panel on farmland is not less than 2.5 m from the ground.
2. The column spacing of pile foundation shall be greater than 4 m, and the row spacing shall be greater than 6 m.
3. The lowest edge of the PV module on the water body should be 0.6 m higher than the highest water level.

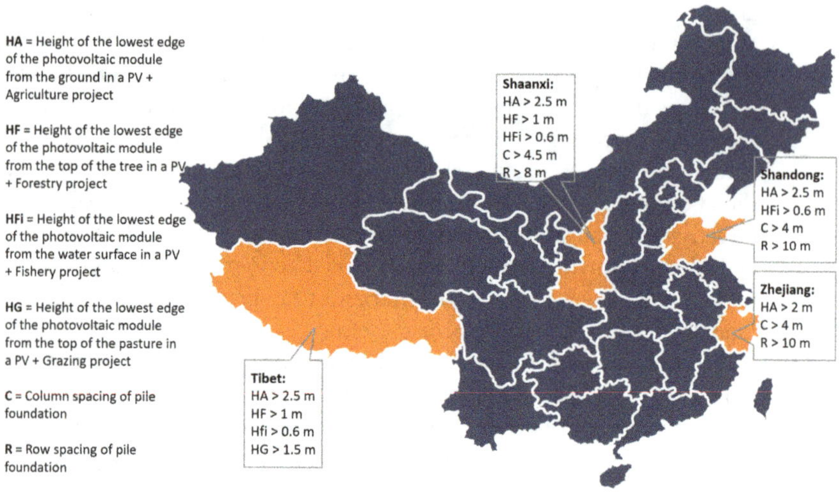

Figure 10.1 Regulations on the installation of PV modules of integrated solar projects in selected provinces in China [4]

4. The lowest edge of the photovoltaic panel on the forest floor should not be lower than the highest point of the trees by more than 1 m.

Taking Jiangshan 200 MW agricultural photovoltaic project in Zhejiang Province as an example, the height of photovoltaic panels is set according to the height of plants [4]. Position of crops (under or between panels) is determined by shade tolerance of crops. According to the height of crops, crops are planted in multiple layers under the PV panel to make full use of vertical space. For example, Chinese herbal medicines are planted under photovoltaic panels, kiwifruit is planted between photovoltaic panels and domestic animals such as chickens, ducks and pigs are also cultivated under photovoltaic panels.

Hainan province requires the construction of plastic greenhouses under photovoltaic panels to realise the function of rain and wind protection and ensure the large-scale production of vegetables in photovoltaic power plants [5]. At present, there is no requirement for agricultural technical standards or data acquisition of Agri-PV power station in China.

10.2.4 Tax regulation

In recent years, the problem of land taxes and fees has made the development of photovoltaic power plants in an embarrassing situation [6]. This limits the healthy development of AGRI-PV industry to some extent [6].

1. According to Article 2 of the Provisional Regulations on Urban Land Use Tax in People's Republic of China (PRC), 'Units and individuals who use land in cities, counties, towns and industrial and mining areas are taxpayers of urban land use

tax (hereinafter referred to as land use tax) and shall pay land use tax in accordance with the provisions of these regulations'.

2. According to the provisions of the current farmland occupation tax in China, Article 2 of the People's Republic of China (PRC) Farmland Occupation Tax Law stipulates: 'Units and individuals that occupy farmland to build buildings, structures or engage in non-agricultural construction in People's Republic of China (PRC) are taxpayers of farmland occupation tax and shall pay farmland occupation tax in accordance with the provisions of this law. If farmland is occupied to build irrigation and water conservancy facilities, farmland occupation tax shall not be paid. The term "cultivated land" as mentioned in this Law refers to the land used for planting crops' [7].
3. Article 14 stipulates that 'those who occupy forest land, grassland, farmland water conservancy land, aquaculture water surface, fishery water beach and other agricultural land for building houses or engage in non-agricultural construction shall be subject to farmland occupation tax according to the provisions of these regulations. If the production facilities directly serving agricultural production occupy the agricultural land specified in the preceding paragraph, the farmland occupation tax shall not be levied' [7].

10.2.5 Development trends

To achieve the goal of carbon emission capping in 2030 and carbon neutrality before 2060, the author believes that the 'photovoltaic+' model will continue to be more widely promoted. The development of Agri-PV in China needs to be considered in different regions. The 14th Five-Year Plan points out that photovoltaic integrated with desertification control and ecological restoration projects should be vigorously developed in the central and western regions. It is foreseeable that this mode will become one of the main modes for developing Agri-PV in the central and western regions of China. For eastern China, local governments tend to increase the proportion of local renewable energy, and the co-location of photovoltaic and open-air agriculture projects and photovoltaic greenhouse projects will be promoted. The local government will introduce more detailed norms and supervision mechanisms for the construction of AGRI-PV projects.

In the future, it is possible for China to develop photovoltaic land planning with the goal of carbon neutrality, to make more effective use of land resources. Considering the local per capita arable land area, land demand for energy consumption, energy self-sufficiency rate, water resources conditions and agricultural self-sufficiency rate, there are regional differences in the formulation of AGRI-PV policies.

AGRI-PV, as an important means to promote rural revitalisation, may continue to be widely promoted in China, and the high-quality development of local agriculture and related industries will be promoted through the coupling of photovoltaic and agriculture. The first stage of AGRI-PV development in China focused on the dual use of energy and agriculture. The next stage is to use the green electricity generated by AGRI-PV project to promote agricultural electrification, such as photovoltaic greenhouses, smart plant factories, agricultural cold storage and agricultural product processing industries, which can all be powered by AGRI-PV system.

10.3 Legal framework of agrivoltaics in France

10.3.1 General context

In 2020, around 20% of France's renewable energy production came from the agricultural sector, which represents 3.5% of France's global energy production. Certain products, such as biofuels, are more heavily represented in this sector [1]. A distinction needs to be made between energy production where the agricultural area hosts the renewable energy installation without participating in it (e.g. wind and solar energy), potentially leading to a conflict of land uses, and energy production where the agricultural sector participates directly in energy production (e.g. biofuel and methanisation), potentially leading to competition between food and non-food production [2]. Agri-PV is an ingenious way of reducing conflicts of land use between agricultural production and the renewable solar energy production. A number of innovative projects have recently been launched in France [3].

Since 2010, a number of incentives have been introduced to encourage energy production in the agricultural sector, for instance, financial support mechanisms for electricity produced by methanisation, photovoltaic, biomass, biomethane and various European, national and regional subsidies for this type of project. These factors have led to an increase in the number of Agri-PV projects, although until 2023 there was no clear legal framework in place to reconcile the interests of preserving agricultural land in a global context of reduced availability, while at the same time encouraging the production of renewable energy. Very often, Agri-PV projects consisted of ground-mounted solar power projects that included, or at least allowed for, ancillary agricultural activity (e.g. grazing sheep or even beehives for honey production).

Through various public policies, France has set itself ambitious targets, with plans to double the installed capacity of renewable energies in 2028 compared with 2017, with installed capacities of 101–113 GW in 2028 (73.5 GW in 2023) and 36% renewable energy in electricity production in 2028 [4]. In this context, the public authorities have sought to clarify the legal definition of Agri-PV to secure and promote the development of such projects [5]. However, it was not until March 2023 that Agri-PV installations were granted a legal framework. The Law to Accelerate the Production of Renewable Energies dated 10 March 2023 (the 'Acceleration Law') has provided a legal framework for Agri-PV installations, bringing legal clarity to the development of such projects even though, the complexity of the legal framework could slow down the development of such projects.

In any case, the dialogue between the agricultural sector and the renewable energy sector is more than ever open to bringing the two sectors closer together with the aim of jointly working on the development of Agri-PV installations. A memorandum of understanding between the Fédération National des Syndicats d'Exploitants Agricoles (FNSEA, the main union defending the interests of farmers) and the Syndicat des Energies Renouvelables (SER, a union defending the interests of renewable energy industrials) to develop joint initiatives in the renewable energy sector was signed on 3 March 2023. Such a memorandum of understanding between two major unions of the farming and renewable energy sector shows that both unions are committed to sharing a common approach based both on the absolute need to lead the fight against climate change by preserving farmland, production and the economic activities of the agricultural sector and the need to develop renewable energies.

10.3.2 The new framework for the development of agrivoltaic installations

The Acceleration Law provides for a general definition of Agri-PV installation as an electricity production facility using the sun's radiative energy, whose modules are located on an agricultural plot where they make a sustainable contribution to the establishment, maintenance or development of agricultural production [6]. Several criteria must be met for an installation to be considered as Agri-PV. More specifically, *at least one* of the following services must be provided directly to the agricultural parcel, guaranteeing significant agricultural production and a sustainable income for farmers: (i) improving agronomic potential and impact, (ii) adapting to climate change, (iii) protecting against hazards and (iv) improving animal welfare. An installation cannot be considered as 'Agri-PV' if it substantially impairs one of the above-mentioned services or if it impairs to a limited extent two of the above-mentioned services. These services therefore imply a technical approach and a good understanding of the farming issues, to ensure that radiant energy is available to the panels without preventing crops from accessing it [7]. The Acceleration Law also sets out exclusion criteria, preventing an installation from being considered as Agri-PV if such an installation does not allow agricultural production to be the main activity on the farm plot or is not reversible. Whether farming is considered to be the main activity can be assessed on the basis of the volume of production, the level of income or the amount of land occupied (the application of these criteria will be specified by decree).

Such a definition demonstrates the search for a balance between the development of the sector and the protection of agricultural production. This quest for balance is clearly stated in France's national energy policy, which is enshrined in the Energy Code: it provides that one of the objectives is to encourage the production of electricity from Agri-PV installations, while reconciling this production with agricultural activity, maintaining the priority given to food production and ensuring that there are no negative effects on agricultural land and prices [8].

A decree is expected to specify the application of the Acceleration Law. Such a decree will notably specify the above-mentioned services ((i)–(iv)) to be provided to qualify as an Agri-PV installation. It will also define the significant agricultural production and a sustainable income for farmers, determine the conditions for deploying and supervising Agri-PV and provide for the monitoring and control of facilities. Such a decree is currently under discussion. A particular point of uncertainty is the definition of agricultural production as the main activity. One of the criteria set out in the draft decree is the ratio of the projected surface area of the photovoltaic modules on the ground to the maximum surface area of the agricultural plot. A maximum projected surface area of the photovoltaic modules will be allowed, beyond which agricultural production will not be considered as the main activity. This threshold, which has a substantial impact on the design of projects, is a source of much debate.

10.3.3 Town planning constraints

One of the main constraints in France is obtaining the necessary permits for the construction of an energy production unit located on natural or agricultural land. In agricultural zones, buildings and facilities necessary for farming or community facilities (depending on the applicable town-planning document) are authorised if they are not incompatible with farming activity and if they do not compromise the preservation of natural areas and landscapes [9]. Before the Acceleration Law, case law and legal doctrine determined what should be considered compatible with agricultural activity, supplementing the relevant provisions of the Planning Code.

On one hand, ground-mounted solar installations have been qualified by case law as necessary for community facilities as they help to meet a public need by producing electricity that is sold to the public [10]. The concept of constructions and installations necessary for community facilities only applies to electricity production installations whose main use is the production of electricity and not to a shed with solar panels, even if the energy produced in this way is intended for resale [11].

The criterion of compatibility of the installation with farming activity is, however, more difficult to assess. The administrative courts considered in a 'Photosol' case that, in order to determine whether the criterion of the compatibility of the project with the agricultural activity is satisfied, it must be assessed 'whether the project allows the exercise of a significant agricultural, pastoral or forestry activity on the land where the project is located, in the light of the activities which are actually carried out in the area concerned [. . .] or, as the case may be, would be intended to be developed there, taking into account in particular the surface area of the plot, the area of the project, the nature of the soil and local uses' [12]. Prior to the 'Photosol' case, the government had specified in a 2011 ministerial response from the Ministry of Agriculture, Food, Fisheries, Rural Affairs and Regional Planning that ground-mounted photovoltaic power projects can be authorised on agricultural, natural or forest areas if they are located 'mainly on land that does not perform these functions, such as wasteland and uncultivated land for which a return to cultivation is unlikely, old quarries, landfills, polluted soils, etc' [13]. Some solar projects have been considered compatible with the exercise of an agricultural, pastoral or forestry activity to a certain extent (with elevated solar panels for instance) [14]. However, this has not always been the case [15] and has given rise to a degree of uncertainty that is difficult to reconcile with the development of photovoltaic projects.

On the other hand, for solar projects which are not ground-mounted installations, as sheds for instance, it is more difficult to assess their compatibility in agricultural zones; if such installations are not considered necessary for community facilities, they would need to be considered necessary for farming. This condition is assessed on the basis of a number of factors [16]. For instance, photovoltaic greenhouses may be deemed necessary for a farm if they are used to produce fruit or vegetables and the economic profit from this operation helps to keep the farm operating [17]. It is also the case when the greenhouses are used to shelter crops that would otherwise be subject to the various natural risks [18].

Since the Acceleration Law, Agri-PV facilities are explicitly considered necessary for farming, which circumventing legal debates on this point [19]. Such an addition, while welcome, could also have specified that Agri-PV installations are compatible for farming activity for the avoidance of doubt, even if it seems that the very definition of Agri-PV implies that the installation is compatible with farming activity. Nevertheless, this compatibility implies that the Agri-PV project falls precisely within the legal definition, as it will be specified by the decree to be issued.

The Acceleration Law specifies that the installation of greenhouses, sheds and shades for agricultural use supporting solar panels must correspond to a need linked to the effective exercise of a significant agricultural, pastoral or forestry activity [20]. The aim of this provision, added by amendment to the bill [21], is to ensure that buildings that are 'unnecessary or even harmful to farming' (in the words of the members of Parliament) are not built solely to support solar panels, and thus to avoid the numerous 'alibi' projects, particularly under greenhouses. Besides, the Acceleration Law provides for higher development constraints for solar projects as a framework document will define the agricultural and forestry areas open to solar projects. This means that no solar project may be installed on agricultural or forestry land not identified in this framework document [22]. Fortunately, Agri-PV installations are excluded from such provisions and can therefore be installed on any agricultural or forested zones, although local planning constraints may apply. A case-by-case study is therefore necessary to take account of local planning constraints.

10.3.4 Environmental and agricultural constraints

Unlike wind projects, solar projects (including Agri-PV projects) are usually only subject to a planning authorisation and, as the case may be, environmental-related authorisations (such as tree-clearing authorisations, Water Law declarations/authorisations [23] or derogations related to protected species). However, they should not be subject to an 'environmental authorisation' (with exceptions to be analysed on a case-by-case basis) but could require a modification notice in relation to ICPE regulations if the farm is covered by such regulations.* Solar projects may however be subject to an environmental impact assessment (systematically above

*ICPE means an industrial facility (*Installations Classées pour la Protection de l'Environnement*) which may present dangers or inconvenient for the environment or human health and for which the operator is required to file either an ICPE Prior Declaration, an ICPE Registration or an ICPE Authorisation (and as from 1 March 2017, an Environmental Authorisation).

1 MWp), with the exception of installations on shaded areas located on parking areas and installations on roofs [24]. Administrative authorities [25] consider that the term 'installations on roofs' refers to the covering of enclosed buildings, including greenhouses used for agricultural or arboricultural production that are enclosed on all sides and the covering of sheds. Therefore, Agri-PV installations meeting these criteria are exempt from environmental impact assessment. Such an environmental impact assessment identifies and assesses the risks of environmental impact arising from a project, to guide the authorities in their decision-making and, in particular, in setting the relevant requirements. When a project consists of several installations, it must be considered as a whole, including where it is split up over time and space and where there are several project owners, in order to assess its environmental impact as a whole [26].

In addition, Agri-PV installations are explicitly the subject of an agricultural impact assessment since the Acceleration Law. This document deals with agricultural issues to be taken into account during the project examination phase. Such an agricultural impact assessment is also required for projects which, by their nature, size or location, are likely to have a significant negative impact on the agricultural economy [27].

Since the Acceleration Law, Agri-PV installations (and any solar project located on agricultural, natural or forest areas) are authorised for a limited period, subject to dismantling at the end of this period or at the end of operation of the structure if this occurs earlier [28]. The characteristics of these structures must ensure that their installation is reversible. The owner of the site on which the structure is built must remove the structure within a reasonable time and restore the site to its original state when the structure is not used or is no longer used or when it is established that the conditions of compatibility with agricultural, pastoral or forestry activity are no longer met or, at the latest, at the end of a period determined by regulation. Such provisions will be specified by decree.

Moreover, since the Acceleration Law, the administrative authorities may require Agri-PV facilities to provide the financial guarantees needed to dismantle and restore the site [29].

10.3.5 Support mechanisms

Support mechanisms in France can take the form of feed-in premium contracts (*contrats de complément de rémunération*) or feed-in tariffs contracts (*contrats d'obligation d'achat*). Such contracts are awarded on the basis of an open-window mechanism or a call for tenders. An open-window mechanism is a process mainly based on the fact that the project meets the criteria set by a ministerial order under which certain projects are, given their characteristics, entitled to obtain a financial support mechanism, subject to compliance with certain conditions, the eligibility being automatic once the criteria are met. A call for tenders is a competitive process based on eligibility criteria and the reference tariff price proposed by the applicant, obtaining the financial support mechanism being subject to the designation of the project following the call for tenders.

Currently, order dated 6 October 2021 [30] provides for an open-window mechanism pursuant to which certain PV installations on buildings, sheds or roofs using photovoltaic solar energy and with a peak installed capacity of less than or equal to 500 kilowatts may be eligible to a feed-in tariff mechanism. The order explicitly targets agricultural-type installations, although it does not refer to the definition of Agri-PV installations provided for by the Acceleration Law.

Before the Acceleration Law, calls for tenders for which Agri-PV installations were eligible had been launched. For instance, in 2017, even though the definition of an Agri-PV installation did not exist at the time, these projects were eligible for contract for difference installations that make it possible to combine secondary photovoltaic production with main agricultural production, creating a demonstrable operating synergy with innovations focusing in particular on photovoltaic systems and control tools and services to optimise electricity production and agricultural production (e.g. better sharing of light between photovoltaic systems and crops using mobile structures in open fields or integrated into greenhouses) [31]. Since the adoption of the Acceleration Law, Agri-PV installations may be subject to specific calls for tenders [32].

Moreover, agricultural land is eligible for support under the Common Agricultural Policy, subject to various conditions. However, under French law, plots of land used for solar installations were not eligible for such aid. Since the Acceleration Law, the presence of Agri-PV facilities on agricultural land does not prevent the same land from being eligible for Common Agricultural Policy aids [33].

10.3.6 Real estate aspects to be considered

As a general rule, land rights are secured by the conclusion of a long-term lease granting in rem rights (*bail emphytéotique*) which, when the land is farmed by a third party, is signed in the presence of the holder of the rural lease, the conclusion of the lease being concomitant with the partial termination of the rural lease and the compensation of its holder. Agri-PV calls into question this usual scheme. In particular, the question of the long-term future of the farm is at stake (an issue that will have to be addressed in the light of the Acceleration Law and its decree of application). Operators are already looking at alternative schemes to ensure agricultural continuity or to integrate the provision of agricultural services into the contract signed with the owner of the land.

10.3.7 Conclusion

The legal framework for Agri-PV installations was enshrined in the Acceleration Law. Acceleration Law provides a precise definition of the concept of Agri-PV installations (it should be noted that the debate is not over insofar as application texts are still awaited), clarifies the compatibility of these installations with agricultural zones and conditions for project development facing various constraints (reversible character of the installation, constitution of financial guarantees for dismantling). The clarification of the legal framework for Agri-PV installations is very welcome, especially as the development of solar projects is going to be more constrained as a framework document, provided for in the Acceleration Law, will in the coming months define the agricultural and forestry zones open to solar projects. Fortunately, Agri-PV installations are excluded from such provisions and may therefore be installed on any agricultural or forested zones, although local planning constraints may still apply.

The application decree and its practical implementation are likely to raise legal debates and could reveal diverging interests between the various stakeholders There is no doubt that the Acceleration Act will give new momentum to Agri-PV projects and that this sector will consolidate over the next few years.

10.4 Legal framework of agrivoltaics in Germany (Jens Vollprecht)

The main regulations of Agri-PV in Germany are presented below: Firstly, the agricultural regulations are explained, followed by construction law and energy law, and then tax law is presented. Finally, an outlook on current legal developments is given.

10.4.1 Agricultural regulation: direct payments of the EU

Within the framework of the European agricultural policy, farmers are granted direct payments for land that is primarily used for agricultural purposes, provided that the relevant conditions are met. An important question is therefore whether an agricultural area loses this eligibility due to the use of Agri-PV.

For the implementation of the numerous European guidelines for the Common European Agricultural Policy, there is, among others, the GAP Direct Payments Regulation (GAPDZV). According to this regulation, the loss described does not occur if the installation is an Agri-PV project [1]. What is such an Agri-PV project, is defined in § 12 (5) sentence 1 GAPDZV. According to this, an Agri-PV system is a system for the use of solar radiation energy that is erected on an agricultural area and does not exclude the cultivation of the area using standard agricultural methods, machines and equipment and reduces the agriculturally usable area by a maximum of 15% based on DIN SPEC 91434. If the requirements are met, 85% of the area is considered eligible [2].

This new regulation is to be welcomed. It resolves uncertainties that arose from the previous regulation. There were very good arguments that the flat-rate loss of subsidies for the construction of a solar installation on the land was not in conformity with European law. Rather, subsidies were to be granted if agricultural use was not severely restricted by the operation of the solar system [3]. This understanding is also to be 'read into' § 12 (5) sentence 1 no. 1 GAPDZV, so that, for example, the grazing of the area by sheep can also be understood as 'cultivation of the area using usual agricultural methods'. Since the usable area is often far below 15%, especially in the case of high-rise systems, the operator should be given the opportunity to receive subsidies for more than 85% of the area if he can prove this. This could be implemented by adding the following sentence 3 to § 12 (5) GAPDZV: 'If the operator proves that the plant reduces the usable agricultural area by less than 15 percent based on DIN SPEC 91434:2021-05, the eligible area according to sentence 2 increases accordingly'.

10.4.2 Building regulation

Agri-PV systems are structural installations within the meaning of the building code, so a building permit is generally required. A building permit is granted if the project does not conflict with public law regulations. Even installations that do not require a permit must comply with the provisions of public law. The public law requirements that must be observed include, in particular, those of building planning law. Admissibility under public law depends on whether the property is located within the scope of a development plan (§ 30 Baugesetzbuch, BauGB), in the so-called 'unplanned inner area' (§ 34 BauGB) or in the so-called 'unplanned outer area' (§ 35 BauGB).

10.4.2.1 Outside the scope of a development plan

If the property is not located within the scope of a development plan, the permissibility under planning law depends on whether the project is in the unplanned inner area as defined by § 34 BauGB or in the unplanned outer area as defined by § 35 BauGB. Typically, the areas in question for the realisation of Agri-PV projects will be in the unplanned outer area. Therefore, § 34 BauGB will not be addressed here.

An explicit privilege for Agri-PV plants was recently introduced in § 35 (1) sentence 1 no. 9 BauGB: It is noteworthy here that the installation must meet all requirements for financial support of so-called special solar systems also arable land, crops or grassland according to the EEG [4]. Thus, the stipulations of the BNetzA and, above that, also the already mentioned DIN SPEC 91434 play an important role. In addition, the project must be spatially-functionally related to a farm according to § 35 (1) sentence 1 no. 1 or 2 BauGB, the base area of the special solar system must not exceed 25,000 m^2 and only one system is operated per farm or farm location.

§ 35 (1) sentence 1 no. 8 lit. b BauGB includes projects that serve the use of solar radiation energy and are located on an area along highways or railroads of the superordinate network within the meaning of § 2b of the General Railway Act with at least two main tracks and at a distance from these of up to 200 m, measured from the outer edge of the roadway. Because Agri-PV systems are projects that serve the use of solar radiation, this privilege can also be applied to Agri-PV systems.

If the Agri-PV installation is to be built on a different area, the justification effort for the classification as a privileged project can increase significantly. However, depending on the design of the project, various other privileged statuses can be considered.

A privileged project is, among other things, a project that serves an agricultural or forestry operation and only occupies a subordinate part of the operating area (§ 35 (1) sentence 1 no. 1 BauGB). The term agriculture is legally defined in § 201 BauGB. Special attention is to be paid here to the term 'serving' [5]. In principle, the facilities necessary for the energy supply of the farm's buildings fulfil this requirement, if the electricity portion used for self-supply primarily serves the farm and there is a spatial connection to it. The operation-related share of energy generation must be significant in relation to the total capacity of the installation. If the operational share does not clearly outweigh the share intended for feeding into the public grid, the 'serving' function of the installation may be lacking. The BVerwG has deemed the use of approximately two-thirds of the electricity generated by a wind turbine in an agricultural operation to be sufficient [6]. In view of climate change and the future requirements in agriculture, forestry and horticulture (e.g. protection against hail, heavy rain, strong solar radiation), the serving function within the meaning of § 35 (1) sentence 1 no. 1 BauGB could arise from the protective function for plants, the soil and the water balance if the solar systems are designed accordingly. In other words: A low own consumption of electricity would not prevent the classification as a privileged project in this case. At this point, the privileged status of § 35 (1) sentence 1 no. 2 BauGB should also be mentioned, according to which a project that serves a horticultural production operation is also considered a privileged project. Admittedly, it would have to be critically examined in individual cases with regard to § 35 (1) sentence 1 no. 1 BauGB whether the additionally required subordination is given. Within the framework of § 35 (1) sentence 1 no. 2 BauGB, this additional requirement does not play any role.

So much for the individual privilege conditions. For all privileged projects according to § 35 (1) sentence 1 BauGB, it is also required that no public interests are opposed, and that the development is sufficiently secured. Public concerns are described – not conclusively – in § 35 (3) BauGB. Here, for example, the contradiction of an FNP or an impairment of the interests of nature conservation and landscape management are listed. In the case of Agri-PV projects, sufficient development refers to the accessibility by road, as well as the grid feed-in possibility [7].

10.4.2.2 Creation of a legal basis for planning

If the permissibility of a project in the unplanned external area is not possible, the area can be 'pulled' from the unplanned area into the planned area. This is achieved by drawing up a development plan for the area. The planning authorities have several instruments at their disposal.

When issuing a simple or qualified development plan, the strict classification according to § 9 BauGB must be observed. According to this, the municipality is not entitled to make determinations that cannot be based on § 9 BauGB or the BauNVO. For example, a basis under planning law can be created by defining a special area within the meaning of § 11 (1), (2) BauNVO. According to this, other special areas are areas for facilities that serve the research, development or use of renewable energies, such as wind and solar energy. The problem is that defining the area solely as a special area for solar energy may not sufficiently consider the agricultural use of the area. It therefore seems advisable to designate the area simultaneously as an area for agriculture according to § 9 (1) no. 18 lit. a BauGB. It is recognised that regulations according to § 9 BauGB for individual areas can overlap and thus complement each other. Different regulations for one and the same area are excluded if they are mutually exclusive. However, Agri-PV installations ensure the implementation of the different specifications on the same area, so that these 'double' specifications are possible.

With the enactment of a project-related development plan in accordance with § 12 BauGB, planning law leeway could also be used, since the authority can permit the project here without strict consideration of the type constraint.

10.4.3 Energy regulations: Renewable Energy Source Act 2023

Agri-PV systems generate electricity from solar radiation energy and are thus to be classified as 'quite normal' as systems for the generation of electricity from renewable energies within the meaning of § 3 no. 1 Renewable Energy Source Act (EEG 2023).

10.4.3.1 Grid connection and cost bearing

Operators of such installations therefore have a claim against the operator of the grid for general supply (grid) for priority connection to the grid in accordance with § 8 (1) EEG 2023. EEG 2023 also specifies the point at which the installation is to be connected to the grid. The guiding principle here is the minimisation of economic costs: The grid connection variant with the lowest total economic costs must be determined. Only when this variant has been found is it determined who must bear which costs. The grid connection costs are borne by the operator of the installation, the grid expansion costs by the grid operator [8].

10.4.3.2 Purchase of the electricity

In addition, the operator of the Agri-PV system is entitled to priority purchase of the generated electricity from the grid operator in accordance with § 11 (1) EEG 2023. This means that the grid operator may only prevent the feed-in to the grid in exceptional cases. This applies, for example, to redispatch 2.0; in this case, however, the operator of the Agri-PV is entitled to compensation for the electricity not fed into the grid [9].

However, the plant operator does not have to feed the electricity into the grid but can also use it himself 'before' the grid or supply it to a third party 'before' the grid [10].

10.4.3.3 Financial support

The situation becomes more complex regarding financial support under the EEG 2023 for the electricity fed into the grid. Here, the EEG 2023 has resulted in numerous positive changes for Agri-PV installations.

10.4.3.3.1 Obligatory direct marketing, participation in tenders and funding period

First of all, it should be noted that operators of installations with an installed capacity of more than 100 kW [11] are obliged to market the electricity to a third party (so-called 'mandatory direct marketing') [12]. However, this does not mean that the grid operator does not have to make any payments in this case: In the case of the so-called 'subsidised direct marketing', the system operator is entitled to the so-called 'market premium' from the grid operator in accordance with § 20 EEG

2023. The market premium is the difference between the so-called 'value to be applied' [13] and the annual market value for solar energy [14]. Since the electricity is sold to the third party and the third party pays the agreed price to the operator, the operator of the installation receives the market premium and the price agreed with the direct marketer for the electricity fed into the grid.

If the system does not exceed the 100 kW-limit the operator of the installation can claim the so-called 'feed-in tariff' from the grid operator. The feed-in tariff is calculated from the value to be applied, from which 0.4 € ct per kWh is deducted in the case of fluctuating renewable energies such as Agri-PV [15]. However, if he wishes, he can also switch to another form of sale, such as subsidised direct marketing.

Operators of Agri-PV systems with an installed capacity of more than 1,000 kW [16] must successfully participate in a tender for solar systems of the so-called 'first segment' or the so-called 'second segment' according to §§ 29 ff. in conjunction with §§ 37 ff. or 38c ff. EEG 2023. The first segment includes ground-mounted systems and solar systems that are to be erected on, at or in structural facilities that are neither buildings nor noise barriers. In contrast, the second segment is defined as solar installations that are to be erected on, at or in a building or noise barriers.

An exception to the obligation to tender may apply if the Agri-PV system is operated by a so-called 'citizen energy company' as defined in § 3 no. 15 EEG 2023: In this case, the limit above which participation in a tender is required increases to 6,000 kW [17]. Thus, the establishment of a citizen energy company may not only make sense with a view to increasing the acceptance of the Agri-PV project. Without a surcharge or, in the case of installations in the first segment, without a so-called 'payment entitlement', operators of installations cannot claim any financial support from the grid operator under the EEG [18].

§ 27a has been abolished [19]. This deletion is to be judged positively. Because of its integration into agricultural operations, Agri-PV is in principle well suited to cover the electricity needs of farms. The electricity can be fed into the grid and financial support can be claimed for it in the future, even if the operator uses a part of the produced electricity himself.

In the case of ground-mounted systems, the '20-MW-limit' stipulated in § 38a (1) no. 5 EEG 2023 must also be observed: If the plant is larger, no financial support can be claimed for the exceeding part of the plant. Exceptionally, this limit will be increased to 100 MW for a bid deadline in 2023 [20]. For plants in the second segment, the bid quantity per bid may not exceed 20 MW [21].

The financial support will be paid for 20 years from the date of commissioning of the installation [22]. For installations whose financial support is fixed by law, the payment is extended until 31.12. of the 20th year. It should be noted at this point that the entitlements to connection to the grid and to the purchase of electricity are not limited in time – in this respect, they are 'perpetual rights'.

10.4.3.3.2 Prerequisites for financial support

In addition to the general requirements for financial support under the EEG 2023, the special requirements for solar energy in particular must be met. These are outlined below and are therefore not exhaustive:

10.4.3.3.3 Installation on a building or other structural facility

As can be seen from § 48 (1) no. 1 EEG 2023 [24], there is an entitlement to financial support if the system is installed at, on or in a building or other structural facility and this has been erected primarily for purposes other than the generation of solar electricity. In the final analysis, this means that the solar system is installed on an area that is used 'anyway' ('dual use').

In the first case, the PV system can also be installed as a roof [23]. In the case of PV systems on greenhouses, for example, it must be ensured that the use of the greenhouse in its actual function is in the foreground. The subsidy does not prevent the building structure from being optimised to a certain extent, particularly with regard to its stability and durability, in order to accommodate and operate the power generation systems and also with regard to any remuneration to be obtained under the EEG, even if such measures would not otherwise have been absolutely necessary to achieve the main purpose pursued by the building structure [24]. As a result, a precise examination of the individual case is required. In the case of 'non-residential buildings' in unplanned outdoor areas pursuant to § 35 BauGB – such as greenhouses – the restriction pursuant to Paragraph 3 of § 48 EEG 2023 must also be observed. For rooftop systems with an output of up to 1,000 kW, the financial support provided increases if – in short – the entire electricity generated is fed into the grid and not just the 'surplus' [25].

If it is not a building, it must be checked, if it is another structural facility – the second above-mentioned case. If, for example, it is a scaffolding system for hops on which the module is exclusively installed, this can be well justified. This is because the scaffolding is erected to enable the hop plants to grow upwards, i.e. for purposes other than solar power generation. The aforementioned considerations of the BGH on optimisation, especially regarding stability and durability, are likely to be transferable to this constellation. This means that a certain reinforcement of the scaffolding system would not prevent financial support according to § 48 (1) sentence 1 no. 1 EEG 2023. The same applies to trellises in orchards and vineyards. But here, too, the individual case must be carefully considered.

10.4.3.3.4 Ground-mounted PV

If the requirements of a building or another structural facility are not met, the financial support for ground-mounted systems comes into consideration.

In the case of ground-mounted systems, the payment entitlement may pursuant to § 38a (1) no. 5 lit. b EEG 2023 only be issued if, among other things, the system is not located on an area that has been legally designated as a nature reserve within the meaning of § 23 BNatSchG or a national park within the meaning of § 24 BNatSchG at the time of the decision to draw up or amend the development plan.

The prerequisite in these cases is that there is usually [26] an adopted development plan for the area and that the area is not drained, agriculturally used moorland [27]. If for example this development plan has been established or amended after 01.09.2003 with the purpose of erecting a solar system, the Agri-PV installation must be located on certain areas, e.g. according to § 37 (1) no. 2 lit. c EEG 2023 and § 48 (1) no. 3 lit. c aa EEG 2023 along motorways or railroads within a strip of 500 m width, measured from the outer edge of the paved roadway [28].

Newly included in the EEG 2023 are, among other things, three different eligibility criteria in § 37 (1) no. 3 lit. a to c EEG 2023 and § 48 (1) no. 5 lit. a to c EEG 2023 for certain Agri-PV systems.

According to letter a, installations on arable land with simultaneous crop cultivation on the same area (Arable Agri-PV) are eligible for subsidies. It should be noted at this point that the bidder must submit according to § 37 (2) no. 3 EEG 2023 a self-declaration when submitting the bid that it has checked that the land is not arable land relevant to nature conservation. It is unclear which requirements can be derived from this with regard to the requirements for financial support. In the case of installations whose value to be applied is determined by law, according to § 48 (1) no. 5 lit. a EEG 2023 the area must not have been legally designated as a nature reserve within the meaning of § 23 BNatSchG or as a national park within the meaning of § 24 BNatSchG.

In order to qualify for support under letter b, the systems must be installed on land with simultaneous agricultural use in the form of cultivation of permanent or perennial crops on the same land (Cultivation-Agri-PV). It should be noted at this point that the bidder must submit according to § 37 (2) no. 3 EEG 2023 a self-declaration when submitting the bid that it has checked that the land is not arable land relevant to nature conservation. It is unclear which requirements can be derived from this with regard to the requirements for financial support. In the case of installations whose value to be applied is determined by law, according to § 48 (1) no. 5 lit. b EEG 2023 the area must not have been legally designated as a nature reserve within the meaning of § 23 BNatSchG or as a national park within the meaning of § 24 BNatSchG.

Letter c was added at the last minute of the legislative process : This provides for the subsidisation of installations on grassland with simultaneous agricultural use as permanent grassland, if the area is not located in a Natura 2000 site as defined in § 7 (1) no. 8 BNatschG and is not a habitat type listed in Annex I of Directive 92/43/EEC (Grassland-Agri-PV). In the case of installations whose value to be created is determined by law, according to § 48 (1) no. 5 lit. c EEG 2023 the area must not have been legally designated as a nature reserve within the meaning of § 23 BNatSchG or as a national park within the meaning of § 24 BNatSchG, is not located in a Nature 2000 site as defined in § 7 (1) no. 8 BNatschG and is not a habitat type listed in Annex I of Directive 92/43/EEC.

The areas must not be classified as moorland. It should also be emphasised that the areas do not have to be in the area of an adopted development plan [29].

If these Agri-PV systems are elevated with a clear height of at least 2.10 m, the value to be applied is increased by a technology bonus, which amounts to 1.2 € ct per kWh if the surcharge is granted in 2023 and gradually decreases to 0.5 € ct per kWh if the surcharge is granted in 2026 to 2028 [30].

This technology bonus is to be welcomed but should be increased significantly. This is because the additional costs for the elevations are so high that the resulting additional costs can generally not be covered with this bonus level. It is therefore feared that these systems will not be realised. It would also be conceivable to give the BNetzA the authority to increase this value. A comparable competence is granted to the authority in § 85a EEG 2023. Another possibility would be to create a separate bidding segment for these Agri-PV installations with a significantly higher maximum value than currently provided for solar systems in the first segment. This would determine the additional costs in a competitive process.

Furthermore, the technology bonus is only granted if the operator of the installation has successfully participated in the tenders for solar systems of the first segment. This results from the reference to 'special solar plants according to § 37 (1) no. 3 lit. a, b or c' in § 38b (1) sentence 2 EEG 2023. Thus, this bonus cannot be claimed for installations for which the value to be applied is determined by law. This is not comprehensible. Because even in this size class, the additional costs for the elevation arise. There is no apparent objective reason for such unequal treatment. Therefore, the following sentence 3 should be added to § 48 (4) EEG 2023: '*§ 38b (1) sentence 2 shall be applied with the provision that it is not the time of the surcharge that is to be taken into account, but the time of commissioning of the system*'.

Another category of special solar plants is the so-called Moor-PV in § 48 (1) no. 5 lit. e EEG 2023 [31]. The prerequisite for financial support here is that the plants are erected on moor soils that have been drained and used for agriculture and that the areas are permanently rewetted with the erection of the solar installation. Here, too, the area does not have to be within the scope of a development plan. § 38b (2) sentence 3 EEG 2023 provides for a 'moorland bonus' of 0.5 € ct per kWh. However, this bonus can also not be claimed for installations for which the value to be applied is determined by law. There is also no apparent objective reason for this. Therefore, in § 48 (4) sentence 1 EEG 2023, the addition '*and sentence 3*' should be inserted after '*sentence 1*' and the '*is*' should be replaced by '*are*'.

The details regarding these special solar installations will be determined in determinations by the BNetzA pursuant to § 85c EEG 2023. Regarding the Arable Agri-PV and the Cultivation-Agri-PV, the determination of the BNetzA dated 01.10.2021 is to be applied [32]. With regard to the Grassland-Agri-PV and the Moor-PV, the BNetzA has made the details in a further determination [33]. In the definition, the possibility is opened to use the land for agricultural purposes even in the case of Moor-PV, so that this category can also be called Agri-PV.

The respective determinations can be issued on 01.10 of a year with effect from 01.01 of the following calendar year [34].

It is positive that additional paludiculture can be regulated in the case of Moor-PV. Because it makes sense to provide for this additional use as well. However, the technology bonus should be granted for high-rise moorland systems in addition to the 'moorland bonus'. This can be ensured by inserting the following half-sentence in § 38b (1) Sentence 3 EEG 2023: '*; Sentence 2 applies accordingly if the additional agricultural use of the land (paludiculture) is regulated in accordance with § 85c (1) Sentence 3*'.

An extension of this range of areas results only for installations that participate in tenders: Financial support can then also be considered for areas that have been used as arable land or grassland at the time of the decision to draw up or amend the development plan, in short, cannot be assigned to any other land category than that specified in § 37 (1) no. 2 EEG 2023 and are also located in a disadvantaged area [35]. It should be noted that the definition of the disadvantaged area in § 3 no. 7 EEG 2023 has been extended to include the areas referred to in letter b of the definition. However, this extension only applies if and to the extent that the state government has issued a legal ordinance for bids on the corresponding areas. On the basis of the EEG 2021 – i.e. still with the current definition of the disadvantaged area – this has so far only been done in Bavaria, Baden-Württemberg, Hesse, Lower Saxony, North Rhine-Westphalia, Saxony, Saxony-Anhalt, Saarland and Rhineland-Palatinate.

In principle, there is no exclusion or superior/subordinate relationship between the regulations on rooftop installations, installations on other structural facilities, the 'classic' ground-mounted installations and the special solar installations. Such a relationship does not result from the wording of the provision, nor is it required by the purpose of the subsidy. Rather, the respective payment entitlement is merely linked to different area-related prerequisites and the bidder or operator of the installation can decide which subsidy category he would like to claim if the prerequisites for several subsidy statuses are met [36]. Only the areas in the disadvantaged areas according to § 37 (1) no. 2 lit. h or i EEG 2023 may not be areas listed under one of the letters a to g or j of § 37 (1) no. 2 EEG 2023.

10.4.3.3.5 Innovation tenders [37]

Only system combinations as defined in the InnAusV are eligible to participate in the innovation tenders. This means that the Agri-PV system must be coupled with a storage system or another renewable energy system and feed in the electricity via the same grid interconnection point [38]. Bidders who prevail in the tender receive the market premium from the grid operator for the electricity fed into the grid in accordance with § 19 (1) no. 1 EEG 2023 [39]. The so-called fixed market premium has been abolished, as have the special regulations for special solar plants, as these have been transferred to the tenders of the first segment under the EEG 2023 as of 01.01.2023 [40].

It is questionable whether the technology bonus can also be claimed if a high-mounted Agri-PV system is part of the system combination. The same applies regarding the moorland bonus for Moor-PV.

In § 3 (3) InnAusV old version, reference was made to § 38b EEG 2021 and its corresponding application was ordered. This corresponding application was necessary because § 38b EEG 2021 was based on the value to be applied and § 38b EEG 2021 should be 'read' with reference to this fixed market premium. This reference has been deleted with the Easter amendment. The reason given is that the repeal is a consequential amendment to the abolition of the fixed market premium, so that 'the cited standards are to be applied directly' [41]. The connecting factor for this direct application is § 3 (1) InnAusV. Thus, there is a lot to be said for the fact that § 38b (1) sentence 2 or 3 EEG 2023 is also to be applied to the electricity from a high-mounted Agri-PV system or a Moor-PV system that is part of the system combination in the case of tenders from 01.01.2023. If a storage facility is part of such a combination, this will ultimately also apply to the electricity from the storage facility that is fed into the grid. This is because § 8 (2) InnAusV states, in short, that the electricity fed out receives the same financial support as the electricity fed in if it had been fed directly into the grid. In order to avoid legal uncertainties, it is nevertheless advisable to include a corresponding clarification in the InnAusV and to add the following sentence 2 to § 8 (1) InnAusV: '*The entitlement may be increased in application of § 38b (1) sentences 2 and 3 of the Renewable Energies Act*'.

At baseline, it will be important for states to define and standardize agrivoltaic practices statutory and regulatory language. Increased utilization of dual-use land terminology as well as the promotion and funding of agrivoltaic research through legislation will allow states to understand how to properly incorporate agrivoltaic practices into their land use schematics. Once this understanding of the benefits and viability of establishing agrivoltaic practices in a particular state or region, legislatures will then be able to address taxation and zoning barriers that might otherwise inhibit agrivoltaic funding and project development.

10.4.3.3.6 Municipal participation in ground-mounted systems

Regarding acceptance, § 6 of the EEG 2023 is interesting: This regulation provides that operators of ground-mounted solar plants can grant subsidies of up to 0.2 € ct per kWh fed into the grid to affected municipalities. For systems that receive financial support, the subsidies are reimbursed by the grid operator. For other installations (so-called PPA installations), the operator of the installation must pay this amount out of his own pocket.

10.4.4 Tax regulation: inheritance-, gift-, real estate- and real estate transfer tax

If ground-mounted systems are built on an area, there is a risk that the area will no longer be assigned to agricultural and forestry operations, but to real estate. This means that the preferential treatment of agricultural and forestry assets regarding inheritance, gift, land and real estate transfer tax will cease to apply – and this may even be the case retroactively if the so-called 'retention periods' have not yet expired after the transfer of the farm.

According to the identical decrees of the supreme tax authorities of the federal states on the attribution and valuation of Agri-PV systems of 15.07.2022 [43], the following regulation applies for the purposes of land tax, inheritance and gift tax and land transfer tax: Land on which solar systems are located, which are category I or II Agri-PV systems according to DIN SPEC 91434, is to be attributed to agricultural and forestry assets. Areas on which PV systems are located that are not category I or II Agri-PV systems according to DIN SPEC 91434 are to be allocated to real estate. This makes it clear that these advantages do not apply to all ground-mounted systems, but only to certain (!) Agri-PV systems.

It should be noted at this point that there is a special regulation in Bavaria. In accordance with Article 9 (3) Bavarian Land Tax Act (BayGrStG), taxation according to the more favourable property tax A is possible despite a transfer of the land previously assigned to agricultural and forestry assets for the operation of ground-mounted photovoltaic systems, if a contractual obligation to deconstruct with subsequent continuation of agricultural and forestry use has been contractually agreed.

However, it should be pointed out that there are design options to avoid the disadvantages described, even if the plants do not meet the requirements of DIN SPEC 91434.

10.4.5 Outlook

In the federal government's cabinet resolution of 16.08.2023 (BT-Drs. 20/8657), the so-called 'Solar Package I', only changes to the EEG 2023 are addressed.

Bids for ground-mounted systems that are to be installed on agricultural land may not be submitted if ground-mounted systems with an installed capacity of more than 80 GW have been commissioned and registered on agricultural land three months before the respective bid date after 31.12.2022. After 31.12.2030, the capacity threshold increases from 80 GW to 177.5 GW. This is also relevant for ground-mounted systems, whose value to be applied is determined by law.

Agri-PV systems with a clear height of at least 2.10 m incur, as mentioned, higher costs compared to conventional ground-mounted systems due to the more complex installation. These have so far been compensated with the technology bonus. In the Solar Package, the additional costs for the tendering systems are to be offset using a different format: The tenders for the first segment are to be carried out in two stages. In the first stage, only the special solar installations – the Agri-PV with a clear height of at least 2.10 m on the one hand and the moor-, floating- and parking lot-PV on the other – will initially compete with each other. The volume in this first stage will be 500 MW in 2024 and is set to increase gradually to a total of 3,000 MW by 2029. To reflect the aspect of higher costs, the maximum value is to be significantly increased to 9.5 € ct per kWh in 2024. In subsequent years, the maximum value will be determined based on the tender results from previous years but may not exceed 9.5 € ct per kWh. Once the tender for the first stage has been completed, the group of systems eligible to participate in the second stage will be opened and all systems that have not yet been awarded a contract will compete with each other – including those installations that were not considered in the first stage. In addition, the additional costs for Agri-PV systems with a clear height of at least 2.10 m, whose value to be applied is determined by law, are now also to be covered by the technology bonus: The additional remuneration amounts to 2.5 € ct per kWh in 2024. In the years thereafter, the bonus is basically the difference between the maximum value for the first stage of the tenders and the value to be applied for conventional ground-mounted systems.

In addition, a new facet of Agri-PV is to be introduced into the EEG, the so-called 'extensive Agri-PV'. Agri-PV systems within the meaning of § 37 (1) no. 3 lit. a to c EEG 2023 can receive a bonus of 0.3 € ct per kWh if the operators implement appropriate extensification measures such as reduced use of herbicides and fertilisers, thereby ensuring greater biodiversity.

With the planned financial support for so-called 'biodiversity-PV', the area for ground-mounted systems will be expanded. Biodiversity-PV are installations that are constructed and operated on areas and in a manner that particularly promotes the conservation and expansion of biodiversity. The ordinance required for the implementation of the new eligibility criteria is to be issued by 31.03.2024 at the latest and may also provide for a biodiversity bonus. Agricultural use of the land will probably not be required here, but it should not be prohibited either.

Penalties will also be introduced. These are intended to ensure that the entitlement to remuneration for Agri-PV systems is not permanently forfeited in the event of certain infringements – as the BNetzA's determinations might suggest – but only temporarily.

It should also be noted that in future, financial support for solar installations in disadvantaged areas will no longer be dependent on the respective federal state government allowing this by issuing a legal ordinance. Instead, the areas will be opened up and the respective federal state government must prohibit financial support in the disadvantaged areas if they wish to exclude financial support. However, this will only be possible if a certain proportion of the agricultural land in the respective federal state is already occupied by ground-mounted systems, or the land is in landscape conservation areas or nature parks. It is also interesting to note that this area was previously only intended for installations that had to take part in a tendering process. According to the Solar Package I, the disadvantaged areas are now also to be opened to ground-mounted systems, the value to be applied of which is to be determined by law.

The planned changes are very welcome and should come into force on 1 January 2024. However, it remains to be seen whether, to what extent and when these regulations of the Solar Package I will become part of the EEG.

10.5 Legal framework of agrivoltaic in Italy (David Moser)

Italy has recently updated the National Energy and Climate plan (NECP) [1] to accelerate the country's growth path to achieve the European objectives for 2030 and 2050. One of the key legislative tool to accelerate the deployment of renewable energy is represented by the legislative decree of 8 November 2021, n. 199 (199/2021) [2], which is the transposition of the RED II directive [3]. Another key element with a particular impact on the introduction of innovative technologies and relevant financial support is represented by the National Recovery and Resilience Plan (NRRP) [4].

In this context of accelerated deployment and introduction of innovative PV system design, it is of particular importance that the sustainable transition pathways combine the need of achieving the ambitious decarbonisation targets with the needs related to environment and landscape preservation. Among the various points to be addressed, a key issue is represented by the integration of renewable source systems, in particular photovoltaic, built on agricultural land.

One of the emerging solutions is to create so-called 'agrivoltaic' systems (or agro-photovoltaics or Agri-PV), i.e. photovoltaic systems that allow the continuity of agricultural and pastoral cultivation activities while ensuring energy production. The attention level to this topic in Italy is very high where, in this regard, a specific measure has for example been envisaged as part of the NRRP, with the aim of promoting and supporting the most advanced methods and technologies in the field of Agri-PV.

The topic is highly relevant as, depending on the agreed definition for Agri-PV, the potential can be vast. The discussion is currently impacted also by the process of identifying the so-called 'suitable areas' for the installation of renewable energy systems, provided for by the Legislative Decree 199/2021 and, therefore, at the different possible levels of deployment of photovoltaic systems in agricultural areas, including that envisaged by the PNRR. It is thus possible that some elements of this chapter might become obsolete or no longer valid depending on the direction the norms and legislative effort will take.

Some definitions that so far have contributed to the internal debate are:

1. Agrivoltaic system (or agrovoltaic, agro-photovoltaic or Agri-PV): photovoltaic system that adopts solutions aimed at preserving the continuity of agricultural and pastoral cultivation activities on the installation site.
2. Advanced agrivoltaic system: agrivoltaic system which are in compliance with the provisions of article 65, paragraph 1-quater and 1-quinquies, of the legislative decree of 24 January 2012 [5].

   The definition of advanced agrivoltaic system relates to systems that:
3. Adopts innovative integrative solutions with installation of modules elevated from the ground, also involving the rotation of the modules with trackers, in any case in such a way as not to compromise the continuity of agricultural and pastoral cultivation activities, also possibly allowing the application of digital agriculture tools and precision;
4. Provides for the simultaneous creation of monitoring systems that allow verifying the impact of the photovoltaic installation on crops, water saving, agricultural productivity for the different types of crops, the continuity of the activities of the agricultural companies involved, the recovery of soil fertility, microclimate, resilience to climate change.

Regarding the experience in Italy with Agri-PV demonstrator, notable examples are the elevated Agri-PV system, around 40 kWp at 6 m heigh in Laimburg, South Tyrol, Italy installed over an apple orchards in 2008. Other example as reported by Toledo and Scognamiglio [6] is from the University of Piacenza that together with REM Tec patented an agrivoltaic solar tracking system named Agrovoltaico®. It was examined for maize crop production by Amaducci *et al.*, while Agostini *et al.* assessed economic and environmental performance. The first two Agrovoltaico systems were installed in 2012 in Castelvetro Piacentino (1.3 MWp) and Monticelli d'Ongina (3.2 MWp) in the North of Italy covering an area of 7 ha and 20 ha, respectively [6], also had an influence on the development of the guidelines which will be presented in the next section.

As a sign of the level of interest in the topic, the main Italian trade association are closely monitoring the market, market potential and are active in the normative and legislative tables. A notable experience is represented by the Associazione Italiana Agrivoltaico Sostenibile (AIAS) [7] which have grouped together several stakeholders active along the Agri-PV value-chain.

Finally, Italian research and industry organisations are present in the three projects funded by the European Commission with a dedicated Horizon Europe call. The projects all started in 2023 and more information can be found on their respective websites: SYMBIOSYST [8], PV4PLANTS [9] and REGACE [10].

10.5.1 Legislative and normative framework

Various documents deal with definitions of Agri-PV in Italy with the following three documents as the most impactful:

1. Guidelines published in June 2022 by the Ministry of the Energy Security previously known as the Ministry of the Ecological transition (Linee Guida in materia di Impianti Agrivoltaici) [11].
2. In January 2023, the Italian Chapter of the IEC (CEI) published the 'Public Available Specification PAS 82-93 Ed.1 Agrivoltaic Systems', developed by the Working Group (WG) 15 'Agrivoltaics', which aims to satisfy the need of operators to define the characteristics of the various types of such systems. The group was made up of 45 experts from photovoltaic operators, agricultural operators, research centres, as well as from the CEI.
3. In August 2023 UNI published the 'Reference Practice 148 – Agrivoltaic systems – Integration of agricultural activities and photovoltaic systems', promoted by ENEA, Università Cattolica del Sacro Cuore, REM TEC and UNI and which also saw the contribution of RSE, CEI and KB Development srl. This document aims to define the necessary requirements for a photovoltaic system to be defined as an 'agrivoltaic system', indicating the conditions for its optimal integration into the landscape, in accordance with current guidelines.

The guidelines introduced the following requirements to differentiate between different Agri-PV typologies:

REQUIREMENT A: The system is designed and built in such a way as to adopt a spatial configuration and appropriate technological choices, such as to allow the integration between agricultural activity and electricity production and enhance the production potential of both subsystems.

At least 70% of the surface is intended for agricultural activity, in compliance with good agricultural practices (GAP) with

$\boldsymbol{S_{agri} \geq 0.7 \quad S_{tot}}$ and $\boldsymbol{LAOR \leq 40\%}$ where LAOR is land area occupation ratio and the Sagri is calculated by subtracting the impeded area from the total area.

LAOR is the ratio between the area of the modules and the area of land that they occupy, expressed in percentage. Depending on the definitions on how the modules' area is calculated, low tilt angles and larger row distance will lead to higher LAOR values which increase the radiation to allow the photosynthesis of the crop, but to the detriment of electricity generation. Therefore, finding the trade-off between energy and agricultural demands will require a careful analysis of the implications of selecting the design criteria.

REQUIREMENT B: The agrivoltaic system is operated, during its technical life, in such a way as to guarantee the synergistic production of electricity and agricultural products and not compromise the continuity of agricultural and pastoral activity.

(B.1) the continuity of agricultural and pastoral activity on the land subject to the intervention;

(B.2) the electrical producibility of the agrivoltaic system, compared to a standard system and its maintenance of efficiency with the following low threshold:

$$PV_{agri} \geq 0.6 \cdot PV_{standard}$$

REQUIREMENT C: The agrivoltaic system adopts innovative integrated solutions with modules elevated from the ground, aimed at optimising the performance of the agrivoltaic system both in energy and agricultural terms. A concept of minimum height is introduced with 2.1 m for agricultural purposes and 1.3 m in case of livestock.

REQUIREMENT D: The agrivoltaic system is equipped with a monitoring system that allows the verification of the impact on crops, water saving, agricultural productivity for the different types of crops and the continuity of the activities of the agricultural companies involved.

(D.1) water saving;

(D.2) the continuity of agricultural activity, that is: the impact on crops, agricultural productivity for the different types of crops or livestock and the continuity of the activities of the agricultural companies involved.

REQUIREMENT E: The agrivoltaic system is equipped with a monitoring system which, in addition to complying with requirement D, allows the recovery of soil fertility, the microclimate and resilience to climate change to be verified.

The guidelines states that:

- Compliance with requirements A, B is necessary to define a photovoltaic system built in an agricultural area as 'agrivoltaic'. Compliance with requirement D.2 should also be required for these systems.
- Compliance with requirements A, B, C and D is necessary to satisfy the definition of 'advanced agrivoltaic system' and, in compliance with the provisions of Article 65, paragraph 1-quater and 1-quinquies, of Legislative Decree 24 January 2012, n. 1, classify the plant as deserving of access to state incentives based on electricity tariffs.
- Compliance with A, B, C, D and E are pre-conditions for access to NRRP contributions; however, further criteria may be defined in terms of subjective or technical requirements, reward factors or priority criteria.

As previously mentioned and also included in guidelines, the definition of 'advanced agrivoltaic system' is not new in Italy as it was already included in the legislative decree in 2012 which was subsequently modified in 2021, before the publication of the guidelines.

The 'Public Available Specification PAS 82-93 Ed.1 Agrivoltaic Systems' [12], developed by the Working Group (WG) 15 'Agrivoltaics', took the guidelines as inspiration and defined for each typology how to calculate the geometrical parameters needed to comply with the A-E requirements. This PAS (Public Available Specification) has an experimental nature and provides information regarding the characterisation of agrivoltaic systems, also with respect to photovoltaic systems. It also deals with the classification of the various types of agrivoltaic systems and the related basic requirements, as well as the monitoring and evaluation of electricity production. Finally, it provides elements for electrical safety in the exercise of electrical and agricultural activities, for O&M activities and for checks on agrivoltaic systems.

The document reports installation typologies, which at present are considered feasible, but it is not intended to be restrictive with respect to other possible types of agrivoltaic plants and systems.

The debate created by the publication of these two documents is ongoing and at the time of publication there is no clear consensus on the modifications needed. The key elements from the guidelines and from the PAS which are deemed as critical by the (PV) operators are mainly linked to the definition of elevated systems (impact on requirement C) and to the calculation of the area impeded for agricultural purposes for various Agri-PV plants typologies (impact on requirement A). The current PAS Ed.1 currently defines the impeded area for non-elevated tracked system as the area with the tracking system in the horizontal position while the height of the system is calculated with the trackers at the maximum technical tilt angle as provided by the datasheet. Discussion on how not to penalise non-elevated tracked system is currently ongoing in the view of reaching an agreement for an Ed.2 of the PAS. The definition of impeded area is important for the achievement of any requirement related to the reduction in the effective agricultural areas over the total area. What is impeded by the presence of structures and PV module shading and rotation will also depend on the type of crop but most importantly on the health and safety of the operators. Another element included in the PAS is related to the role of the agronomic expert who will take part in the assessment process of Agri-PV projects. The role and the details on their duties is still under discussion.

In the meantime, in absence to a clear decree or document with normative or legislative validity with clear definitions and to avoid speculative approaches, regions have started developing their own guidelines (e.g. Piemonte and Campania) with clear indications on their preference towards elevated systems as the only Agri-PV typology that can be authorised with simplified procedures.

10.5.2 Permitting and simplification

For the authorisation of agrivoltaic systems, the regulations for ground-based systems apply, subject to some simplifications. If such systems are located within 3 km of areas for industrial and commercial use, it is possible to authorise agrivoltaic systems of any power by filing the SAP (Simplified Authorization Procedure) governed by Article 6 of Legislative Decree 28/2011. It consists in the deposit of the technical project to the relevant Municipality and allows the start of the works 30 days after the submission (unless additional requests or suspensions by the Municipality). If the selected project site does not interfere with the so-called 'unsuitable areas' (listed in Annex 3 letter f of the Ministerial Decree 10/09/2011), no further environmental procedures are envisaged for Agri-PV plants with nominal power up to 10 MW. In case the PV plant AC power exceeds 10 MW and up to 20 MW, in addition to the SAP, a procedure for verifying eligibility for an Environmental Impact Assessment (so-called EIA Screening) might be carried out at a regional level, while for higher AC power connection requests, the authorisation process requires a national EIA.

Agrivoltaic systems built directly by agricultural entrepreneurs (possibly in joint ventures with energy producers) at heights exceeding 2 m can be freely installed, provided that they are in compliance with possible requirements from landscape authorisations and that there are no interferences with protected areas or those forming part of the Natura 2000 network.

Outside of the two cases mentioned, agrivoltaic systems follow the same regulations applied for ground-mounted photovoltaic systems.

10.5.3 Incentives

The NRRP has committed around 1.1 billion euros to incentivise around 1 GWp of Agri-PV system with innovative features, namely elevated Agri-PV system integrating innovative components. Although several drafts have been circulated it is still unclear the level of contribution to the systems: it is likely that the measure will both cover initial capital investment as well as providing incentives on the generation as feed-in-tariff.

While waiting for the European Commission to return the ministerial decree regulating the incentives, the GSE (Gestore dei Servizi Energetici) has launched a public consultation process in order to acquire useful elements for the definition of future rules. Specifically, the public consultation aims to delve deeper into some aspects that will be regulated in the Rules:

1. Subjective requirements and verification methods.
2. Requirements for agrivoltaic systems worthy of the resources provided by the PNRR. Design and construction characteristics.
3. Operation requirements of agrivoltaic systems with access to the resources provided by the PNRR. Monitoring systems.
4. Characteristics of the main monitoring system (continuity of agricultural and livestock activity and related assessment methods).

At the stage of publication, no feedback has been provided by the GSE following the public consultation.

10.5.4 What is missing? Recommendations

An important document for the deployment of renewables in Italy is linked with the 'suitable areas' decree. At the stage of publication, only drafts are available which can only lead to speculations. The drafts seem to introduce no constraints for elevated Agri-PV systems while any other form of Agri-PV is limited by given percentages of the total agricultural area. In any case the clear feedback from the parties involved is that it is necessary to clearly clarify the technical characteristics that justify these choices, and, to safeguard agricultural production, it is necessary to specify that in any case all types of agrivoltaic systems must guarantee the maintenance of the agricultural production or, possibly, the transition to new crops of higher economic value, in line with requirement B of the aforementioned 'Guidelines on Agrivoltaic Systems'.

It is also necessary to make this criterion explicit in the Ministerial Decree as the aforementioned Guidelines are not currently binding from a normative and legislative viewpoint. Particular attention must be paid to the definition of agrivoltaics, as, at the moment, it includes a vast range of possibilities, some of which are not very functional for an effective synergistic integration with agriculture. Many stakeholders and policy makers have noticed that some systems in the authorisation process defined as 'agrivoltaic' in reality were nothing more than traditional photovoltaics combined with cultivation of little value. Since the 2022 MITE guidelines are not a binding document, the current drafts of the decree do not introduce legally imposed limits. Therefore, it is necessary to either reach a consensus for the definitions in a legally / normative binding document or to include definitions in the 'suitable area' decree to guarantee an interpretation and application that leaves no margin for doubts. To this extent, the ed. 2 of the TC 82 PAS could fill the gap.

10.6 The legal framework of agrivoltaics in Japan

10.6.1 Background information

Japan's pioneering role in Agrivoltaics started with the implementation of the first 'solar sharing' project by Akira Nagashima in 2004. Over the next 20 years, more than 3,474 additional Agri PV projects were added after the initial introduction in Chiba prefecture [1].

In recent years, Agri-PV is mentioned in key strategy papers by the national government and relevant ministries. Agri-PV is seen as an additional instrument to achieving carbon neutrality by 2050 and reaching the 2030 renewable energy targets. According to the Basic Energy Plan, a road map for the future energy sector, the current capacity of solar photovoltaic (PV) must be doubled by 2030 [2]. Japan's geographical context makes an expansion of large-scale ground-mounted solar PV projects challenging since only 34% of the land area is flat and agricultural land is scarce Ground-mounted solar PV projects are generally allowed on forested mountainsides in Japan, but the negative impact on the environment and the landscape raises opposition from local stakeholders [3,4]. Agrivoltaics can increase the available areas to install solar PV on already utilised farmland and therefore, reduce the land use conflict.

Japan does not only struggle to become energy independent but has similar problems in the food sector. Japan's food self-sufficiency on a calorie supply basis was only 38% in 2021, one of the lowest rates among developed countries. The goal of the government is to raise this number to 45% by 2030 [5]. However, the average age of farmers in Japan is over 68 years and the number of abandoned farmlands is increasing [6]. Especially small-scale farmers are struggling to make a living from farming alone, making the agricultural sector unattractive for successors. Agrivoltaics can diversify and increase the income of farmers in Japan, help with rural revitalisation, and eventually raise the food self-sufficiency of Japan.

Even though agrivoltaics have a high potential to solve pressing societal problems and agrivoltaics are generally recognised by the government, a clear vision or numerical targets specifically for agrivoltaics are missing [7].

Besides supporting policies by the government, a suitable legal framework is a decisive factor in the expansion of any technology [8]. In 2013, Japan was one of the first countries to set up regulations for the implementation of agrivoltaics on farmland. Other countries that are starting to implement their legal framework for agrivoltaics can learn from over ten years of regulations, their amendments and remaining legal barriers in Japan.

10.6.2 Relevant laws and regulations

10.6.2.1 Agricultural regulation: Cropland Act

The Ministry of Agriculture, Fisheries and Forestry (MAFF) is the relevant authority in the agricultural sector in Japan. The ministry's tasks include the overall policy-making, planning and implementation of regulations related to the protection and management of farmland. It also promotes the reuse of abandoned farmland and the revitalisation of rural communities.

The Cropland Act is the key law related to agricultural land use and stipulates the possibility of land use conversion to non-agricultural purposes, such as the installation of renewable energy projects. The purpose of the Cropland Act is to protect the limited high-quality farmland in Japan.

Farmland is classified into five different categories based on its quality for agriculture and the status of the surrounding land use, e.g. its vicinity to urban areas. There are different regulations for the specific types of farmlands. For example, ground-mounted solar PV plants are generally not permitted on the highest classes of farmland. On the other hand, it is possible to install agrivoltaics on all types of farmlands (Table 10.1).

10.6.2.2 Directives on agrivoltaics

The Agricultural Promotion Bureau of MAFF clarified the requirements for the temporary land use conversion for agrivoltaics in its first directive on agrivoltaics in March 2013 [9]. A directive is issued by a ministry or director-general of an agency to clarify the implementation of a specific law but does not have the legal strength of an actual law.

Table 10.1 Farmland classification with the description of the possibility of temporary land use conversion for ground-mounted solar PV (GM-PV) and agrivoltaics

Farmland classification	Description	Possibility of temporary land use conversion	
		GM-PV	Agrivoltaics
Farmland in agricultural land use zone	Agricultural land within the agricultural promotion area development plan established by the municipalities	No	Yes
Type A farmland	Agricultural land with particularly good farming conditions in urbanisation control areas	In principle, no	Yes
Type 1 farmland	Agricultural land with good farming conditions.	In principle, no	Yes
Type 2 farmland	Low-productivity farmland under 10 ha	On conditions, yes	Yes
Type 3 farmland	Agricultural land located within an urban area or within an area with a denoted tendency toward urbanisation	In principle, yes	Yes

It is deemed sufficient to only convert the area needed for the mounting frame of the agrivoltaics system to non-agricultural land use. The following requirements for the approval of land use conversion were decided:

1. The mounting frame must be only temporary and easily removable.
2. The shading rate must allow sunlight for crop growth.
3. The minimum panel height is 2 m.
4. The agrivoltaics systems should not impede the agricultural practices of the surrounding farmland nor should it negatively affect the implementation of the Agricultural Promotion Plan.
5. The agricultural yield reduction must be under 20% compared to the average level of the surrounding farmland.

Annual reporting to the local agricultural councils by the farmers is necessary to prove that the requirements are fulfilled. MAFF provides a template for the report that can be adapted by the local agricultural councils [10]. Besides the unit yield, photos showing the growth of crops below the agrivoltaics system just before harvesting and any sales documents should be included. Moreover, quality-related indicators, such as the grade of the crops or sugar content should also be reported. If such indicators do not exist, information about the suitability of the crops for shipment or any significant difference to crops without agrivoltaics systems in the region can be reported instead.

If the farmers do not fulfil the requirements, the local agricultural council will provide guidance on how to improve agricultural production. The farmer must make efforts to follow the instructions. Otherwise, the agrivoltaics system must be removed if the efforts are considered not sufficient. However, the practical enforcement of the sanctions is doubtful since it is questionable if the directives would hold up in court. So far, no agrivoltaics system was removed because of legal reasons.

In MAFF's first directive, the duration of the temporary land use conversion was limited to a maximum of three years. However, it soon became obvious that such a short permit duration brings challenges. Financial institutions are reluctant to provide loans for projects that face the risk of being dismantled before the loans can be paid back. Therefore, the possibility of a permit duration of a maximum of ten years was given in the second directive of the MAFF in May 2018 [11]. Agrivoltaics systems that fulfil one of the following requirements can receive the longer permit of ten years:

1. The farmer can demonstrate competence in agricultural practices and management.
2. The agrivoltaics system is installed on 'devastated farmland'.
3. The system is installed on Type 2 or Type 3 farmland.

Devastated land is defined as 'land that is not currently being used for cultivation, has become dilapidated due to its abandonment, and is objectively impossible to cultivate through normal agricultural work' [12]. The devastated land is determined by municipalities and agricultural committees based on a guideline by MAFF.

In their last directive in March 2021, MAFF decided to lower the requirements for devastated land even further [13]. A temporary land use conversion and achieving the yield requirements are no longer mandatory for agrivoltaics in areas that are labelled as devastated. Moreover, the height requirement was removed for vertical agrivoltaics systems.

A re-permit is necessary after the initial permit duration of three or ten years. A comprehensive judgement is to be made on the best available information on the agricultural situation during the previous period. Leeway should be given if proper farming according to the requirements was not conducted because of unavoidable circumstances such as natural disasters [14].

10.6.2.3 Energy regulation: Feed-In Tariff Act

In 2012, the Ministry of Economy, Trade and Industry (METI) introduced the Act on Special Measures Concerning Procurement of Electricity from Renewable Energy Sources by Electricity Utilities, the so-called Feed-In Tariff (FIT) Act. At its introduction, solar PV projects with a capacity of 10 kW or more received 42 JPY/kWh (~0.31 USD/kWh) for 20 years. The FIT Act was one of the decisive factors for a rapid increase not only in solar PV projects but also in agrivoltaics. The generous FIT allowed farmers to install agrivoltaics systems that had a higher cost compared to ground-mounted solar PV projects. In 2023, the FIT for solar PV projects, including agrivoltaics, between 10 and 50 kW only receive a rate of 10 JPY/kWh (0.075 USD/kWh) [15].

The second amendment of the FIT Act which was enforced in April 2022 and includes relevant changes for small-scale agrivoltaics projects. The revised FIT Act mandates solar PV projects with a capacity between 10 and 50 kW to fulfil certain 'regional use requirements' to obtain a FIT. Among these requirements, a self-consumption rate of at least 30% must be achieved. However, this self-consumption requirement can be waived for agrivoltaics projects if they obtained a land use conversion permit with a duration of ten years.

10.6.2.4 Environmental Impact Assessment (EIA) Law

The Environmental Impact Assessment (EIA) Law was implemented in 1999 and mandates the assessment and reporting of the potential environmental impacts of certain proposed projects. An amendment in April 2020 added large-scale solar PV projects with a capacity of over 30 MW to the list of projects that need to be assessed. Currently, agrivoltaics projects in Japan are not reaching this capacity. However, stricter regulations that dictate EIA for projects with a smaller capacity can be enforced by ordinances on a local level.

10.6.2.5 Building Standards Act

The Building Standards Act regulates the minimum standards for the construction of buildings to ensure proper construction of buildings and determines which buildings need building permits. The handling of agrivoltaics was clarified by the Ministry of Land, Infrastructure, Transport and Tourism in their notification No. 3762 [16]. Agrivoltaics are not classified as buildings if the following requirements are met:

1. A permit for temporary land use conversion under the Cropland Act was obtained.
2. A specific person continues to farm on the agricultural land in question.
3. The agrivoltaics project will not have walls and the design allows for suitable sunlight for the growth of crops.

However, solar PV projects, including agrivoltaics, must follow the Japanese Industrial Standards (JIS), which, for example, regulate to which extent solar panels must withstand impacts caused by natural disasters such as earthquakes or high wind loads [17].

10.6.2.6 Discussion and conclusion

A legal framework for agrivoltaics started to evolve ten years ago in Japan. Directives and amendments in relevant laws and regulations were introduced during this time to react to the dynamic development of this new technology. Nevertheless, many legal barriers remain that hinder the faster expansion of agrivoltaics in Japan [18].

Especia, the current procedure and requirements for land use conversion to non-agricultural land are problematic. Local agricultural councils have the final say in approving or rejecting agrivoltaics projects in their communities. The awareness and knowledge about agrivoltaics differ widely between different regions in Japan. Agricultural councils in Chiba prefecture, where the first pilot project was built and nowadays more than 500 agrivoltaics projects are implemented, are generally more willing to approve projects compared to councils in regions with less experience. A survey in 2018 showed that 60% of members of agricultural councils think that adequate agricultural production is not possible under solar panels [19]. Financial institutions are especially concerned about the short duration of the permits for land use conversion and the theoretical risk that the agrivoltaics projects need to be dismantled if farmers cannot achieve the yield requirements and cannot follow the prescribed guidance by the agricultural councils. Consequently, their willingness to provide loans for projects is low [20]. For farmers, the yield requirement is a point of great concern. The lack of research on crop growth in agrivoltaics settings in the diverse geographical regions throughout Japan means that farmers are uncertain if they will be able to achieve the high yield requirement of 80%. In addition, many find it difficult to commit to a long time of proper farming, as the average age of farmers in Japan is over 68 and there is a lack of suitable successors [19].

The feed-in tariff was a decisive factor in the first expansion of agrivoltaics. However, the current low FIT rate is seen as not sufficient to cover the still higher costs of agrivoltaics compared to ground-mounted solar PV plants [21]. New business models, such as Power Purchase Agreements, are necessary to guarantee the further advancement of agrivoltaics. A suitable legal framework that supports these kinds of new business models is essential.

Overall, the legal framework for agrivoltaics in Japan is already advanced compared to many other countries. Nevertheless, a continuous discussion with relevant stakeholders and revisions of regulations that take the high dynamic of the agrivoltaics sector into account is necessary to reduce the aforementioned legal barriers and unleash the full potential of the technology.

10.7 The legal framework of agrivoltaics in the United States

10.7.1 The developing US legal landscape of agrivoltaics

Regulations impacting the agrivoltaic industry in the United States are newly emerging and rapidly developing. Over the past five years, both federal and state regulations addressing the potential of agrivoltaic technology to further both agricultural and renewable energy goals have come before legislatures, with some bills finding receptive minds. The current US legal regulatory structure for agrivoltaics is developing such that federal entities are the primary sources of funding for agrivoltaic research and development, while more detailed regulation is left to states and localities. With no large-scale permitting incentives in place, local regulation and zoning laws [1] play a significant role in determining the siting of agrivoltaic installations within communities.

10.7.2 Federal

Existing federal regulation of agrivoltaics primarily finances research and development projects with some pending legislation recently introduced that would more formally designate agrivoltaics as an emerging industry worthy of investment. Limited existing legislation presently funds some agrivoltaics research: the Foundational Agrivoltaic Research for Megawatt Scale (FARMS) grant administered by the US Department of Energy's Solar Energy Technologies Office (SETO) and the Sustainably Co-locating Agricultural and Photovoltaic Electricity Systems (SCAPES) grant funded by the US Department of Agriculture's National Institute of Food and Agriculture (NIFA) [2,3]. In 2022, the FARMS grant provided $8 million in funding to support programs exploring how agrivoltaics can provide new economic opportunities for all stakeholders in the agrivoltaics industry, while the SCAPES grant provided $10 million in 2021 to expand research over the course of four years on which crops are best suited for agrivoltaic systems through analysis of microclimate and plant responses to solar panels at three different agrivoltaic research sites [2,3]. Research efforts for the projects funded by FARMS and SCAPES are ongoing.

US senators have also proposed three bills in 2023 that would impact the agrivoltaics industry: The Pollinator Power Act of 2023; The Agrivoltaics Research and Development Act of 2023; and The Protecting Future Farmland of America Act of 2023. At the time of this writing, these bills are still pending, but if passed, would direct robust funding and resources toward agrivoltaic projects.

10.7.2.1 The Pollinator Power Act of 2023

The Pollinator Power Act of 2023 [4], if passed, would direct the US Secretary of Agriculture to prioritise solar projects funded by the Rural Energy for America Program that creates habitat for pollinators beneath solar panels. The act specifies that 'not less than 35 percent of new solar area created under the project shall be pollinator habitat', and 'not less than 50 percent of the pollinator habitat ... shall be planted and maintained as a native cover composed only of native plants' [4]. As for the funding provided, the act notes that grant amounts 'shall not exceed 55 percent of the cost of the activity carried out using funds from the grant, including the cost of the creation of a pollinator habitat' [5]. As such, the Pollinator Power Act would aid subsidisation of agrivoltaic installation incorporating or focused on pollinator-friendly operations.

10.7.2.2 The Agrivoltaics Research and Demonstration Act of 2023

The Agrivoltaics Research and Demonstration Act of 2023 [6], if passed, would direct $15 million per year from 2024 to 2028 toward agrivoltaics research and demonstration projects. Under the act, agrivoltaics is defined as 'a system under which solar energy production and agricultural production, including crop or animal production, occurs in an integrated manner on top of the same piece of land through the duration of the project' [6]. If codified, this definition of agrivoltaics would represent the first official US federal definition of agrivoltaics and one of the only definitions of agrivoltaics at any level of US legislation. The funding would be utilised for studying several questions aimed at determining where, when and how agrivoltaics may be an appropriate technology to implement. Some of these questions include:

1. 'A risk-benefit analysis of agrivoltaic systems in different regions of the United States';
2. An assessment 'of the economic scalability of agrivoltaic systems across different agricultural land types, production systems, and regional markets';
3. An assessment 'of the types of agricultural land best suited and worst suited for agrivoltaic systems';
4. An assessment 'of how to best develop agrivoltaic systems on a national and local scale consistent with–
   (a) Maintaining or increasing agricultural production;
   (b) Increasing agricultural resilience;
   (c) Retaining prime farmland;
   (d) Increasing economic opportunities in farming and rural communities;
   (e) Reducing nonfarmer ownership of farmland; and
   (f) Enhancing biodiversity';
5. An assessment 'of the unique risk management and crop insurance needs of agrivoltaic systems'; and
6. An assessment 'of appropriate modifications to better incorporate agrivoltaic systems into existing Federal–
   (a) Agricultural conservation programs;
   (b) Agricultural risk management programs, including federal crop insurance;
   (c) Renewable energy programs; and
   (d) Investment tax credit' [6].

Further, the act also directs the Secretary of Agriculture and Secretary of Energy to consult with farm and conservation groups to develop a regulatory definition of the term 'agrivoltaic system' [6]. This would add further refinement to federal level definitions of agrivoltaics. In doing so, the act notes that the definition should consider regional needs and market access, existing state and local policies and definitions, and other factors as deemed appropriate.

10.7.2.3 The Protecting Future Farmland of America Act of 2023

If passed, the Protecting Farmland of 2023 Act would amend the Farm Security and Rural Investment Act and Rural Energy for America to incorporate the definition of the term 'agrivoltaic system' into all programs of the United States Department of Agriculture [6]. The act would define 'agrivoltaic system' the same as the Agrivoltaics Research and Demonstration Act, and further directs incorporation of any refined definition into federal agricultural and renewable energy programs [8]. Most notably, the act notes that in the selection of solar projects and their edibility to receive loan guarantees, 'the Secretary shall given priority to projects for which a plan for soil, water, and vegetation management and conservation, including a plan that includes an agrivoltaic system' [7]. The includes collecting data at the federal level from applications on benefits and impacts of agrivoltaic projects as well as, for proposed projects:

1. Whether the proposed project involves ground disturbance;
2. Whether energy generation is primarily for on- or off-farm use;
3. The percentage of farmland in the proposed ... area; and
4. Whether an agrivoltaic system is planned [7].

While it remains to be seen whether this legislation will pass, greater attention to the details of incorporating solar and agricultural production is reaching a wider audience who seem optimistic about the potential of agrivoltaics. Ultimately, the passage of this type of legislation would allow for clarity in developing national and regionally appropriate guidance.

10.7.3 State regulation of agrivoltaics

With much of the regulatory power retained by states and localities, it remains up to various state legislatures to determine whether and how they will support agrivoltaic research and development. Only a handful of states have so far taken steps toward the targeted regulation of agrivoltaics. As for the remaining states, agrivoltaic installations are determined by complex solar siting regulations that likely did not anticipate the type of dual land use the agrivoltaics employs. Thus, several legal hurdles exist for agrivoltaic development in states that have yet to address the promise of colocation of renewable energy and agriculture. In particular, the issues of how solar operations on agricultural land should be zoned (if permitted) and taxed have raised questions among stakeholders pursuing establishment of an agrivoltaic operation.

10.7.3.1 Dual-use for agrivoltaics in state legislation

States are taking a variety of approaches, from encouraging agrivoltaic projects to issuing regulations that protect prime farmland from solar development of any kind. A review of the state approaches follows.

Agrivoltaics-focused state legislation

At the time of writing, four states have passed legislation directly pertaining to agrivoltaics: Massachusetts, New York, New Jersey and Colorado. Massachusetts' Act Driving Clean Energy and Offshore Wind clarifies that agrivoltaics projects are to be treated as agricultural use for property tax purposes [8]. The Massachusetts Act creates commission to identify obstacles to agrivoltaic projects and strategies to overcome those obstacles and contains a provision specifying that 'Land used primarily and directly for agricultural purposes ... or land used primarily and directly for horticultural purposes ... shall be deemed to be in agricultural or horticultural use ... if used to simultaneously site a renewable energy generating source' [8]. This represents one state effort to rectify the daunting property tax penalties that might befall agrivoltaics operators installing solar on agricultural land. Legislation in New York would amend the agriculture and markets law to establish agrivoltaics research programs that will 'develop innovative science-based solutions to facilitate the co-location of crops and photovoltaics while promoting the biodiversity of endemic flora and fauna' while also considering socio-economic factors [8]. In 2021, New Jersey passed the Dual-Use Solar Energy Pilot Program that granted $2 million to research farms to allow for the 'construction, installation, and operation of dual-use solar energy projects on unpreserved farmland, while maintaining the affected land in active agricultural and horticultural use' [10]. The program is to last three years and encourages dual-use solar projects eligibility for farmland assessment [10]. Lastly, in 2023, Colorado passed the Agricultural Producers Use of Agrivoltaics Act that authorises grants for new or ongoing research projects studying the potential, benefits and trade-off of agrivoltaics [11]. The Colorado bill includes that 'solar energy facility' definitions will now 'include[] facilities for agrivoltaics', in an similar effort to mitigate property tax penalties of changed land use [11].

State legislative limitations of agrivoltaics

States that have not yet directly addressed agrivoltaics, like Oregon, still have existing legislation that may hinder the agrivoltaics industry in that state. For example, In 2019, the Oregon Land Conservation and Development Commission approved a rule that bans solar development on Class 1 and 2 soils but allows developments up to 12 acres on class 3 and 4 soils and developments of up to 20 acres if the development includes agricultural uses [12]. Other states take alternative approaches to regulating solar that all the same limit the potential of agrivoltaic research and development. Ohio, for example, passed a law in 2021 allow localities to prohibit the development of solar power facilities in certain unincorporated areas, strictly constraining the size of any agrivoltaic installation and exploration of its utility-scale application [13]. These states represent a stricter regulatory regime for agrivoltaics simultaneously emerging alongside the incentivising legislation in other states. Without additional federal guidance, a variety of approaches to regulating agrivoltaics in the United States at state and local levels is likely to continue.

10.7.4 State-level legal issues facing US agrivoltaics: zoning and taxes

Zoning regulations in the United States are inherently local because land use is generally governed by state and municipal authority [14]. Localities establish land use plans, and within each zoning district, each parcel of land is assigned at least one (although not necessarily the only) allowed land use [15]. When landowners can prove that zoning standards impose unnecessary hardships, variances may be granted to accommodate agrivoltaic installations [15]. Zoning plans govern where operators may locate agrivoltaic installations within the jurisdiction, whether through explicit regulation of agrivoltaics or through the impact of existing regulations (that likely did not consider agrivoltaics at the time of enactment) [1].

Agrivoltaic designs do not fit neatly into existing zoning policies premised upon single-use land, fit for only agricultural use or solar development. Although agrivoltaic systems have intrinsic agricultural function, 'agrivoltaic systems are [still] subject to the permitting and regulatory process of a conventional solar PV installation, with the added condition of placement on agricultural land' [16]. Given that zoning rules incentivise particular land uses and 'state-provided incentives for agricultural or solar energy production can, in some cases, be the determining factor in the decision to invest in solar or agriculture development', local governments hold the unique power to influence agrivoltaic development [17].

Taxation structures also heavily impact decisions to engage in agrivoltaic practices. A shift in tax burdens sometimes follows the installation of solar energy on agricultural land [18]. Agricultural land usually benefits from current use policies that disincentivises activity that would take the land out of agricultural production. With regulations ranging from the explicit prohibition of solar installation on agricultural land to the removal of tax penalties for land use changes in limited instances, inconsistency and uncertainty abound [5]. These varied approaches to regulation create concern amongst the agrivoltaics industry over the financial viability of the operation and the possible legal consequences that might follow from any breach [1].

Beneficial taxation programs, like current use policies, 'create an incentive for private landowners to keep their land undeveloped by providing some relief from market pressure to convert agricultural, open space, and forest land to economically "best uses" through development' [18]. Farmland enrolled in current use taxation programs may complicate the installation of solar arrays, though some states like Pennsylvania have recognised this legal barrier [19]. In states where solar arrays are prohibited, local governments may revoke current use enrolment and/or assess land-use change penalties for converted agricultural land [19].

Some regulatory remedies, like the implementation of overlay districts, could promote agrivoltaics through zoning 'because they entail conditional or special permit uses that are permissive of solar in certain zones, which gives local governments opportunity for strategic siting of agrivoltaics in their jurisdiction' [20]. Additionally, zoning regulations addressing excessive administrative requirements, the inclusion of decommissioning plans and removal procedures, varying temporary and long-term land-use standards, and removal of zoning barriers to its implementation would all potentially incentivise agrivoltaic systems.

References

China

[1] Fan, Tienan; Trommsdorff, Max; Gölz, Sebastian; and He, Jijiang. (2022). Agrivoltaics in China: A Study of the Current State of Agrivoltaics Development, Governmental Support Schemes, and Stakeholder Groups' Perspectives and Acceptance based on Expert Interviews. *Oral Presentation at Conference & Exhibition AgriVoltaics2022*, Piazenca, Italy.

[2] Fan, Tienan. (2022). *Agrivoltaics in China: Governmental Support Schemes, Analysis of Existing Systems, and Lessons Learned* [Master Thesis]. Technical University of Berlin [7 Jan 2022].

[3] General Office of the Ministry of Natural Resources, the Office of the State Forestry and Grassland Bureau and the General Department of the National Energy Administration. (2023). *Notice on Supporting the Development of Photovoltaic Power Generation Industry and Standardizing Land Management.*

[4] Xiao, Yuanyuan. (2022). *Ecological Impact Assessment of Landscape-scale Solar Parks Integrated with Eco-restoration in China* [Master Thesis]. Technical University of Dortmund [11 Jul 2022].

[5] The Specific Office for Fair Price for Vegetable Supply Guarantee and People-center Consideration Actions of Hainan Provinces. (2022). *Letter on Accelerating the Review of Photovoltaic Vegetable Greenhouse Design Scheme.*

[6] *AIP Conference Proceedings 2635, 020002 (2022);* https://doi.org/10.1063/5.0106454. Published Online: 6 Dec 2022.

[7] The Seventh Session of the 13th NPC Standing Committee. (2018). *People's Republic of China (PRC) Farmland Occupation Tax Law.*

France

[1] Courteau, R. and Fugit, L., *L'agriculture face au défi de la production d'énergie, Rapport no. 646 (2019–2020), 16 July 2020. Available (in French) on:* https://www.senat.fr/rap/r19-646/r19-646_mono.html.

[2] Ibid.

[3] For instance, the installation of semi-transparent modules to protect a crop from the sun (*En Alsace, les panneaux tout en transparence de BayWa r.e. protègent les framboises*, Plein Soleil, 26 October 2023, available (in French) on: http://www.plein-soleil.info/actualites/en-alsace-les-panneaux-tout-en-transparence-de-baywa-r-e-protegent-les-framboises/)) or a solar power plant cohabiting with cattle (REDEN, pionnier de l'essai en matière d'élevage bovin sur parc agrivoltaïque, Plein soleil, 2 October 2023, available (in French) on: http://www.plein-soleil.info/actualites/reden-pionnier-de-lessai-en-matiere-delevage-bovin-sur-parc-agrivoltaique/).

[4] *Stratégie française pour l'énergie et le climat, Programmation pluriannuelle de l'énergie, 2019–2028.* Available (in French) on: https://www.ecologie.gouv.fr/sites/default/files/20200422%20Synthe%CC%80se%20de%20la%20PPE.pdf.

[5] *ADEME, I Care & Consult, Ceresco, Cétiac. Caractériser les projets photovoltaïques sur terrains agricoles et l'agrivoltaïsme, Résumé exécutif de l'étude. 2021.* Available on: https://librairie.ademe.fr/energies-renouvelables-reseaux-et-stockage/4992-caracteriser-les-projets-photovoltaiques-sur-terrains-agricoles-et-l-agrivoltaisme.html.

[6] Article L. 314-36 of the Energy Code.

[7] Radiguet R., *L'agrivoltaïsme dans la loi relative à l'accélération de la production d'énergies renouvelables : "je veux du soleil"*, Energie-Environment-Infrastructures, LexisNexis, April 2023.
[8] Article L. 100-4 of the Energy Code.
[9] Articles L. 111-4, L. 151-11 and L. 161-4 of the Planning Code.
[10] Administrative Court of Appeal of Bordeaux, 3 April 2018, no. 16BX00674 (point 16); see also Conseil d'Etat, 13 July 2012, no. 343306 (for wind farms, point 3).
[11] Administrative Court of Appeal of Bordeaux, 22 June 2015, no. 13BX01623; see also J. Maillot, *Le devenir des zones agricoles à l'aune de la jurisprudence sur les permis de construire de "centrales photovoltaïques"*, Construction – Urbanisme no. 9, LexisNexis, September 2019.
[12] Conseil d'Etat, 31 July 2019, Société Photosol, no. 418739.
[13] Rép. Min. no. 107780: JOAN 5 July 2011.
[14] Administrative Court of Appeal of Bordeaux, 15 March 2018, no. 16BX02223; see also very recently Administrative Court of Rennes, 25 July 2023, no. 2303539.
[15] Conseil d'Etat, 31 July 2019, Société Photosol, no. 418739.
[16] Duval J., *Aspects juridiques du développement de projets d'installations photovoltaïques*, JurisClasseur Géomètre expert – Foncier > V° Environment, September 2023.
[17] For examples: Administrative Court of Appeal of Marseille, 3 April 2015, no. 13MA02539; Administrative Court of Appeal of Bordeaux, 14 November 2013, no. 12BX00465.
[18] Administrative Court of Appeal of Bordeaux, 18 February 2020, no. 18BX00809.
[19] Article L. 111-27 of the Planning Code.
[20] Article L. 111-28 of the Planning Code.
[21] Amendment no. 388 rect, available on https://www.senat.fr/amendements/2022-2023/83/Amdt_388.html.
[22] Article L. 111-29 of the Planning Code.
[23] Please note that PV projects may be subject to IOTA regulations which is a set of legal requirements regulating certain facilities or activities (named 'IOTA') likely to have an impact on water. The IOTA regulations provide for a nomenclature of impact on water that may require either an IOTA prior declaration or an IOTA authorisation (i.e. an environmental authorisation as of 1 March 2017). Practice shows that the main categories concerned are 2.1.5.0 (discharge of rainwater into surface freshwater or onto the ground or subsoil), 3.3.1.0 (draining, impounding, filling of wetlands or marshes) and 3.2.2.0 (installations, structures, embankments in the major bed of a water-course). Concerning category 2.1.5.0, administrative authorities consider that ground-mounted PV projects are not systematically subject to this category but that it 'applies in some cases' and that 'in general the panels are spaced to allow rainwater to infiltrate into the soil' (*Installations photovoltaïques au sol – Guide de l'étude d'impact, Ministère de l'écologie,*

du développement durable, des transports et du logement, April 2011). The administrative courts have issued few decisions on the question of the submission of ground-mounted PV projects to category 2.1.5.0. In some decisions, it has seemed to take sides in favour of submitting the projects to category 2.1.5.0. From what we understand, this could be an issue in terms of authorising PV projects.

[24] Article R. 122-2 of the Environmental Code.

[25] *Commissariat général au développement durable, Evaluation environnementale des projets, Guide de lecture de la nomenclature, Mars 2023.* Available (in French) on: https://www.ecologie.gouv.fr/sites/default/files/guide_nomenclature_evaluation_environnementale_des_projets_mars2023.pdf

[26] Article L. 122-1 of the Environmental Code.

[27] Article L. 112-1-3 of the Rural and Maritime Fishing Code.

[28] Article L. 111-32 of the Planning code.

[29] Article L. 314-40 of the Energy Code.

[30] *Arrêté du 6 octobre 2021 fixant les conditions d'achat de l'électricité produite par les installations implantées sur bâtiment, hangar ou ombrière utilisant l'énergie solaire photovoltaïque, d'une puissance crête installée inférieure ou égale à 500 kilowatts telles que visées au 3° de l'article D. 314-15 du code de l'énergie et situées en métropole continentale.*

[31] *Appel d'offres portant sur la réalisation et l'exploitation d'Installations de production d'électricité innovantes à partir de l'énergie solaire.* Available on: https://www.cre.fr/documents/Appels-d-offres/appel-d-offres-portant-sur-la-realisation-et-l-exploitation-d-installations-de-production-d-electricite-innovantes-a-partir-de-l-energie-solaire

[32] Article L. 314-37 of the Energy Code.

[33] Article L. 314-38 of the Energy Code.

Germany

[1] § 12 (4) no. 6 Regulation on the Implementation of CAP Direct Payments (hereafter GAPDZV), 24.01.2022, last amendment 29.08.2023.

[2] § 12 (5) sentence 2 GAPDZV.

[3] *Vollprecht/Kather, InfrastrukturRecht 2021 p. 266 (270 et seq.).*

[4] See below 10.4.6.

[5] On the concept of 'serving': Federal Administrative Court, judgement of 3.11.1972 – 4 C 9.70.

[6] Federal Administrative Court, decision of 4.11.2008 – 4 B 44.08.

[7] *BeckOK BauGB/Söfker*, 56th Ed. 1.8.2021, BauGB, § 35 Rn. 54.

[8] § 16 (1) and § 17 Renewable Energy Sources Act (hereafter EEG) 2023, 21.07.2014, last amendment 26.07.2023.

[9] *Lamy/Lehnert, Zeitschrift für das gesamte Recht der Energiewirtschaft 2021 p. 208 et seq.*

[10] Since the grid is not used in these constellations, grid fees do not apply. The same applies to the grid-connected charges and levies (e.g. CHP levy, Section 19 (2) StromNEV levy, offshore grid levy, concession fee). Therefore, the system operator can, for example, offer the supplied party a more favourable price for the electricity than a supplier who handles the supply via the grid. The EEG levy will be abolished on 1 January 2023, so that this levy will no longer play a role in the future. However, the electricity tax must be taken into account, whereby the exemptions according to § 9 (1) Nos. 1 and 3 StromStG can open up optimisation possibilities.

[11] When determining the size of the installation here and in the following, the 'bracketing provisions' of § 24 (1) and partly also of (2) EEG 2023 must generally be observed.

[12] This can be derived from section 21 subsection 2 no. 2 EEG 2023. According to this, the system operator can sell the electricity to the grid operator. However, he then only receives the low feed-in tariff. This low feed-in tariff is intended to encourage the system operator to switch back to direct marketing as quickly as possible.

[13] The value to be applied is an amount in cents per kilowatt hour.

[14] No. I.2. Annex 1 EEG 2023.

[15] § 21 (1) no. 1 in conjunction with. § 53 (1) no. 2 EEG 2023.

[16] § 22 (3) sentence 2 lit. a EEG 2023.

[17] § 22 (3) sentence 2 lit. b EEG 2023.

[18] § 22 (3) sentence 1 EEG 2023.

[19] In principle, electricity from 'tendering plants' could not be used for self-supply until now. If this was violated, the value to be applied was reduced to zero for the entire calendar year of the violation, cf. section 52 (1) no. 4 in conjunction with p. 3 EEG 2021. S. 3 EEG 2021.

[20] § 100 (13) sentence 1 EEG 2023.

[21] § 38c (2) EEG 2023.

[22] § 25 (1) EEG 2023; see also § 51a EEG 2023, if applicable.

[23] In the case of 'tender plants', this results from § 38c (1) EEG 2023 and § 37 (1) no. 1 EEG 2023.

[24] On the EEG 2004: Federal Court of Justice, judgement of 17.11.2010 – VIII ZR 277/09.

[25] § 48 (2), (2a) EEG 2023.

[26] Exceptionally, this does not apply according to § 37 (1) no. 2 lit. c EEG 2023 if the prerequisites specified in § 35 (1) no. 8 lit. b BauGB are fulfilled (cf. above).

[27] According to § 3 no. 34a EEG 2023, moorland soil is any soil that fulfills the requirements of § 11 (2) of the GAP Conditionalities Ordinance and can be used as a basis for the creation of the area map according to § 11 (3) of the GAP Conditionalities Ordinance. In this respect, it is problematic that the classification of an area on the basis of the criteria listed may not be legally

certain. It would therefore be desirable if the BNetzA were obliged to designate these areas more legally as moorland in a determination.

[28] Up to now, the strip was limited to 200 m and a 15 m wide corridor had to be kept free with-in this strip, cf. § 37 (2) no. 2 EEG 2021 or § 48 (1) no. 3 EEG 2021.

[29] This is a significant difference to the requirements for financial support for 'normal' ground-mounted systems according to § 48 (1) no. 3 EEG 2023 or § 37 (1) no. 2 lit. c EEG 2023.

[30] § 38b (1) sentence 2 EEG 2023.

[31] In the case of 'tender plants', this results from § 37 (1) no. 3 lit e EEG 2023.

[32] Determination of 1.10.2021 – Ref.: 8175-07-00-21/1, available at: https://www.bundesnetzagentur.de/SharedDocs/Downloads/DE/Sachgebiete/Energie/Unternehmen_Institutionen/Ausschreibungen/Innovations/GezeichneteFestlegungOktober2021.pdf;jsessionid=2CCC48CFBAEEAB49D559C0D002E2107D?__blob=publicationFile&v=3.

[33] Determination Ref.: 4.08.01.01/1#4, Available at: https://www.bundesnetzagentur.de/SharedDocs/Downloads/DE/Sachgebiete/Energie/Unternehmen_Institutionen/Ausschreibungen/Solar1/BesondereSolaranlagen/Festlegung.pdf?__blob=publicationFile&v=1

[34] § 85c (2) sentence 2 EEG 2023.

[35] § 37 (1) no. 2 lit. h. and i. EEG 2023.

[36] So also, Clearingstelle EEG|KWKG on the regulations on conversion areas (§ 37 (1) no. 3 lit b EEG 2017) and on other structural facilities (§ 37 (1) no. 2 EEG 2017) in the arbitration award of 11.01.2019 – Az. 2018/39, available at https://www.clearingstelle-eeg-kwkg.de/sites/default/files/Schiedsspruch_2018_39.pdf).

[37] The tenders for innovative concepts with hydrogen-based electricity storage according to § 39o EEG 2023 or for plants for the generation of electricity from green hydrogen according to § 39p EEG 2023 are not addressed here. It is true that Agri-PV plants could also play a role in this context. However, the corresponding ordinances have not yet been issued, so that no concrete statements can be made in this regard at present.

[38] § 2 (1) Regulation on innovation tenders (hereafter InnAusV).

[39] § 8 (1) InnAusV.

[40] On the original regulations Vollprecht/Kather, IR 2021, 266 (268).

[41] Explanatory memorandum, Bundestag printed matter (BT-Drs.) 20/1630, p. 261.

[42] On the similar regulation for wind power plants Baur/Lehnert/Vollprecht, EnWZ 2021, 341 ff. A model contract for community participation was developed by the Bundesverband Neue Energiewirtschaft e.V. (Federal Association of the New Energy Industry), which can be found at https://sonne-sammeln.de/mustervertrag/.

[43] Federal Tax Gazette (BStBl) I 2022 p. 1226.

Italy

[1] National Energy and Climate Plans [Internet]. [cited 2023 Oct 23]. Available from: https://commission.europa.eu/energy-climate-change-environment/implementation-eu-countries/energy-and-climate-governance-and-reporting/national-energy-and-climate-plans_en
[2] Gazzetta Ufficiale [Internet]. [cited 2023 Oct 23]. Available from: https://www.gazzettaufficiale.it/eli/id/2021/11/30/21G00214/sg
[3] Directive (EU) 2018/2001 of the European Parliament and of the Council of 11 December 2018 on the Promotion of the Use of Energy from Renewable Sources (Text with EEA relevance.) [Internet]. 328, 32018L2001 21 Dec 2018. Available from: http://data.europa.eu/eli/dir/2018/2001/oj/eng
[4] Italy's Recovery and Resilience Plan [Internet]. [cited 2023 Oct 23]. Available from: https://commission.europa.eu/business-economy-euro/economic-recovery/recovery-and-resilience-facility/country-pages/italys-recovery-and-resilience-plan_en
[5] Gazzetta Ufficiale [Internet]. [cited 2023 Oct 23]. Available from: https://www.gazzettaufficiale.it/eli/id/2012/03/24/12A03524/sg
[6] Toledo C and Scognamiglio A. Agrivoltaic systems design and assessment: A critical review, and a descriptive model towards a sustainable landscape vision (three-dimensional agrivoltaic patterns). *Sustainability*. 2021;*13* (*12*):6871.
[7] AIAS – Associazione Italiana Agrivoltaico Sostenibile [Internet]. [cited 2023 Oct 23]. Associazione Italiana Agrivoltaico Sostenibile. Available from: https://www.associazioneitalianagrivoltaicosostenibile.com/
[8] Symbiosyst – Create a Symbiosis where PV and agriculture can have a mutually beneficial relationship [Internet]. [cited 2023 Oct 23]. Available from: https://www.symbiosyst.eu/
[9] https://www.pv4plants.eu/ [Internet]. [cited 2023 Oct 23]. PV4Plants. Available from: https://www.pv4plants.eu/
[10] REGACE Project – Using Co2 for Green Energy [Internet]. [cited 2023 Oct 23]. Available from: https://regaceproject.com/
[11] Impianti Agri-voltaici: pubblicate le Linee Guida | Ministero dell'Ambiente e della Sicurezza Energetica [Internet]. [cited 2023 Oct 23]. Available from: https://www.mase.gov.it/notizie/impianti-agri-voltaici-pubblicate-le-linee-guida
[12] CEI Norme [Internet]. [cited 2023 Oct 23]. CEI -CEI PAS 82-93. Available from: http://mycatalogo.ceinorme.it/cei/item/0000019155

Japan

[1] Ministry of Agriculture, Forestry and Fisheries, Results of Permission for Conversion of Agricultural Land to Install Photovoltaic Power Generation Facilities. 2022.

[2] Ministry of Economy, Trade and Industry, Strategic Energy Plan. 2021.
[3] Yamashita, N. and Y. Maruyama, 太陽光発電の地域トラブルと自治体の対応 [Local Problems with Solar Power and Municipal Responses], in どうすればエネルギー転換はうまくいくのか [How can the Energy Transition Work?], Y. Maruyama and M. Nishikido, Editors. 2022.
[4] Doedt, C. and Y. Maruyama, The mega solar Twitter discourse in Japan: Engaged opponents and silent proponents. *Energy Policy*, 2023;**175**: 113495.
[5] Ministry of Agriculture Forestry and Fisheries, *Summary of the Basic Plan for Food, Agriculture and Rural Areas*. 2020.
[6] Ministry of Agriculture Forestry and Fisheries, *Census of Agriculture and Forestry, Survey of Agricultural Structure and Dynamics*. 2022.
[7] Doedt, C., M. Tajima, and T. Iida., The socio-technical dynamics of agrivoltaics in Japan. In *AgriVoltaics2023*. Daegu, Korea; 2023.
[8] FAO, Legal framework analysis for rural and agricultural investment projects: Concepts and guidelines. In *FAO Legal Papers Online*. 2000.
[9] Director-General, Rural Development Bureau, Ministry of Agriculture, Forestry and Fisheries, *Notification No. 24 Noushin Article 2657*. 2013.
[10] Ministry of Agriculture Forestry and Fisheries, *Agricultural Crop Situation Report [Japanese]*, Rural Development Bureau, Editor. 2023.
[11] Director-General, Rural Development Bureau, Ministry of Agriculture, Forestry and Fisheries, *Notification No. 30 Noushin Article 78*. 2018.
[12] Ministry of Agriculture Forestry and Fisheries, *Current Status of Devastated Farmland and Countermeasures*. 2020.
[13] Director-General, Rural Development Bureau, Ministry of Agriculture, Forestry and Fisheries, *Notification No. 2 Noushin Article 3854*. 2021.
[14] Ministry of Agriculture Forestry and Fisheries, *About Farming Solar Power Generation Facilities*. 2021.
[15] Ministry of Economy, Trade and Industry, *Renewable Energy Purchase Prices, Surcharge Rate, and Other Details Related to FIT and FIP Schemes from FY2022 Onward to Be Determined*. 2022.
[16] Ministry of Land, I., Transport and Tourism, *Treatment under the Building Standards Act of Solar Power Generation Facilities Installed with Poles on Farmland (Technical Advice)*. 2014.
[17] Japanese Industrial Standards Committee, *JIS C 8955:2017 Load Design Guide on Structure for Photovoltaic Array* 2017.
[18] Doedt, C., M. Tajima, and T. Iida. Agrivoltaics in Japan: A legal framework analysis. In *AgriVoltaics2022*. Piacenza, Italy; 2022.
[19] Kurasaka, H., From the national survey on solar sharing in Japan. *Journal on Public Affairs*, 2019;**15**(1):280–297.
[20] Magami, T., Agenda for diversification and pervasiveness of agrivoltaic systems' facilities and business scheme. *Journal on Public Affairs*, 2018;**14** (1):375–397.
[21] Sung, K. and K. Suzuwa, The possibilities and prospects of farming photovoltaics (solar-sharing) by the model of power purchase agreement. *Research and Studies in Education*, 2019;**3**:55–87.

USA

[1] Guarino J. and T. Swanson. Emerging agrivoltaic regulatory systems: A review of solar grazing. *Chicago-Kent Journal of Environmental and Energy Law* [Internet]. 2022 Nov 16 [cited 2023 Nov 11]; 1. Available from https://studentorgs.kentlaw.iit.edu/ckjeel/v12i1-2022-2023-1-guarino_swanson/.

[2] *Funding Notice: Foundational Agrivoltaic Research for Megawatt Scale* (FARMS) [Internet]. Energy.gov. [cited 2023 Nov 11]. Available from https://www.energy.gov/eere/solar/articles/funding-notice-foundational-agrivoltaic-research-megawatt-scale-farms.

[3] Osti.gov. 2022 [cited 2023 Nov 11]. Available from: https://science.osti.gov/-/media/sbir/pdf/Market-Research/SETO—Agrivoltaics-August-2022-Public.pdf.

[4] Congress.gov. [cited 2023 Nov 11]. Available from https://www.congress.gov/bill/118th-congress/senate-bill/1555/cosponsors?s=1&r=3&overview=closed.

[5] *Farmland Solar Policy State Law Database* [Internet]. Farm & Energy Initiative; 2020 [cited 2023 Nov 11]. Available from https://farmandenergyinitiative.org/projects/farmland-solar-policy/state-law-database/.

[6] Govinfo.gov. [cited 2023 Nov 11]. Available from https://www.govinfo.gov/content/pkg/BILLS-118s1778is/pdf/BILLS-118s1778is.pdf.

[7] GovInfo [Internet]. Govinfo.gov. [cited 2023 Nov 11]. Available from https://www.govinfo.gov/app/details/BILLS-118s2931is.

[8] Massachusetts H.B. *5060* [Internet]. 2023. Available from: https://malegislature.gov/Bills/192/H5060.

[9] *New York S.B. 2023-S7081* [Internet]. Available from: https://www.nysenate.gov/legislation/bills/2023/S7081.

[10] *N.J. stat. § 48:3-87.13* [Internet]. 2021. Available from: https://casetext.com/statute/new-jersey-statutes/title-48-public-utilities/chapter-483-unjust-or-unreasonable-discriminations-or-classifications-of-rates-board-defined/section-483-8713-rules-regulations-for-dual-use-solar-energy-pilot-program.

[11] *Agricultural Producers Use of Agrivoltaics* [Internet]. Available from https://leg.colorado.gov/bills/sb23-092.

[12] *Oregon Administrative Rules* [Internet]. 660-033-0130(38) 2019. Available from: https://secure.sos.state.or.us/oard/viewSingleRule.action?ruleVrsnRsn=280453

[13] *O.R.S* [Internet]. Section 303.58 2021. Available from: https://codes.ohio.gov/ohio-revised-code/section-303.58.

[14] Hall P.K. Land use conflicts between wind and solar renewable energy and agriculture uses [Internet]. Wvu.edu. [cited 2023 Nov 11]. Available from: https://researchrepository.wvu.edu/cgi/viewcontent.cgi?article=1104&context=law_faculty.

[15] Chalifour N.J., P. Kameri-Mbote, L.H. Lye and J.R. Nolon. *Land Use Law for Sustainable Development* [Internet]. 2007 [cited 2023 Nov 11]. Available from https://portals.iucn.org/library/node/9064.

[16] Pascaris AS. Examining existing policy to inform a comprehensive legal framework for agrivoltaics in the U.S. *Energy Policy* [Internet]. 2021 [cited 2023 Nov 11]; 159(112620): 112620. Available from: https://ideas.repec.org/a/eee/enepol/v159y2021ics0301421521004869.html.

[17] Cleveland T. and D. Sarkisian. *Balancing Agricultural Productivity with Ground-Based Solar Photovoltaic (PV) Development* [Internet]. [cited 2023 Nov 11]. Available from https://nccleantech.ncsu.edu/wp-content/uploads/2019/10/Balancing-Agricultural-Productivity-with-Ground-Based-Solar-Photovoltaic-PV-Development-1.pdf.

[18] Schelly C, Bessette D, Brosemer K, *et al.* Energy policy for energy sovereignty: Can policy tools enhance energy sovereignty? *Solar Energy* [Internet]. 2020 [cited 2023 Nov 11]; 205:109–12. Available from: https://digitalcommons.mtu.edu/michigantech-p/2066/.

[19] Byrne G. *Farmland Solar Policy Design Toolkit: Current Use Taxation.* 2020.

Chapter 11

A visionary outlook on agrivoltaics

Constantin Klyk[1], Stephan Schindele[2] and Christian Dupraz[3]

In this chapter, we present a visionary outlook on the way forward for agrivoltaics. We derive this vision from the results contained in this book. Therefore, the first sub-chapter contains a summary of the key findings of the diverse explorations undertaken from Chapters 1 to 10. It describes the status quo of agrivoltaics and a distillation of our in-depth investigation into the realm of agrivoltaics. Then, as an interlude, the second sub-chapter revisits the debate on the definition of agri-voltaics. We contrast two positions: One from Christian Dupraz that is more in favor of a narrower definition and calls for restrictive quality criteria. And the other from Constantin Klyk and Stephan Schindele, which advocates a broader definition of agrivoltaics embracing as many agricultural activities as possible.

Guided by the knowledge and insights accumulated thus far, in the third sub-chapter, which is the closing chapter of this book, we extend our perspective toward the future, contemplating the potential trajectory of agrivoltaics. In order for the Agri-PV market segment to continue to grow and thrive healthily, it is essential that we are equipped with a robust knowledge base that supports a fact-based policy orientation, that we try to provide throughout the book. We highlight the role of leading industrialized countries and key-markets in agrivoltaics diffusion, present options to enhance sustainable land use and current research trends in agrivoltaics and describe where there is room for improvement in the diffusion of agrivoltaics worldwide.

11.1 The status quo of agrivoltaics: summary of key findings in the book

Chapter 1—The future of utility-scale photovoltaics on farmland is agrivoltaics describes the Political context and Agri-PV Innovation Process. It endeavors to address two central inquiries. First, it explores the rationale behind advocating for the

[1]Agri-PV Consulting, Jörg-Zürn-Str. 22 , 88339 Bad Waldsee, Germany
[2]BayWa r.e. AG, Arabellastraße 4, Germany
[3]National Research Institute for Agriculture, Food and the Environment (INRAE), University of Montpellier, UMR ABSYS, France

proliferation of agrivoltaics technology. Second, it examines the current state of global agrivoltaics market expansion. In doing so, it presents a comprehensive review of the political, legal, and technological advancements in agrivoltaics. It also provides a detailed examination of the evolution of agrivoltaics, tracing its inception from Goetzberger's idea in 1981, through its invention and innovation stages, and finally forecasting the likely future diffusion of this technology.

Chapter 2—Definition and potentials of agrivoltaics delves into the regulatory evolution of agrivoltaics, using the German DIN Spec 91434 as a case study to illustrate how such a preliminary standard can evolve. The chapter scrutinizes various definitions by different authors through a comprehensive literature review, culminating in a proposed definition that takes into account previously established definitions and current trends. Beyond this, the chapter provides an overview of global potential for agrivoltaics and contrasts this with the existing installed Photovoltaics (PV) capacity within specific regions

Chapter 3—Technological insights into photovoltaics in agrivoltaic systems is a segment devoted to PV technologies, extensively probed into the technical facets of PV within the agrivoltaic system. It revisited the classifications of Agri-PV elucidated in Chapter 2, with beacon case studies acting as model demonstrations. Significant attention was allocated to the technical interplays between mounting frames and PV module technology. The chapter also presented and contemplated on appropriate PV cell technologies for agrivoltaics, many of which had already showcased their effectiveness.

Chapter 4—Adoption of agricultural processes with agrivoltaic applications provides a detailed examination of the diverse agricultural technologies employed within Agri-PV systems, focusing particularly on farming machinery and their operation in Agri-PV projects. It offers an in-depth analysis of technologies designed to foster ideal plant growth, including water and light management systems. The chapter also highlights the importance of overhead Agri-PV, emphasizing its protective role and outlining the advantages it offers in terms of promoting plant growth and conserving resources.

Chapter 5—The agronomy of crops in agrivoltaic systems is an exploration of ecological impacts aimed to balance the ecological pros and cons of agrivoltaic systems. It focused on identifying suitable plants and essential management strategies for successful implementation. Yield changes for respective crops were discussed and the influence of shading on the plants is examined.

Chapter 6—Crop and power yield modelling: The current status of the predictions is examined at different models. The primary focus is on agricultural yield and the electrical yield, key parameters for predicting the economic viability of a planned agrivoltaics projects. Particular attention is given to the prediction of shadow frequency for areas between or under the solar modules.

Chapter 7—Operations and maintenance considerations at agrivoltaic sites: The chapter on operation and management sheds light on the challenges encountered in the operation of agrivoltaic systems, emphasizing the compatibility of agricultural machinery and the effects of field cultivation on the PV system. The cleaning of modules, cable management systems, and harvesting strategies is also considered.

Chapter 8—The importance of social aspects on Agri-PV diffusion discusses social aspects and examines the impact of agrivoltaic systems on society. It considered the importance of general acceptance for the success of agrivoltaic technology, analyzing perceptions of agrivoltaics compared to other renewable energy sources and considering how to maintain and achieve high social acceptance of agrivoltaics.

Chapter 9—Economic performance of agrivoltaic systems: A comprehensive analysis studies factors influencing the economy of agrivoltaic projects. Capital expenditure (CAPEX), operational expenditure (OPEX), and Levelized Cost of Electricity (LCOE) values were discussed in detail. Exemplary calculations were made, comparing agrivoltaic projects with single-use systems like common agriculture or ground-mounted PV-systems. The chapter also explored factors affecting profitability, economic risks, and the impact of global subsidies on the economic viability of the project.

Chapter 10—Legal frameworks in agrivoltaics provides a comparative analysis of the legal landscapes pertaining to agrivoltaics across various nations. It delves into countries with progressive agrivoltaic legislations, namely Germany, France, Italy, China, Japan, and the United States. Each country's legal context is elucidated by an author familiar with the respective jurisdiction. This not only includes an exploration of the existing laws but also the incentives available, and in some instances, the approval process required for initiating an agrivoltaic project.

This chapter, **Chapter 11—A visionary outlook on agrivoltaics**, concludes the book. Beginning with a recap it moves to the differentiation between a conventional GM-PV project and an agrivoltaic project. The second sub-chapter is a debate between Christian Dupraz, who takes on the more restrictive position, and Constantin Klyk as well as Stephan Schindele, that argue in favor of a more liberal definition. Finally, we look toward the future of agrivoltaics. Research trends are presented as well as the obstacles to agrivoltaic diffusion and how the development is hindered in developing countries. The chapter as well as the book is finalized with a recap of the potentials of Agri-PV diffusion from Chapters 1 and 2.

11.2 From narrow to broad: defining the boundaries of agrivoltaics—a discussion

Before we enter the debate on a narrow or broad definition of agrivoltaics, let's first review what the key distinctions are between conventional GM-PV and agrivoltaics.

Distinction between GM-PV and agrivoltaics: In a conventional GM-PV project, there is no farming activity, and no farmer is involved in the operation of the project over the lifetime of 30 years. Under these conditions, the project area is used for maximum solar power generation, which in turn prevents long-term economically viable farming on the project area due to the technical implementation of the PV system. Figure 11.1 show two extremes of GM-PV project. On the left, we see the PEG system, that comes with 1.7 MWp installed PV capacity per hectare or a GCR

Figure 11.1 PEG system with a GCR close to 100% (left) [1] and a GM-PV system in the coal pit in Hambach by © RWE

of almost 100%. The picture on the right shows GM-PV system in the Hambach coal pit, where farming is excluded due to the missing soil quality. Optimal GCRs for GM-PV systems should avoid self-shading losses (Figure 11.1).

In a region or project area where no agricultural activity is possible anyway, there is no land use competition with agricultural production. These types of GM-PV projects are therefore not bad or reprehensible per se and have their legitimacy, especially against the background of social justice and the most cost-effective energy supply possible. Areas that come into question for GM-PV implementation are, for example, desert areas, agricultural areas that can no longer be cultivated due to climate change, polluted areas such as former military training areas, landfills or former coal mining areas with very low soil quality.

In an agrivoltaics project, there is always an agricultural activity on the project area and a farmer is involved in the operation of the project during its 30-year lifetime. Under these conditions, the project area must be suitable for agricultural activity and the PV generator must technically adapt so that the farmer can carry out long-term economically viable farming on the project area, for example by raising the PV modules higher so that farming can take place underneath (overhead) or by installing the PV module rows further apart so that farming can take place between the module rows. Figure 11.2 shows one overhead and one interspace agrivoltaics project on agricultural land.

Agrivoltaics projects involve a farmer in project implementation, increase land use efficiency in terms of the agricultural yield in a multifunctional configuration compared to a mono-functional land use on the same area, thereby mitigating land use competition between food production and PV development. Against this background, agrivoltaic projects can only take place where agricultural activity can also take place, i.e., on agricultural land. According to the FAO Land Use Matrix of 2021 [2], agricultural land can be divided into two categories. Agricultural land and land under protective cover, which includes farm building, farmyards, but also greenhouses.

Figure 11.2 Overhead Agri-PV system developed By Krinner Solar(left) and Interrow Agri-PV system showing a Zim Tracker © Zimmermann PV Stahlbau

Agricultural land, on the other hand, comprises the three main categories arable land, land under permanent crops and land under permanent pastures and meadows.

Agrivoltaics projects are most of the time more costly than conventional GM-PV due to their technical adaptation to enable agricultural activity and the lower solar power yield per area. The higher costs can be justified by the relative advantages of agrivoltaics compared to GM-PV, e.g., higher land use efficiency, higher food/feed production compared to mono-functional land use, enabling techno-ecological synergies (TES) along the food-energy-water-biodiversity nexus, strengthening farming sector and rural areas economically, mitigation and adaptation of agricultural sector to climate change, or access to land units that are prohibited from being set aside from agriculture production (when this ban exists in the legislation such as in France).

The differences between GM-PV and agrivoltaics and the integration of an agricultural activity on the project area in combination with an appropriate land category can be identified in the United States. The DOE of the US government funds the Agrivoltaics research programme InSPIRE. In this context, NREL created an agrivoltaics map for the United States, covering all known agrivoltaics projects. As of October 2023, 371 agrivoltaics projects are recorded, including 296 SolarGrazing with sheep projects, 74 SolarPollinator Habitat projects, 2 SolarGreenhouses, and 24 Cropvoltaics projects, which are also implemented as hybrids, e.g., with SolarGrazing or SolarPollinator. The total agrivoltaics capacity of these projects is 1.25 GWp (DC) and the projects are mainly located in the north-east and east of the United States. The distribution of the projects can be seen in the agrivoltaics map in Figure 11.3 [3].

A total of approx. 80 GWp (DC) of utility-scale PV has been implemented in the United States, most of which is from GM-PV in the south and south-west of the United States, see Figure 11.4 [4].

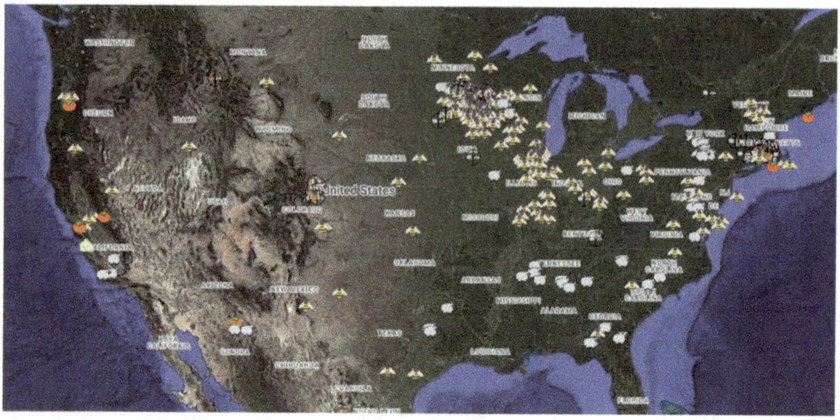

Figure 11.3 *Map of the United States showing all known Agri-PV projects. The map was created in the InSpire project.*

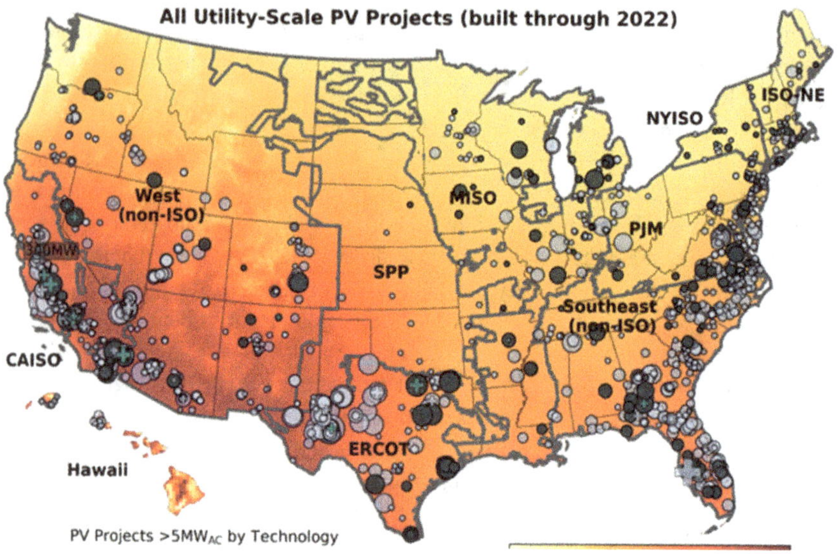

Figure 11.4 *Distribution of utility-scale PV projects in the United States. Most of the large utility PV projects are located in the deserted south-west [4].*

In the west and south of the United States, i.e., in the states of New Mexico, Arizona, southern Texas and southern California, the climate is very hot and dry. These regions are in the arid and semi-arid climate zone and intensive agriculture is possible mainly along river courses and, in the case of livestock farming, on steppes. Most of the land in these regions is not usable for agricultural production.

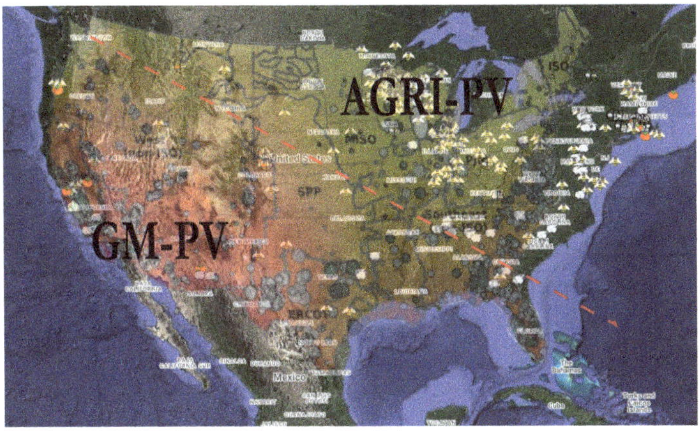

Figure 11.5 *Superimposed map of the two above (Figures 11.3 and 11.4) showing the distinction between large utility-scale GM-PV projects in the south-west and a stronger accumulation of Agri-PV projects in the north-east*

The rural areas are relatively sparsely populated, and most people live in larger metropolitan areas, such as Los Angeles, Phoenix, Santa Fe, Austin, and Houston.

If the agrivoltaics map of the United States from NREL and the utility-scale PV map of the United States are superimposed, it becomes clear that the largest PV projects are not installed on agricultural land and accordingly not with a farmer and thus not as agrivoltaics (Figure 11.5). In fact, the agrivoltaics market share of the utility-scale PV market segment is only 1.5% and of the total 135.6 GWp (DC) PV capacity installed in the United States by the end of 2022, only about 1%.

For countries with no desert areas, such as most European countries, there is hardly any opportunity to locate GM-PV on unused areas. The conflict for the use of agricultural land is more serious, and often the law simply prohibits any GM-PV on agriculture land and forest land. This is the case in France where the recent (March 2023) law on the acceleration of the adoption of renewable energies has banned any GM-PV activity on agricultural land. In that case, the only option to install PV capacity on farmland is to go for Agri-PV.

There are two lessons learned from these market observations in the United States. On the one hand, it is reasonable to implement GM-PV in desert regions, on areas of intensive desertification, and/or on areas with very low soil quality, as farming in these areas is not possible (anymore). On the other hand, wherever farmland is still productive, it is important to protect it. On these agricultural land areas, Agri-PV is either the preferred option, or, if GM-PV is prohibited, the only option. There will be agricultural land areas in transition, meaning climate change is threatening the continuation of the farming process. In such case, there might be a small, but realistic chance that Agri-PV projects can support the farmer and local community to keep the farmland active, as the project will bring a different micro-climate and additional benefits to the rural community, such as rainwater collection and storage, desalination of groundwater, water irrigation, local food production, clean energy, jobs, and tax income. In such case the impacted farming sector may learn from cultivation options and crops produced in sunnier regions, such as millet and yams in the sub-Saharan Africa. The strongly promoted energy transition is a chance to positively impact the agricultural transition through Agri-PV, which has the potential to become an adaptation strategy for the farming sector and increase the resilience of soil fertility and thus the local community.

11.2.1 Sharpening the focus: defining legal agrivoltaics by limiting the GCR of Agri-PV projects to avoid the dropout of agriculture

The economical margins of agriculture in Europe are small and any reduction in crop or pasture yields in agrivoltaics would threaten the profitability of the agriculture activity under the panels, and, as a consequence, the durability of the agrivoltaic systems. If the farmer can't make a profit with the crop production, he will logically decide to stop cultivating under agrivoltaic arrays. This happened already in the past. Many recent agrivoltaic greenhouses are void of crops today in the South of France. This could be predicted: most agrivoltaic greenhouses were designed with all the south facing slopes of the rooves covered with PV panels. As a result, the shade is very dense, especially in winter, when greenhouses are very useful and efficient. As the crop yields in these very dark greenhouses dropped too much, many farmers stopped cultivating them. Some research programmes are now addressing this critical issue: what can be cropped with profit under a very shaded greenhouse? The answers are slow in coming.

This dropout of agriculture under agrivoltaic systems is a risk as the electricity producers often grant the farmer and/or the landowner with a high rent. Why bother to crop under panels when a large income is provided automatically, independently of the cultivation process? This issue is even more critical when the farmer is the landowner and gets a high compensation. Agriculture net margins are very low in Europe, and any significant drop in yield usually results in negative net margins. In Europe, the loss of the Common Agriculture Policy subsidies may also deter the farmer from cropping. Assuming CAP subsidies will remain available in Agri-PV projects is optimistic, but even in that case, we may end up with "CAP subsidies hunting" crops in Agri-PV, which is not desirable. If farmers crop only to get the CAP subsidies, irrespective of any production, what is the benefit for society? If farmers crop only to reduce taxation, irrespective of any production, what is the benefit for society?

There will always be a temptation for the landowner to become an annuitant, cancelling the farming lease if necessary. This is why the contracts between landowners, farmers and electricity producers should be carefully designed to avoid any temptation of agriculture abandonment.

We all dream of biting twice the solar cherry. But the rules of photosynthesis are unsurmountable. The more shade from PV, the less agricultural yield. This rule was evidenced by a synthesis of all published yield measurements available so far under agrivoltaic systems [5]. Ignoring this issue and allowing very shaded agrivoltaic systems will put the farmers under a very high pressure: if the farmer fails to meet the yield required by law, the system will be disqualified, with ominous consequences. In Japan, where an 80% field relative yield is required in agrivoltaics, controls have evidenced that the yields were sometimes not meeting this target. As a consequence, the feed-in tariffs for electricity could be cancelled, inducing the bankruptcy of the controlled projects. Several projects were already dismantled for this reason in Taiwan. But the consequences are even more tragic: the investment funds that finance the agrivoltaic facilities are now very reluctant in Japan, fearing that farmers will not meet the required yield targets. Some countries like Germany are less demanding and accept a 33% decrease in crop yield. There should be a consistency in the regulations between the accepted loss of agriculture yield and the maximum area of panels per hectare that can be installed. If not, farmers will face an impossible challenge, and will be taken as responsible for the failure of the project. This is why several voices call for a strict regulation of agrivoltaic systems.

The ground coverage ratio of the system has been proposed as an integrative indicator of the shade cast by the system. It needs to be adjusted to take into consideration the fine details of any AV installation, but these calculations are easy (Table 11.1).

Table 11.1 Calculation of the adjusted ground coverage ratio of an agrivoltaic system as an indicator of the radiation availability for the crops and pastures

	Equations	**Comments**
Calculation of the gross GCR	GCR_{gross} = *Area of PV panels/ Area of the array*	The area of the array may include a headland
Correction for semi-transparency of panels	$GCR_{adjusted} = GCR_{gross} * (1-Tr)$	Tr: average transmissivity of the panels (%)
Correction for unoptimal positioning of the panels (fixed and mono-axial trackers)	$GCR_{adjusted} = GCR_{gross} - alpha/10$	alpha: angle of the azimuth of PV panels with the South (°)
Correction for mobile panels with solar tracking	$GCR_{adjusted} = GCR_{gross} * (1+CG)$	CG: Capture gain allowed by the panel mobility (%) Usual rates: 25% for mono-axial trackers 40% for bi-axial trackers
Correction for mobile panels with agronomical tracking	$GCR_{adjusted} = GCR_{gross} * (1-beta*BT)$	BT: Backtracking rate to provide full sun to the crop during the crop cycle (%) Beta: efficiency coefficient of the irradiation provided to the crops
Composition of the corrections	$GCR_{adjusted} = GCR_{gross} * [(1-Tr) * (1+CG) * (1-beta*BT)] - alpha/10$	

Adapted from [6].

Table 11.2 Adjusted ground coverage ratios required to meet some thresholds of crop and pasture relative yields under AV systems

		Expected Field Relative Yield				
		100%	90%	80%	70%	60%
Uncropped Land area	10%	6.8%	10.1%	15.5%	25.3%	44.5%
	20%	4.4%	6.5%	10.1%	16.4%	28.9%

Source: [6], Chapter 5, this book.

When the regulations include some levels of relative yields to achieve, they always consider the field relative yield: a field with an AV system is compared to a field with no AV system. The rate of uncropped areas must be included in the relative yield calculation. This has a strong impact on the maximum GCRs that can allow to achieve expected field relative yields (Table 11.2). From this table, and assuming that the uncropped land area is 20%, it is clear that in Japan, where an 80% relative crop yield is demanded by law, the maximum GCR of the systems should be 10% while most projects use a 30% GCR. In Germany, where a minimum 66% relative crop yield is expected, GCRs should not exceed 21% to ensure that this target is met.

When the rate of unused land area is high in an agrivoltaic system (such as projects with low elevated panels and crops or pastures only in the alleys between the panel rows), the agricultural yield reduction is always large. If only 50% of the area is cropped or grazed, the agriculture production per hectare will drop. In many projects, this is even compounded by the change in agriculture production. When you replace productive crops with sheep grazing, the farm income decreases sharply. Sheep will not save agriculture in agrivoltaic projects where they replace more lucrative productions, but the additional income for the farmer from the electricity side may cover for the loss of agriculture revenues.

Two strategies are currently proposed to qualify and control agrivoltaic systems: the first one suggests to control agrivoltaic systems *a posteriori*, by measuring the crop or pasture yields under the system and in a control open area. The second option is to limit the density of panels of agrivoltaic systems *a priori*, as a security option to provide enough light to the crops.

A mix of the two systems can be proposed (Table 11.3) and would leave a large freedom of designs: agrivoltaic projects with low GCR would be exempted of the crop productivity control, while projects with high GCRs would need to prove their compatibility with the agriculture component. Such a regulation would not hinder creativity in the design of AV systems, as all options are still available.

Ceiling the GCR of agrivoltaic systems will not only secure the agriculture component of the system; it will stimulate the creativity of both researchers and electricity companies. If there are allowed to call "agrivoltaics" any standard ground-mounted system that is simply elevated, the need for research and innovation is null. If they are allowed to call "agrivoltaics" a system where the crop is only

Table 11.3 A qualification and control strategy for agrivoltaic systems leaving full liberty for the choice of panel density in AV systems

Adjusted ground coverage	Control options	Practicalities
$GCR_{adjusted} \leq 20\%$	No yield control Evidence of cultivation and harvest	Camera with pictures shot at regular time intervals (weekly)
$20\% \leq GCR_{adjusted} \leq 30\%$	Simple yield control	One Control Zone without PV panels five replications Duration: 5 years Beyond 5 years, evidence of cultivation regime
$GCR_{adjusted} \geq 30\%$	Reinforced yield control	Two Control Zones without PV panels five replications Duration: 10 years Beyond 10 years, evidence of cultivation regime

cultivated between widely spaced row of PV panels (with nothing under the panels resulting in a high rate of uncropped area, and hence a very high reduction of crop yield), there is no need of any research either. Let's think twice: the people who will design an agrivoltaic system where the crop yield is guaranteed will have access to an unlimited area for producing electricity. This is the ultimate goal of agrivoltaics, and to achieve this we need to adopt a definition of legal agrivoltaics where the crop yield is preserved. Designing a system on twice the area with half the panel density (as compared to a standard ground-mounted system) will meet the electricity production needs and secure the crop production. The cost of the electricity produced will of course be slightly higher as compared to standard ground-mounted PV systems, but this is the price to pay for accessing an unlimited area of land. The game is worth the candle.

To conclude, it is not a definition of agrivoltaics per se that we need. The simple definition as a co-located production of both electricity and agriculture products on the same parcel is always valid. What we need is a definition of legal agrivoltaic systems, that will meet the needs of the society. And if this is the target, a restricted definition of legal agrivoltaics is required, to stimulate creativity, secure crop yields, and meet the society expectation of protecting of the agricultural land for the future.

11.2.2 Embracing the big picture: advocating for a broad definition of agrivoltaics for greater inclusivity, innovation, and investment

To supply a growing demand for food and energy, to use water in agricultural production as efficient as possible, and to combat the loss of biodiversity in rural areas, Agri-PV applications that consider the food-energy-water-biodiversity nexus may contribute to an increase of today's sustainability levels in the farming and PV sector alike. There is hardly any ambivalence on Agri-PV diffusion amongst the sectors. The low conflict level of Agri-PV is a great strength of this inter-disciplinary land use configuration. And yet, there is ambiguity on how a technical standard can guarantee a certain level of minimum agricultural output on the Agri-PV project area. There are concerns that farmers and Agri-PV project developers will cheat and implement alibi- or so-called pseudo-Agri-PV systems. The fear is that if the farmer's income is much higher from land-lease or service contracts to the Agri-PV system, and if the farming activity itself is not profitable anymore, the farmer will stop the agricultural production on the Agri-PV project area. This concern is purely theoretical. In reality, the consequence of stopping agricultural production would eliminate the eligibility of agricultural subsidies. The project area would lose its status of farmland and transit to industrial area. With this, the landowner would need to pay much higher inheritance and income tax when passing on the project area to ancestors. The landowner usually has an intrinsic motivation to maximize the economic output of the land. If a farmer decides to stop farming in an Agri-PV project, the landowner will search for another farmer who is willing to receive access to land with relatively low land lease cost (as most of the land lease is paid by the PV generator) and also the agricultural subsidy. If the landowner is also the farmer in the Agri-PV project, besides the tax impact also the contractual framework with the Agri-PV project will enable to more than double the income of the project area in Agri-PV compared to no Agri-PV. In fact, the Agri-PV sector cannot (politically) afford on the long-run to implement "pseudo-Agri-PV" projects, as sooner or later more restrictive regulations will be introduced to command and control a minimum output of agricultural yield on the Agri-PV project area. And yet, especially at the beginning of Agri-PV diffusion, it is of utmost importance to follow a liberal approach in Agri-PV market regulation. There must be political willingness to allow inefficiencies in the market introduction because it must be more attractive compared to GM-PV to develop new, innovative, and creative Agri-PV land use configurations jointly with the farming sector. A too restrictive market regulations, such as the proposed GCR quality criteria, will ultimately lead to divestment. The GCR criterion is the attempt to create a template that has to fit all applications. But the farming sector is far too divers as it could fit all into one template. To give you an example: Adolf Goetzberger proposed in 1981 to combine utility-scale PV projects with sheep shepherding. Sheep keeping on permanent pastures is without a doubt an agricultural activity. We need wool to manufacture our cozy jumpers and socks for the wintertime. Shepherds-pie is a delicious British dish. In Europe, hundreds if not thousands of SolarGrazing sheep projects have been executed and operate successfully with over 10 year experience. The GCR ratio of these projects is between 55% and 60%. Christian Dupraz proposes a GCR of below 30% for these types of Agri-PV applications, claiming that there is no sufficient scientific evidence on how fodder production will be maintained if there is not enough light and rainwater distribution on the project area. In reality, most Agri-PV projects are implemented and operated by businesses, such as farmers and independent power producers and not by scientists. The knowledge and competence on how to successfully graze sheep in Agri-PV projects is there, even if little scientist data has been published on this. The unknow in science is no excuse to consult policymakers along the science-policy interface to restrict a successfully tested and implemented Agri-PV application by practitioners.

In general, to install Agri-PV above permanent pastureland in combination with livestock keeping such as sheep, chicken, and cattle is fairly easy compared to other Agri-PV applications implemented on arable land or above permanent crops. Land under permanent pastures has a share of 68% of the total globally utilized agricultural area (UAA) and most commonly is utilized for domesticated livestock grazing, e.g., cattle, horses, sheep, chicken, or pigs. This sub-segment of the agricultural land represents by far the largest area potential for Agri-PV installations. SolarGrazing with sheep and cows is the low-hanging fruit of Agri-PV market introduction and not much command and control appears necessary. The biomass vegetation on permanent pastures if often forage, which consists of grasses, legumes, and other non-grass herbs, whereas grasses are the pre-dominant plants. Compared to other agricultural land outputs, such as cereal, vegetables, or special cash-crops like fruits and nuts, forage is very robust in yield output in Agri-PV microclimates. The temperature, solar radiation, rainfall distribution, and soil conditions seems to be sufficient for forage production and grazing in Agri-PV projects. Consequently, an increase in land-use efficiency is easily doable. At the same time, techno-ecological synergies are more than obvious in these Agri-PV projects: sheep, chicken, and cows find shelter under the PV-modules in case of heavy rain, hail, or heat. The animals take over the green-keeping job, improve the soil-restauration and biodiversity with their droppings, and it is a perfect symbiosis of farming production (wool, meat, and environmental services) and solar PV generation. On the long-run, with high probability, Hay- and Cow-PV hybrid projects will become the most important Agri-PV application on permanent pastures. In all cases, the GCR of SolarGrazing can be between 55% and 60% and its diffusion in the farming sector is very desirable.

Agri-PV is not a single product. Agri-PV applications are just as drivers as the farming sector. This diversity is a challenge to the PV sector, but an opportunity for the farming sector and society. Every Agri-PV project must include a farmer executing a professional agricultural activity on the project area, respecting the standards for good agricultural and environmental conditions of the farmland. Without a doubt, in every Agri-PV project the collaborating farmer must gain an economic surplus from being involved—otherwise there would be no Agri-PV project. Before the project is implemented, it is the job of the local permitting authority to check if the proposed Agri-PV project is offering an increase in land use efficiency compared to a monofunctional land use and to evaluate if the assumptions made by the collaborating partners (farmers and Agri-PV project developer) are realistic or maybe too optimistic. It is the responsibility of the Agri-PV project developer to provide all evidence needed toward the permitting authorities that an objective decision can be made. The 5C-Agri-PV-Assessment-Tool provided by NREL generates a report that information on the (1) conditions of the project, e.g., soil quality, landscape impact, solar radiation, water impact; the (2) configuration describes in detail the Agri-PV design, e.g., engineering plan, height of system, agricultural machinery used under or between the PV modules rows; the (3) crop selection or animal husbandry explains the agricultural concept of the farmer within the Agri-PV project; the (4) compatibility provides all analysis outcomes of the project, e.g., assumed agricultural yields, justifying the quality of the project and the matchmaking of all legal standards and requirements, such as farming activity, Agri-PV specifications, and regulatory framework prerequisites; and (5) finally, all Cs are integrated into a collaboration agreement, capturing the mutual agreement between the farmer, landowner, and Agri-PV project developer to execute the project as proposed. A much better-quality standard than the single GCR quality criteria is a detailed agricultural concept for the Agri-PV project that is verified by a third-party agricultural expert. The 5C-Agri-PV-Report from NREL gives a good structure on such an agricultural concept and coordinates the complexity of the interplay between farming and PV power generation. The main output criteria in the 5C concept are agricultural yields, the environmental impacts, and the economic benefits to the farmer, rather than a simple PV-technical design as proposed with a minimum GCR for Agri-PV applications.

Permanent crops, such as viticulture, olives, citrus, and apple horticulture, represent the smallest area potential for Agri-PV, accounting for only 4% of the global UAA. Despite being the smallest sub-segment, they produce the most valuable crops, often referred to as "cash-crops." These crops are typically grown in protected cultivation systems, like hail nets or plastic rain covers, and often require water irrigation systems for optimized yield. Agri-PV applications in this sub-segment tend to be more expensive but can create the most techno-ecological synergies. The cost–benefit ratio in "Fruit-PV" or "Vino-PV" projects is usually very positive, especially when compared to Agri-PV applications with lower economic value outputs, such as cereals or grass. The key advantages for farmers using Agri-PV applications include water demand reduction, protection from harsh weather events, substitution of existing cultivation protection systems, and reduction of plastic waste. The Ground Coverage Ratio (GCR) for Agri-PV applications in this sub-segment can range from 55% to 100%. However, implementing Agri-PV projects in this sub-segment requires careful planning and detailed configuration, as the crop's economic value is often higher than the solar PV power generated on the same area. The crops, being more complex to cultivate, require a balanced synergy with the Agri-PV systems, making these projects costly. Regarding the GCR, Christian Dupraz proposed a method to adjust the GCR by multiplying it with the average transmittance of the PV module. For instance, in the Fruit-PV projects of GroenLeven/BayWa r.e. in the Netherlands, the GCR of 90% is multiplied by 1–45% light transmittance of the used PV modules, equating to a 49.5% GCR. However, the French draft law for promoting Agri-PV sets a maximum GCR of 40%. This regulation could hinder the implementation of projects like "Fruit-PV" in France, even though these projects have been scientifically validated by Wageningen University. To achieve the 40% GCR, one could consider increasing the pitch between the berry rows or increasing the level of light transmittance. However, increasing the pitch could significantly reduce the agricultural yield. On the other hand, increasing the level of light transmittance would decrease the PV capacity per module, which could increase the levelized cost of electricity by roughly 25%. This approach doesn't consider that berries naturally grow in forests and may not require higher light availability. Unfortunately, such "adjusted" GCR of 40% could unnecessarily increase subsidy costs and limit the number of projects receiving funding due to the increased project cost. This situation highlights the need for better scientific advice and understanding of regulations in agrivoltaic projects.

So far, we have examined Dupraz proposal to use GCR between 20% and 30% as a single quality criterion for Agri-PV projects on the FAO's land use categories "Land under permanent pastures" and "Land under permanent crops." We found out that there is no rational justification to introduce such restrictive criteria to evaluate the successful implementation of Agri-PV projects on these "permanent" land use configurations. Agri-PV projects have been operated with much larger GCR, ranging from 55% (grazing with sheep and cows in Agri-PV) to 90% (raspberry and blueberry cultivation in Agri-PV). We learned, due to this shortfall of the restrictive GCR, the GCR concept can be adjusted, e.g., if the level of light transmittance is considered when calculating the adjusted GCR. The third and final land use category of the FAO's land use matrix that is of relevance to Agri-PV implementation is arable land used for temporary crops and meadows. Globally, 28% of the UAA is used to produce cereals, oil seeds, vegetable, and other temporary crops on arable land. Typically, temporary crops are cultivated within a crop rotational system, which results in different agricultural machinery processes from year to year. Therefore, compared to other agricultural land categories used for Agri-PV implementation, this sub-segment is more complex. The technical adaptation to enable PV integration into a farming process is relatively higher against Agri-PV applications on permanent pastures or permanent crops. The pitch between the PV rows is wider, as the ag-machinery used in arable farming are much larger and wider compared to the ones used in hay production, animal husbandry or special cultures production. Furthermore, the temporary crops grow naturally higher than grass on permanent pastures, leading to a larger clearance height between the surface and the PV modules to avoid shading from the crops to the PV generator. To answer the complexity of Agri-PV in combination with temporary crops, PV-tracker system appear an attractive solution. The PV modules can be moved out of the optimal PV power production in times when the temporary crop demands much light and temperature within its vegetation period. This principle of adjusting the PV-tracking software according to the plant physiological demands of the crop is often referred to as "dynamic shading." In the logical of Dupraz's GCR this means: a higher GCR is allowed if this "dynamic" function is included in the Agri-PV project. The adjusted GCR is calculated by multiplying the higher GCR with adjustments according to the captured agricultural gain allowed by the "panel mobility/dynamic shading." Dupraz recommends 25% capture gain for single-axis tracker and even 40% for bi-axial PV-trackers.

Assuming a single axis 2p-tracker, as implemented by EWS Sonnenfeld in Austria or Doppelernte in Germany, a pitch/distance between the PV-tracker rows of 10 m, allowing an agricultural zone of 9 m. In the horizontal position of the tracker, two PV modules with the length of 2,30 m, leads to 4,6 m covered, leading to a GCR of 46%. If a dynamic shading capture gain for a single-axis tracker of 25% is considered, the adjusted GCR is 46% × (1%–25%) equals 34.5%, which is higher than the maximum of 30% that Dupraz recommends as a unified GCR quality assurance criteria and the project would not be eligible for funding. As ag-machinery increase in steps of 3 m, and considering a 50 cm security distance between the ag-machinery and the PV-installation, the pitch of 13 m would lead to an agricultural zone of 12 m. GCR adjustment according to dynamic shading results in 4.6/13 × 0.75 = 26.5%. Due to the large width of ag-machinery, on arable farming the GCR is relatively low even with 2p single-axis-PV-trackers. If the adjustments of GCR according with dynamic shading functions are considered, this adjusted GCR request might even work in reality. And yet, we find that the GCR criteria only works if it is adjusted and multiple exceptions according to the agricultural demands and PV technology adaption is included in the GCR concept. These adjustments and exceptions do often not reflect the reality and local conditions. In our opinion, no farmer, not on permanent pastures, not on permanent crops, and not on arable land used for temporary crops will tell and guarantee what he or she will produce on the project area in 20 or 30 years from now (the typical lifespan of an Agri-PV project). Also, farmers will not know what ag-machinery and agricultural robotics or new water irrigation systems will be available in future. Farmers need to have the flexibility to adapt to a fast-changing environment. The focus of an Agri-PV quality standard should always be the agricultural yield assumptions and the evidence of the agricultural concept that needs to be proposed and confirmed by a local authority in a permitting process. The agricultural yield, however, should never be linked to the financial support of implementing the Agri-PV projects, as the agricultural yield cannot be predicted for 20–30 years. Therefore, the hurdle rate for the farmer to participate in the Agri-PV project must be as low as possible to enable Agri-PV projects to become the new industry standard for PV diffusion on farmland. It is clear that for the PV sector it is not enough to only produce green-power, but especially on farmland additional benefits such as techno-ecological-synergies and income diversification and increase of the farmer must be included into the Agri-PV project. Any obligatory must-have criteria like the GCR will hinder the flexibility and investment and constrain the creativity of the farmer. The most important quality criteria is a motivated farmer that sees a financial benefit on providing his or her farmland to Agri-PV project implementation and who can provide in a joined effort with the Agri-PV project developer a consistent and holistic agricultural concept (5C-report) to the local permitting authorities. The technical design parameters must be secondary, as there will be interdependencies. If the national government request a certain maximum GCR, all projects' ideas will be designed around this parameter and creative and innovative approaches will be eliminated. In a worst case scenario, just as the current Agri-PV degree in France, there will be a single GCR criteria requested for all three classifications of agricultural land. This will lead to divestment and inefficient financial supports, as it artificially drives of project cost and thus limits the eligible PV capacity to the constrained budgets available for the financial support. Equity and debt finance investors into Agri-PV projects need legal "bankability." Any regulative constrain adding to the implementation cost of the Agri-PV project, like reducing the installed PV capacity per area due to a limited GCR criteria, will increase the cost, threaten the economic concept of the farmer and the Agri-PV project, and thus increase the risk of failure and rejection. Agri-PV is a chance for the farming sector to reduce CO_2 emissions on land use, while at the same time adapt to climate change, diversify, and increase the income of the farm, and providing the economic flexibility for the farm contributing to an agricultural transformation. The farming sector demands investment, not divestment. The theoretical correlation of a low GCR in Agri-PV projects miss-matches the reality of hundreds of Agri-PV projects that have been already successfully implemented and operational. It constrains the innovation and economic benefit of farmer participating in Agri-PV projects. An example in livestock cattle keeping: This farming sector suffers from high CO_2 emissions, too much nitrate inflow on farmland, and decreasing acceptance for intensive animal stocking. A reduction of animals per area, and thus a decrease in CO_2 emissions, increase in animal welfare and acceptance can be justified by a higher and diversified income from Agri-PV. The GCR is aligned with the agricultural concept and the future objectives of the farm. A larger GCR can lead to more income for the farmer, allowing him/her to reduce stocking density and therefore have ecological and economic benefits. It is this freedom that is required in the project development phase to have an open dialogue between the involved stakeholders.

We argue for a liberal definition of Agri-PV, the merger of agriculture and photovoltaics, addresses both sustainable energy and farming needs. It thrives on a broad definition, incorporating diverse farming practices to foster innovation and productivity, as there is no single quality criteria that fits all the Agri-PV applications thinkable. Whether it's traditional farming or high-intensity methods like hydroponics, agrivoltaics fits all, especially in permanent pastures and meadows. Integration with perennial crops, despite its complexities, can yield significant synergies and potentially replace single-purpose technologies. However, arable farming presents unique challenges and opportunities due to crop diversity, rotation, and machinery. The key is an inclusive agrivoltaic standard, embracing all potential technologies and not stifling innovation. Restrictive criteria may discourage investment and innovation, undermining agrivoltaics' essence. The GCR and its adjustments create insecurity, narrow the freedom to innovate, and destimulate the integration of agricultural subsystems like water irrigation, agricultural robotics, or insect and shading nets into the Agri-PV project. According to social scientist Ortwin Renn, the technology with the highest net benefit will succeed. This highlights the need for a broad agrivoltaic definition, encouraging competition to derive the best solutions. In essence, with farmers' innovative spirit and an expansive agrivoltaic definition, we can create sustainable, efficient, and productive agricultural systems.

11.3 Seeds of tomorrow: visioning the open horizons and future potentials of agrivoltaics

The upcoming segment provides an in-depth analysis of Agri-PV's future, encompassing the latest research trends and the challenges inhibiting its widespread adoption. Finally, the future potentials are scrutinized, offering a glimpse into what lies ahead in this field.

11.3.1 Unveiling the horizon: exploring the latest research trends in agrivoltaics

The field of Agri-PV, the synergistic production of solar energy and agriculture on the same land, has seen a significant surge in research interest in recent years. This trend is driven by the increasing recognition of the potential of Agri-PV to address multiple challenges, including food security, energy transition, and climate change adaptation.

A review of the latest research trends in Agri-PV, reveals several high-frequency keywords that indicate the focus areas of current studies. These include "economic feasibility," "agriculture activities," "Agri-PV design," "performance of crops and soil," "food and water aspects," "climate change adaptation," "biodiversity," and "storage capacity management" [6–8].

Economic feasibility is a critical aspect of Agri-PV research, with studies exploring the cost-effectiveness of integrating photovoltaic (PV) systems with agricultural practices. This includes assessing the potential return on investment, operational costs, and the impact of government policies and incentives on the viability of agrivoltaic projects.

It would be quite intriguing to observe the variations in policy regimes, legislations, and incentives across different nations. Moreover, comparing the evolution of these guidelines and their impact on the propagation of Agri-PV in the corresponding countries would also be insightful. Agriculture activities in Agri-PV projects are another significant area of research. Studies are investigating the effects of solar panel shading on crop growth, yield, and quality. This includes exploring the optimal panel arrangement and spacing to balance energy production and agricultural productivity.

The design of Agri-PV systems is a crucial research area, with studies focusing on optimizing the integration of PV systems with various types of agricultural practices, including crop farming, livestock rearing, and aquaculture. This involves exploring innovative design solutions to maximize the benefits of co-location.

Research on the performance of crops and soil under agrivoltaic systems is examining the impacts of solar panel shading on soil moisture, temperature, and nutrient cycling. This includes studying the potential benefits of reduced evapotranspiration and improved water-use efficiency.

Food and water aspects are also a key focus of Agri-PV research. Studies are exploring the potential of Agri-PV to contribute to food security and sustainable water management, particularly in arid and semi-arid regions.

Climate change adaptation is a significant theme in agrivoltaic research, with studies examining the potential of agrivoltaic systems to enhance the resilience of agricultural systems to climate change impacts, such as increased temperatures, changing precipitation patterns, and eco-system as well as soil restoration.

To combat the loss of biodiversity in rural areas, farmers can alter parts of their intensively used agricultural land to regenerative, organic, or non-productive, but biodiversity enhancing areas. If biodiversity-PV projects are integrated into these set-a-side areas of farmland, these extensive Agri-PV projects promoting flora and fauna on the project area can harmonize the positions of the agricultural, environmental, and PV sectors.

Finally, the management of the storage capacity of Agri-PV systems is a key area of research. This includes exploring solutions for storing the solar energy produced for use during periods of low sunlight, as well as investigating the potential for agrivoltaic systems to contribute to grid stability and reliability.

While there is a lot of research happening in this field, there is still the need for further quality measurements to raise project quality and comparability. NREL suggest a strategy of 5C as a guideline for agrivoltaic projects: Conditions, Configuration, Crop rotation, Compatibility, and Collaboration [9]. Extending this 5C by a Sixth C, Consumption, could further increase quality of Agri-PV projects and react to upcoming needs in the electrified farming sector with net-zero CO_2 emissions.

The self-supply of solar power takes the agrivoltaic concept a step further by integrating electricity directly on the farm. This adds a new dimension to the economic feasibility of the system, providing an onsite, renewable source of power for various farming activities and processes. For instance, the harvested solar energy could be used to power electrified tractors or farming robots. These automated assets can significantly reduce the manual labor required, increasing efficiency and potentially improving crop yield. Moreover, the use of clean, renewable energy in these operations reduces the carbon footprint of the farming process, contributing to overall sustainability.

In addition, this locally produced electricity can be utilized for further processing of farmed goods. One crucial aspect here is cooling, which is often needed to preserve harvested crops and maintain their quality before they reach the market. By powering cooling systems with solar energy, farms can ensure the freshness of their products while reducing their reliance on external power sources. Another promising aspect is the self-production of solar PV fueled fertilizers, as offered by the company Nitricity [10], which has the potential to substitute the Haber-Bosch process and thus reduce the demand for natural gas in the farming sector significantly.

The 6C also relieves the national grid and reduces energy costs for the farmer, as well as strengthens the bond between the farmer and the agrivoltaic system. It creates a self-sufficient, circular system where the farmer is both a producer and consumer of energy. This sense of ownership can encourage more farmers to adopt and invest into Agri-PV systems, helping to spread the benefits of this integrated approach to farming and energy production.

In conclusion, the latest research trends in Agri-PV reflect the multidisciplinary nature of this field, encompassing aspects of economics, agriculture, engineering, environmental science, and climate change adaptation. These trends indicate the growing recognition of the potential of Agri-PV to contribute to sustainable development and climate change mitigation and adaptation.

11.3.2 Obstacles to agrivoltaic development

The main technical obstacles in Agri-PV include the higher upfront cost compared to GM-PV, ongoing expenses like maintenance and insurance, the need for technical expertise, and the reduction of agricultural land. The upfront cost of an agrivoltaic system can be a significant barrier preventing inventions in agrivoltaic systems. Ongoing costs such as maintenance and insurance can also add up over time. Technical expertise is missing with many project developers, which slows down the development process of an agrivoltaic project. Additionally, agrivoltaic setups require some loss of agricultural land, and the crop response is hard to predict for many crops until now, as a solid database is missing. Despite these challenges, incentives and tax benefits can make Agri-PV more feasible in the long run [11].

While there are already solutions for these technical and economic difficulties, Policy and regulatory hurdles pose significant challenges, with many countries having uncertain legal situations around Agri-PV. Addressing this requires a proactive policy-making approach that supports the renewable energy transition.

Lastly and maybe most important, social acceptance and stakeholder involvement are crucial for the widespread adoption of Agri-PV. Studies suggest that support for local solar projects increases when an agricultural function is incorporated. Therefore, engaging with the community and stakeholders from diverse backgrounds in a meaningful way can be an effective strategy in overcoming this challenge. A critical aspect for the social acceptance of Agri-PV is the grabbing of the Agri-PV rent by a very limited group of well-informed farmers and landowners. This may wreak havoc in rural communities, inducing jealousy, inequality, and bribery. A significant progress would be to share the Agri-PV rent among all farmers of the area. There are good examples in France of communities that try to organize this (such as the chamber of Agriculture of Corrèze).

Agrivoltaic development is significantly lagging in many developing nations globally, with a notable gap in areas like Africa, Central Asia, and South America. They possess immense surface potential, and some have not only that but also an urgent need for renewable energy investment. This in combination with extreme weather conditions and water shortages should be a favorable surrounding for Agri-PV to flourish [12–14]. The combination of photovoltaic solar panels with agriculture could revolutionize the situation. It could offer protection against climate extremes, enhance food production, and guarantee income for numerous small-scale farmers who don't have access to large machinery. However, the legal and political landscape surrounding Agri-PV is often ambiguous, leading to a dearth of private investment. Furthermore, corruption sometimes impede the effective use of available funds. The journey to realize this potential is riddled with challenges. The limited spread of knowledge, primarily from US and European research projects, has created a considerable knowledge void in the agrivoltaic field [11]. This knowledge gap, along with the absence of investments stemming from political instability and corruption, has stunted the deployment of Agri-PV in developing countries.

In conclusion, while there are significant challenges in implementing Agri-PV, there are also viable solutions and strategies to overcome these. By addressing these obstacles, we can drive the growth of Agri-PV in a sustainable and efficient manner, thereby contributing to the reduction of greenhouse gas emissions and the mitigation of climate change.

11.3.3 Potentials and developments

Chapter 1 predicts the spread of Agri-PV within the Photovoltaic market to meet the projection of Goldschmidt *et al.* from Fraunhofer ISE and PIK [15], aiming for 80 TW of installed PV capacity. To achieve this, an enormous number of 46 utility-scale projects, each averaging 30 MW, would need to be introduced consistently for 78 years every day on a global level. With the world's population and prosperity, and therefore the demand for food, set to rise over this period, the potential for conflict over PV implementation on farmland will likely rise. The global potentials of Agri-PV are astoundingly vast and transformative. As established in Chapter 2, there's a colossal potential for agrivoltaic systems, with only 1% of the UAA being sufficient to bring about a revolution in renewable energy. This assumption, based on FAOSTAT data [2], focuses on different types of agricultural land - arable land, permanent crops and permanent meadows and pastures.

Interspace solutions are particularly attractive due to their cost-efficiency and minimized resource utilization. The capacities used in the analysis are 0.4 MWp/ha on arable land, 0.8 MWp/ha on permanent crops, 0.5 MWp/ha on permanent meadows and pastures, and 0.9 MWp/ha on natural meadows considering solar grazing and biodiversity PV.

This approach can drastically increase the installed global PV capacity. On only 1% of the global UAA, the 2022 global installed PV capacity could be expanded by a staggering 38 times. To achieve the 1.5 °C goal of the Paris agreement, only 0.5% of permanent meadows would be required to increase the installed capacity to almost 13TW and reach the current goal of 77% renewable energies in the mix for 2050.

In Europe, where arable lands are prevalent, utilizing just 1% of UAAs can enhance the installed capacity more than fivefold, reaching an overall capacity of 1TW – nearly the entire PV capacity required by 2050 to stay in line with the EU contribution to the Paris Agreement. Similar potentials are observed in Asia (10TW), North America (4TW), and Africa (3TW), with just 1% of their agricultural lands.

Consequently, it's expected that the current key markets for Agri-PV, including China, the United States, and major EU agricultural producers

(France, Germany, Italy, the Netherlands), will most likely expand their Agri-PV support regimes rather than limit them.

Simultaneously, new Agri-PV markets are emerging. Since the beginning of 2022, governments have been discussing or have already implemented Agri-PV support instruments, including Israel, Luxembourg, Croatia, the Czech Republic, and Austria. Luxembourg, for instance, has assigned 52.7 MW in an Agri-PV tender. Additionally, pilot projects have been undertaken in Kenya, Spain, and Switzerland. In France, Germany, and Austria, the first interspace "Cropvoltaics" projects have been successfully tested. In Bavaria, Germany, the first "Cowvoltaics" project was launched in 2022.

The diffusion of Agri-PV is in its early stages and still needs considerable policy learning, coordination, and intersectoral cooperation. The legal frameworks often pose significant challenges to Agri-PV diffusion more than the technical and economic ones do. The public's crisis awareness regarding food security and climate protection will influence whether policymakers recognize the potential of Agri-PV in preserving agricultural activities in rural areas and show a willingness to support it.

The GM-PV sector must understand that just producing green energy is not sufficient. The societal demands on PV projects will increase, as will the demand for multifunctionality. Agri-PV increases land-use efficiency and acceptability among affected citizens, considering the food-energy-water-biodiversity nexus.

For successful Agri-PV diffusion, capacity building in PV companies is necessary, and the PV industry must work hand in hand with the Agricultural sector to ensure Agri-PV projects are harmoniously implemented with agriculture and environmental protection.

In 2021, the global market value of Agri-PV was estimated at 3.6 billion USD, projected to increase to 9.3 billion USD in 2031, growing at a CAGR of 10.1% from 2022 to 2031. These figures underline that—**The future of utility-scale PV on agricultural land is Agri-PV**.

References

[1] Tonita, E.M., Russell, A. C. J., Valdivia, C. E., *et al.*, Optimal ground coverage ratios for tracked, fixed-tilt, and vertical photovoltaic systems for latitudes up to 75°N. *Solar Energy* 2023;**258**:8–15. https://doi.org/10.1016/j.solener.2023.04.038.

[2] Belectric. *Peg® Solarkraftwerk: Die Revolution Im Solarkraftwerksbau.*

[3] FAO. Land use statistics and indicators: Global, regional and country trends 1990–2019 [FAOSTAT ANALYTICAL BRIEF 28 Rome]; 2021, 2021.

[4] Inspire. *Agrivoltaics Map.* 2023. Available from https://openei.org/wiki/InSPIRE/Agrivoltaics_Map.

[5] Bolingerl, M. *Utility-Scale Solar.* October 2023.

[6] Dupraz C. Assessment of the ground coverage ratio of agrivoltaic systems as a proxy for potential crop productivity. *Agroforest Syst* 2023. https://doi.org/10.1007/s10457-023-00906-3.

[7] Macknick, J. Agrivoltaics. 2023. Available from https://www.nrel.gov/solar/market-research-analysis/agrivoltaics.html.

[8] Chalgynbayeva A, Gabnai Z, Lengyel P, Pestisha A and Bai A. Worldwide research trends in agrivoltaic systems—A bibliometric review. *Energies* 2023;16(2):611.

[9] Klokov A, Loktionov E, Loktionov Y, Panchenko V and Sharaborova E. A mini-review of current activities and future trends in agrivoltaics. *Energies* 2023;16(7):3009.

[10] Macknick, J, Hartmann, H, Barron-Gafford, G, *et al.* The 5 Cs of agrivoltaic success factors in the United States: Lessons from the InSPIRE Research Study.

[11] Nitricity. *Electrified Climate-Smart Fertilizer*. 2023. Available from Electrified Climate-Smart Fertilizer.

[12] Pascaris AS, Schelly C, Burnham L and Pearce JM. Integrating solar energy with agriculture: Industry perspectives on the market, community, and socio-political dimensions of agrivoltaics. *Energy Res Soc Sci* 2021;75:102023.

[13] Hannachi, A. Exploring the benefits and challenges of agrivoltaic adoption. 2021 [cited 2023 Sep 2]. Available from https://www.linkedin.com/pulse/exploring-benefits-challenges-agrivoltaic-adoption-angeline-hannachi/.

[14] Matalucci, S. The agrivoltaic projects stirring interest in Africa: With record-high temperatures in Northern Africa and worries over food security rampant from Egypt to Morocco, agrivoltaic projects in the region are getting ever more attention. 2022 [cited 2023 Sep 1]. Available from https://www.dw.com/en/climate-change-puts-agrivoltaic-projects-in-northern-africa-in-the-spotlight/a-61972022.

[15] Bingwa, B. Decentralized and with triple benefit – the potential of agrivoltaics in sub-Saharan Africa: New projects in Mali and The Gambia aim to combine solar power with agriculture – merging the management of food, energy and water. Welthungerhilfe; 2023. Available from https://www.welthungerhilfe.org/global-food-journal/rubrics/agricultural-food-policy/agrivoltaics-in-sub-saharan-africa.

Index

www.ingramcontent.com/pod-product-compliance
Lightning Source LLC
LaVergne TN
LVHW011830200926
848484LV00002B/343